Undergraduate Texts in Mathematics

Editors

S. Axler
F.W. Gehring
K.A. Ribet

T0214173

Paul Cull   Mary Flahive
Robby Robson

Difference Equations

From Rabbits to Chaos

With 16 Illustrations

 Springer

Paul Cull
Dept. Computer Science
Dearborn Hall
Oregon State University
Corvallis, OR 97331
USA
pc@cs.orst.edu

Mary Flahive
Dept. Mathematics
Kidder Hall
Oregon State University
Corvallis, OR 97331
USA
flahive@math.orst.edu

Robby Robson
Eduworks
3520 Northwest
  Hayes Ave.
Corvallis, OR 97330
USA
rrobson@eduworks.
  com

Mathematics Subject Classification (2000): 39-01, 39Axx, 68Rxx, 11B37, 11B39

Library of Congress Cataloging-in-Publication Data
Cull, Paul, 1943–
    Difference equations: from rabbits to chaos / Paul Cull, Mary Flahive,
  Robby Robson.
      p. cm. — (Undergraduate texts in mathematics)
    Includes bibliographical references and index.
    ISBN 0-387-23234-6 (alk. paper)
      1. Difference equations. I. Flahive, Mary E., 1948– II. Title. III. Series.
  QA431.C85   2004
  515′.625 — dc22   2004058968

ISBN 1-4899-8823-8 (softcover)   Printed on acid-free paper.
ISBN 0-387-23233-8 (hard cover)

9 8 7 6 5 4 3 2 1   SPIN 10950852 (softcover)   SPIN 10967645 (hardcover)

springeronline.com

Preface

Some years ago we noticed that various seemingly disparate fields were using similar models and techniques to solve similar problems. From mathematics to computer science to engineering to biology, various forms of difference equations were appearing in research papers and in textbooks, but with no common background, the same results were being independently derived over and over again. As we were noticing this, some mathematics curricula were being revised with discrete mathematics replacing calculus as the first college mathematics course. New discrete mathematics courses were created, and several superb textbooks appeared. In some fields a year of discrete mathematics actually replaced a year of calculus, while in other fields students took both discrete mathematics and calculus.

With these changes, what happened to difference equations? Some texts in discrete mathematics ignored them. Others had a few examples of difference equations as applications of proof by induction. Still others devoted a chapter to difference equations, but only solved a few special cases and/or represented generating functions (also called Z-transforms) as the principal or only method for finding solutions. With this lack of common background, texts on algorithms, signal processing, and population biology were still forced to devote chapters to the difference equations used in their areas. Even students who took several of these courses had difficulty seeing that they were working with the same difference equations in different contexts. Many instructors had written notes to flesh out the coverage given in texts, but such notes were of necessity usually so terse that students were led to believe that difference equations were very complicated and hard to understand.

With these problems in mind, we set out to write a book on difference equations that is accessible to undergraduates. As a text, it is meant for undergraduate majors in one of the mathematical sciences, presumably in their junior or senior year. We've written it for the student who likes to compute and is comfortable with mathematical proof, but the book can be profitably read by students who approach the subject from either a computational or theoretical point of view.

We wanted our text to have an algorithmic spirit. In this book, each chapter leads to techniques that can be applied by hand to small examples and also can be programmed for larger examples. In many cases we give explicit algorithms, which we decided to write in pseudocode rather than in a specific programming language for several reasons. First, it is easy to translate from pseudocode into any reasonable programming language. Second, there are many programming languages available, and translating from one language to another is often more difficult than translating from pseudocode. Third, we are not sure that programming these algorithms is worth the effort, because for almost all of our examples there are high-quality implementations readily available on the Web. It probably makes more sense to use one of these programs rather than to cobble together a program that will be used only a few times and/or will be prone to problems when the input is not exactly in the form assumed by the programmer. A number of mathematically oriented computer packages are also available. For example, MATLAB, Maple, and Mathematica all have packages that will solve difference equations and recurrence relations. In many cases these packages give numeric answers as well as symbolic solutions when possible. Using these packages is much simpler than programming from scratch.

In this book we start with the old story of Fibonacci's rabbits and progress through several generalizations, ending with some nonlinear difference equations. We deal with familiar mathematical structures such as the real numbers, the complex numbers, the integers, and the integers modulo an integer. We were tempted to discuss more general structures in order to show, for example, how theories of computation could be represented as difference equations, but we soon discovered that this would result in either a very large book or a very formal book, which would be at variance with our goal of accessibility. After developing the theory and techniques for solving linear difference equations in Chapters 2 to 4, we specialize to equations with nonnegative coefficients in Chapters 5 and 6 and then consider the generalization to matrix difference equations in Chapter 7. Chapter 8 considers equations over other rings, including integers modulo m and finite fields. Chapter 9 considers some issues in computational complexity, including divide-and-conquer algorithms. We end with some nonlinear systems in Chapter 10. Along the way we use linear algebra, develop formal power series, solve some combinatorial problems, visit Perron–Frobinus theory, use graph theory, discuss pseudorandom number generation and integer factorization, and use the FFT to multiply polynomials quickly.

There are four appendices serving different purposes. The first is a collection of worked examples, which are meant to supplement the early chapters of the book. Because the material in Appendices B and C is essential to an understanding of the book, we suggest working through them before beginning Chapter 2. Although many of the difference equations we consider have integer or real coefficients, it is often necessary to consider the coefficients as complex numbers. Appendix B gives the highlights of the complex analysis we use, and no prior experience is necessary to understand this appendix. On the other hand, only the most exceptional student could learn new material at the rate at which linear algebra is presented in Appendix C. One of the aims of this book is to show students that linear algebra is a powerful and coherent subject whose ideas have diverse applications, and we hope Appendix C is a helpful review. Appendix D outlines a method of Morris Marden [105] that can be used to decide when the general solution of a difference equation converges to zero. This appendix is not needed for an understanding of the book.

Computation

Most of our examples work with "small" difference equations, equations that can be completely solved by hand. In particular, for these equations their characteristic polynomials can be found, the roots of these polynomials can be computed exactly, and the associated eigenvector equations can be solved. While the theory we develop applies to both small and large equations, these computations may be difficult or impossible for large equations. For example, actually factoring polynomials is not possible in general, and rational computation of characteristic polynomials may require numbers with very many digits. Numerical approximation methods are often used for these computations, and we refer the interested reader to Acton [1], who gives a good introduction to numerical methods. (More serious users might refer to the compendium [131] or to the classic [170] by Wilkinson.) In general, we do not cover numerical methods. The one exception to this rule is our discussion of the use of Newton's method for finding the positive root of a nonnegative polynomial. We include this method for several reasons: it rapidly finds this root, the proof of its convergence and its speed of convergence are relatively easy, and the method is an example of a commonly encountered nonlinear difference equation.

Notational Preliminaries

In this book, we use the following fairly standard notation:

$\mathbb{Z}$ is the set of **integers**.

$\mathbb{Z}_m$ is the set of **integers modulo m**.

$\mathbb{Z}^k$ is the set of all **k-tuples** with integer coordinates.

$\mathbb{Z}_m^k$ is the set of all **k-tuples** of integers modulo m.

$\mathbb{N}$ is the set of **natural numbers**, including 0; $\mathbb{N} = \{0, 1, 2, \ldots\}$.

$\mathbb{N}^+$ is the set of **positive integers**.

$\mathbb{Q}$ is the set of **rational numbers**.

$\mathbb{R}$ is the set of **real numbers**.

$\mathbb{F}$ denotes a **finite field**.

$\mathbb{C}$ is the set of **complex numbers**.

$\mathbb{R}[x]$ is the set of **polynomials** with real coefficients.

$\mathbb{C}[x]$ is the set of **polynomials** with complex coefficients.

$\mathbb{Z}_m[x]$ is the set of **polynomials** whose coefficients are integers modulo m.

$\mathbb{F}[x]$ is the set of **polynomials** with coefficients from the finite field $\mathbb{F}$.

$\lfloor x \rfloor$ is the **floor** of $x \in \mathbb{R}$, the largest integer n with $n \leq x$.

$\lceil x \rceil$ is the **ceiling** of $x \in \mathbb{R}$, the smallest integer n with $n \geq x$.

$k \pmod{m}$ means the equivalence class $\{k + jm \, : \, j \in \mathbb{Z}\}$, while

$k \bmod m$ means the least nonnegative integer in the class $k \pmod{m}$.

Contents

1
Fibonacci Numbers

This chapter is devoted to the Fibonacci numbers. We start with the familiar definition, move on to some more sophisticated points of view, and then formulate some questions that are typical of those that can be addressed using the material of this book.

1.1 The Rabbit Problem

In the year 1202 the Italian mathematician Leonardo Pisano (which means Leonardo of Pisa) published *Liber Abaci*,[1] a book of problems whose purpose was to illustrate the usefulness of Arabic numerals in arithmetic computations because at that time cumbersome Roman numerals were still being used in Italy. One of the problems discussed in Pisano's book considers pairs of breeding rabbits. Each pair of rabbits matures in two months and produces one new pair each month thereafter, beginning with the last day of its second month. Starting with a single infant pair born at the beginning of Month 0, how many pairs will there be one year after this pair begins breeding? We can find our way to a solution by considering what happens in the first few months:

1. At the end of Month 0 there is only one pair, and they are not yet breeding.

---

[1]The book was reprinted in 1857–1862 by Baldassarre Boncompagni [11]. A translation by L. Sigler [147] has recently been published by Springer-Verlag.

2. At the end of Month 1 the first pair gives birth to a second pair.

3. At the end of Month 2 the second pair is one month old and not yet breeding. The first pair produces another pair, giving three pairs at the end of Month 2.

4. By the end of Month 3, the first pair has produced yet another pair, and the second pair has given birth to its first progeny. Therefore, there are five pairs at the end of Month 3.

5. For Month 4, the two new pairs born the previous month are not bearing young, but all three older pairs give birth to one new pair. Eight pairs are alive at the end of Month 4.

We observe that at the end of the nth month each pair that was alive at the $(n-2)^{\text{nd}}$ month has given birth to a new pair during the month. Therefore, the total number of pairs alive at the end of the nth month is the number from the $(n-1)^{\text{st}}$ month *plus* one new pair for every pair that was alive at the end of the $(n-2)^{\text{nd}}$ month. In other words, the number of rabbit pairs at the end of n months is the sum of the numbers at the end of the two previous months.

The numbers generated by this most famous example of a recurrence relation (which we often simply call a "recurrence") are called the **Fibonacci numbers**, because Pisano is usually referred to as Fibonacci, which means son of Bonaccio. These Fibonacci numbers are given by the sequence

$$1, 2, 3, 5, 8, \ldots,$$

whose first two terms are $1, 2$ and each subsequent term is the sum of the preceding two terms. The Rabbit Problem asks for the thirteenth element of this sequence, which is the number 377. If we are asked for the number of rabbits after one thousand months, could we compute the 1001st Fibonacci number without computing all 1000 numbers that came before? Examination of some slightly more sophisticated ways to view the Fibonacci numbers will provide an answer to this question.

1.2   The Fibonacci Sequence

In calculus we often think of a function such as $f(x) = x^2$ as a simpler object than an infinite sequence, but a **sequence** $s_0, s_1, s_2, \ldots$ of complex numbers is really just a function

$$s : \mathbb{N} \to \mathbb{C},$$

where s_n is the value of the function s at n. In this book we will denote elements of sequences by using subscripts and use $\langle s_n \rangle$ as shorthand for the entire sequence. We will also define a sequence by giving its "generic term," so $\langle n^2 \rangle$ is the sequence $\langle s_n \rangle$ with $s_n = n^2$ for all $n \in \mathbb{N}$.

We will slightly modify the Fibonacci sequence, and from now on define the first two Fibonacci numbers to be 0 and 1, while each subsequent Fibonacci number is still the sum of its two immediate predecessors. Such a rule, which defines the elements of a sequence by a formula involving some fixed number (in this case two) of preceding elements, is called a **recurrence relation**. The Fibonacci sequence is defined by the linear recurrence

$$(1.1) \qquad\qquad f_0 = 0, \quad f_1 = 1,$$
$$(1.2) \qquad\qquad f_{n+2} = f_{n+1} + f_n .$$

The recurrence itself is (1.2), and the two values given in (1.1) are called the **initial conditions**. The system (1.2) and (1.1) is referred to as an **initial value problem**. The equation in (1.2) is called **linear** because the terms of the sequence are connected in a linear manner, using only linear combinations of previous elements of the sequence. (In general, scaling by real or complex constants is allowed.)

As we said above, it's often helpful to think of a sequence as a function f defined on the set of natural numbers $\mathbb{N}$. The Fibonacci sequence is the function $f(n) = f_n$ with

$$(1.3) \qquad\qquad f(0) = 0, \quad f(1) = 1$$
$$(1.4) \qquad\qquad f(n + 2) = f(n + 1) + f(n).$$

Here we begin the sequence at the position $f(0)$ because verifying and developing formulas is often easier when we initialize with $f(0)$ rather than with $f(1)$.

When we define the Fibonacci sequence it's customary to think of the index n in (1.2) and (1.4) as a natural number, but there's no purely mathematical reason to limit ourselves to nonnegative n. From any two consecutive Fibonacci numbers the recurrence (1.2) allows us to compute the preceding element of the Fibonacci sequence using $f_n = f_{n+2} - f_{n+1}$. Therefore, we can proceed backwards as well as forwards, and any integer can be used as an index. Some "center" terms of the associated doubly infinite Fibonacci sequence are

$$\ldots, -8, 5, -3, 2, -1, 1, 0, 1, 1, 2, 3, 5, 8, \ldots .$$

(Exercise 1.6 asks you to verify this pattern.) Extending recurrences to include negative indices can have some advantages, sometimes giving a deeper insight into the theory. It's especially helpful for modeling population dynamics and other time-dependent phenomena.

1.2.1   Computing Fibonacci numbers

The definition of the Fibonacci function given in (1.4) is called a **recursive definition**. It allows us to write a short recursive computer program to compute (at least in theory) the value of the Fibonacci function at any given nonnegative n.

```
PROCEDURE  RECRFIB(n,f)  (* Input n, Output f(n) *)
        IF  n ≤ 1
            THEN   f := n
            ELSE RECRFIB(n − 1,g)
                 RECRFIB(n − 2,h)
                 f := g + h
```

Another style of program for computing $f(n)$ for nonnegative n is given by

```
PROCEDURE FIB(n,f)  (* Input n, Output f(n) *)
        IF  n ≤ 1 THEN  f := n
                  ELSE f := 1
                       FOR  i := 2 TO  n
                            t := f
                            f := f + j
                            j := t
                       ENDFOR
```

The first procedure, RECRFIB, computes values of the Fibonacci function by making calls to itself. Such a program is usually called **recursive**. Program FIB does not call on itself, but computes by repeatedly executing the statements in the FOR loop. Such a program is usually called **iterative**.

   Once we know *some* program that computes a given function, it is natural to ask for a program that computes the function rapidly. Finding such a program is not as easy as it sounds. For instance, for the Fibonacci function, we might want to know how quickly we can produce a *single term* $f(n)$ in terms of the size of the input n, or we might want to know how quickly we can compute *all* of the first n terms of the Fibonacci sequence in terms of the input size. Or we might want to answer these same questions in the context of constraints on the size of the memory of the computer used for the computation. These may be realistic or simply theoretical constraints. More effective procedures for computing Fibonacci numbers are given in [40].

1.2.2 A formula for the Fibonacci numbers

Although the defining recurrence (1.2) *theoretically* allows us to compute $f(n)$ for any n, it's not an especially tidy mathematical expression for $f(n)$, since all previous terms are needed in order to calculate just one term of the Fibonacci sequence. For example, think about the question of quickly estimating the number of rabbit pairs after one thousand months of rabbit breeding. If the function f had a nice formula, it would be easier to answer such a question. Fortunately, and perhaps surprisingly, there is such a formula for values of the Fibonacci function:

$$(1.5) \qquad f(n) = \frac{1}{\sqrt{5}} \left[\left(\frac{1+\sqrt{5}}{2} \right)^n - \left(\frac{1-\sqrt{5}}{2} \right)^n \right].$$

You might recognize the number $\dfrac{1+\sqrt{5}}{2}$; it's often referred to as the **golden mean** or **golden section**. This formula (1.5) is known as **Binet's Formula** and can be derived using the techniques given in the next chapter (refer to Exercise 2.10). Although it is usually attributed to Jacques Phillipe Marie Binet (1786–1856), Donald Knuth [88, vol. 1, p. 82] says that Abraham de Moivre reported the formula in 1730 [48, pp. 26–42] when he considered the general linear recurrence.

For now, we can appreciate the power of Binet's Formula by using it to estimate size of the 1000th Fibonacci number. Since

$$\frac{1+\sqrt{5}}{2} \approx 1.618 \text{ and } \frac{1-\sqrt{5}}{2} \approx -.618.$$

and $(.618)^{1000}$ is very small, then

$$f(1000) \approx \frac{1}{\sqrt{5}} (1.618)^{1000}.$$

If we're interested in estimating the size of $f(1000)$ in terms of the number of its decimal digits, we can use the base-10 logarithm. (Refer to Exercise 1.9.) Since

$$1000 \cdot \log(1.618) - .5 \log(5) \approx 208.629,$$

we expect the number of decimal digits in $f(1000)$ to be about 209. In fact, the 1000th Fibonacci number is the number

$$f(1000) = 43466557686937456435688527675040625802564660517371$$
$$78040248172908953655541794905189040387984007925516$$
$$92959225930803226347752096896232398733224711616429$$
$$96440906533187938298969649928516003704476137795166$$
$$849228875,$$

which does have the predicted 209 digits.

With $\lambda_0 = \frac{1+\sqrt{5}}{2}$, $\lambda_1 = \frac{1-\sqrt{5}}{2}$, and $\alpha = 1/\sqrt{5}$, Binet's Formula can be rewritten as

$$f(n) = \alpha(\lambda_0^n - \lambda_1^n).$$

Since $|\lambda_1| < 1$ and $\alpha < 1/2$, then $|f_n - \alpha\lambda_0^n| = \alpha|\lambda_1|^n < 1/2$, and the fact that f_n is an integer therefore gives the pleasant identity

(1.6) $f_n = \text{Round}\left(\lambda_0^n/\sqrt{5}\right)$ for all $n \geq 0$,

where $\text{Round}(X)$ returns the integer nearest to X.[2] We analyze this notion of **roundability** in Chapter 5. (If you want to look ahead, the main result is Theorem 5.2.2).

1.2.3   Further Fibonacci facts

The beauty and arcana of Fibonacci numbers are studied in a number of places. We would be remiss if we did not mention the *Fibonacci Quarterly*, a journal that has been publishing for about half a century. One of the mainstays of this journal was Brother Alfred. His book [16] is a nice introduction to Fibonacci numbers. From the Russian literature, Vorobev's book [164] is a concise introduction to Fibonacci facts and formulas. A more extensive recent book is *Fibonacci and Lucas Numbers* [90]. For some of the minor arcana, see our paper [22].

1.3   Notation for Asymptotic Analysis

There are often many algorithms for the same problem. For instance, in the case of computing Fibonacci numbers we've already written two types of algorithms for computing a specific Fibonacci number, either using the definition directly or using Binet's Formula. Our usual method of assessing the efficiency of various algorithms for a problem will be to compare

---

[2]Note that $\text{Round}(\frac{k}{2})$ is not defined when k is an odd integer.

their run times, and usually this is done with an asymptotic analysis. For this there are three standard forms of notation, **Big-Oh**, **Big-Omega**, and **Big-Theta**. Each notation removes unimportant details so we can see the size of run time more clearly. This notation was codified for computer scientists by Donald Knuth [85, 86] in 1976, when he drew upon notation already used in analytic number theory. Big-Oh[3] was introduced by Bachmann in 1894, and something very similar to Big-Omega was used by Hardy and Littlewood in 1914. More information on the history of this notation (and that of Big-Theta from the 1960s) can be found in Knuth's article. Although our principal use of this notation is for positive real numbers, our definitions allow $T(n)$ to be a *complex-valued* sequence.

Let $T(n)$ and $f(n)$ be two complex-valued sequences. Then

$T(n) = O(f(n))$ (we say $T(n)$ has order at most $f(n)$) means that there exists a positive constant c such that

$$|T(n)| \leq c|f(n)| \quad \text{for all sufficiently large } n.$$

$T(n) = \Omega(f(n))$ (we say $T(n)$ has order at least $f(n)$) means that there exists a positive constant c such that

$$|T(n)| \geq c|f(n)| \quad \text{for all sufficiently large } n.$$

$T(n) = \Theta(f(n))$ (we say $T(n)$ has order exactly $f(n)$) means that there exist positive constants c_1, c_2 such that

$$c_1|f(n)| \leq |T(n)| \leq c_2|f(n)| \quad \text{for all sufficiently large } n.$$

For instance, from (1.6) we know that the asymptotic size of f_n is $\Theta(\lambda_0^n)$ where $\lambda_0 = (1+\sqrt{5})/2$. To get some practice with this notation, you should verify that for any fixed pair of positive integers $i \leq j$, each of the following holds:

$$n^i = O(n^j), \quad n^j = \Omega(n^i), \quad \Theta(n^i) = \Theta(n^j) \text{ implies } i = j.$$

1.4   Exercises

Ex 1.1. (Taken from *Liber Abaci*, pp. 283 ff.) Translate the following:

``Quot paria coniculorum in uno anno ex uno pario germinentur.''
    Qvidam posuit unum par cuniculorum in quodam loco, qui erat

---

[3]The term Big-Omicron is used by Knuth, but most other authors call this Big-Oh notation.

undique pariete circundatus, ut sciret, quot ex eo paria
germinarentur in uno anno: cum natura eorum sit per singulum
mensem aliud par germinare; et in secundo mense ab eorum
natiuitate germinant. Quia suprascriptum par in primo mense
germinat, duplicabis ipsum, erunt paria duo in uno mense.
Ex quibus unum, scilicet primum, in secundo mense germinat; et
sic sunt in secundo mense paria 3; ex quibus in uno mense duo
pregnantur; et germinatur in tercio mense paria 2 conciculorum;
et sic sunt paria 5 in ipso mense;... Cum quibus etiam additis
parijs 144, que germinatur in ultimo mense, erunt paria 377;
et tot paria peperit suprascriptum par in prefato loco in
capite unius anni. ...

Ex 1.2. Let $x \in \mathbb{R}$. Let $\lfloor x \rfloor$ be the **floor** of x, that is, the largest integer n such that $n \leq x$. Let $\lceil x \rceil$ be the **ceiling** of x, that is, the least integer n such that $x \leq n$. Show that $\lfloor x \rfloor = \lceil x \rceil$ iff $x \in \mathbb{Z}$.

Ex 1.3. Suppose $\langle s_n \rangle$ is any sequence that satisfies the Fibonacci recurrence (1.2) but possibly has different initial values. Show that for any j, any term s_n can be written as a linear combination of s_j, s_{j-1} with *integer* coefficients.

Ex 1.4. Suppose $\langle s_n \rangle$ is any sequence that satisfies the Fibonacci recurrence (1.2) but possibly has different initial values. Let $n_1, n_2 \in \mathbb{N}$ be any fixed indices. Show that any term s_n can be written as a linear combination of s_{n_1}, s_{n_2} with *rational* coefficients.

Ex 1.5. For the Fibonacci sequence, let $n_1, n_2 \in \mathbb{N}$ be any fixed indices with $\gcd(n_1, n_2) = 1$. Show that any term f_n can be written as a linear combination of f_{n_1}, f_{n_2} with *integer* coefficients.

Ex 1.6. Use (1.2) to show that for the Fibonacci function we have

$$f(-n) = (-1)^{n+1} f(n) \text{ for all } n \geq 0.$$

Ex 1.7. Verify Binet's Formula (1.5) for $f(2), f(3), f(4)$.

Ex 1.8. Use mathematical induction to prove that every element of the Fibonacci sequence satisfies Binet's Formula.

Ex 1.9. Show that the base-10 logarithm of each of the integers $1, 2, \ldots, 9$ lies in the interval $[0, 1)$, and that the logarithm of any two-digit integer lies in the interval $[1, 2)$. In general, show the number of decimal digits in an integer n is the ceiling of its logarithm to the base 10.

Ex 1.10. If the base-10 logarithm of an integer begins with 3.53, how many digits does the integer have and what's its first digit?

Ex 1.11. This exercise deals with the sequence of Lucas Numbers. [4]
Consider the sequence $\langle L_n \rangle$ generated by

$$L_0 = 2, \quad L_1 = 1,$$
$$L_{n+2} = L_{n+1} + L_n.$$

(a) Calculate L_2, L_3, L_4.
(b) Find constants $A, B \in \mathbb{R}$ such that for each of $n = 0, 1$ it is true that

$$(1.7) \qquad L_n = A \left(\frac{1 + \sqrt{5}}{2} \right)^n - B \left(\frac{1 - \sqrt{5}}{2} \right)^n.$$

(c) Use mathematical induction to prove that your formula in the previous part holds for all $n \geq 0$.

Ex 1.12.   (a) Write a short program that computes the Fibonacci and the Lucas sequences.
(b) Use your program to calculate f_{30} and L_{30}. Check your answer using the closed forms in (1.5) and (1.7). Compute the ratio $\dfrac{f_{30}}{L_{30}}$.
(c) The ratio $\dfrac{f_n}{L_n}$ has a limiting value. Use your program to calculate this value to ten decimal places.

Ex 1.13. For this exercise, consider the sequence $\langle s_n \rangle$ defined by

$$s_0 = 0, \quad s_1 = 1,$$
$$s_{n+2} = 2s_{n+1} + 2s_n.$$

(a) Calculate the first five terms of the sequence $\langle s_n \rangle$.
(b) Check that the following formula correctly calculates s_2, s_3, s_4:

$$s_n = \frac{1}{\sqrt{12}} \left[(1 + \sqrt{3})^n - (1 - \sqrt{3})^n \right].$$

(c) Show that the general n^{th} term of the sequence $\langle s_n \rangle$ satisfies the formula in the previous part.

Ex 1.14. Show that the variable t in the procedure FIB can be eliminated by using $f := f + j$, $j := f - j$.

---

[4]The French mathematician Édouard Lucas (1842–1891) studied the properties of this and other sequences in the first volume (page 186) of the *American Journal of Mathematics*, which appeared in 1878.

Ex 1.15. Show that every natural number n can be expressed in a "binary" form $n = \sum_{i=2}^{K} b_i f_i$, where each b_i is either 0 or 1 and K depends on the natural number n. Show further that this binary Fibonacci representation is NOT unique. If you impose the additional stipulation that no two consecutive b_i are both 1, show that the binary Fibonacci representation is unique.

Ex 1.16. Define $\log_{FIB}(n)$ to equal the least index i for which $f_i \geq n$.
(a) Calculate some values of $\log_{FIB}(n)$.
(b) For each $i \geq 1$, define $\#(i)$ to be the number of natural numbers n for which $\log_{FIB}(n) = i$. Show that $\#(i) = f_i$ for all $i \geq 3$.

Ex 1.17. Show that any solution to

$$s_n = s_{n-1} + \lfloor \sqrt{n} \rfloor$$

with nonnegative initial conditions satisfies $s_n = \Theta(n^{3/2})$.

2
Homogeneous Linear Recurrence Relations

The simplest type of recurrence relation is the homogeneous linear recurrence with constant coefficients, one in which s_{n+k} is given as a *linear* function of $s_n, \ldots, s_{n+k-1}$. In other words, for all $n \geq 0$,

$$\text{(HL)} \qquad s_{n+k} = c_1 s_{n+k-1} + c_2 s_{n+k-2} + \cdots + c_k s_n,$$

where $c_1, \ldots, c_k$ are complex constants and $c_k \neq 0$. This is called a kth order homogeneous linear recurrence with constant coefficients. The purpose of this chapter is to analyze these recurrences using tools from linear algebra. Appendix C contains a review of the basic linear algebra that we will assume here. Even if we were only interested in *integer* recurrences, we would still need to consider recurrences whose coefficients are complex numbers because this is more or less forced on us by the algebra. In addition, more compact formulas can often be obtained by using more general number systems.

Equation (HL) with given values $s_0, \ldots, s_{k-1}$ is called an **initial value problem**. For any set of k initial values, the recurrence (HL) can be successively applied to compute the infinite sequence $\langle s_n \rangle$ that satisfies this recurrence. From $s_0, \ldots, s_{k-1}$ the s_k term is specified by $s_k = c_1 s_{k-1} + \cdots + c_k s_0$, the s_{k+1} term by $s_{k+1} = c_1 s_k + \cdots + c_k s_1$, and so on. What this informally shows is that every initial value problem has a solution, and the solution is unique. The first k terms determine the rest of the sequence. In Exercise 2.1 we ask you to supply a formal verification of this fact, which is fundamental to the remainder of this chapter.

Arithmetic Operations on Sequences

We can add and scale sequences in exactly the same way as functions. This means that $\langle s_n \rangle + \langle t_n \rangle = \langle s_n + t_n \rangle$. Viewed as functions, this is the rule $(s+t)(n) = s(n) + t(n)$. A sequence is multiplied by the scalar $\lambda \in \mathbb{C}$ using the rule $\lambda \langle s_n \rangle = \langle \lambda s_n \rangle$. The set of all functions from $\mathbb{N}$ into $\mathbb{C}$ (which is the same as the set of all complex sequences) is a complex vector space under these operations of addition and scalar multiplication.

2.1   The Solution Space of (HL)

We begin by analyzing the second–order initial value problem

$$(2.1) \qquad s_{n+2} = 3s_{n+1} - 2s_n, \quad s_0 = 2, \ s_1 = 3,$$

which generates the sequence

$$2, 3, 5, 9, 17, 33, 65, 129, 257, 513, \ldots,$$

and has the closed formula $s_n = 2^n + 1$. We might ask what the sequence looks like under other initial conditions. Here are two such sequences:

$$2, 5, 11, 23, 47, 95, 191, 383, 767, 1535, \ldots,$$
$$3, 1, -3, -11, -27, -59, -123, -251, -507, -1019, \ldots.$$

It's not quite as easy to guess a formula for the nth term of the sequence starting with $2, 5$, but you might be lucky and notice that

$$23 = 24 - 1,$$
$$47 = 48 - 1,$$
$$95 = 96 - 1,$$

and so arrive at the formula $s_n = 3 \cdot 2^n - 1$. By thinking negatively, we can find the formula $s_n = 5 - 2^{n+1}$ for the third set of initial conditions $s_0 = 3, s_1 = 1$. (In Exercise 2.2 you're asked to prove these formulas by an inductive argument.) Since we used the same recurrence (2.1) to generate all three sequences, it's not surprising that there are similarities among the formulas. For instance, all three formulas involve powers of 2. One of our goals is to understand how and why the sequences corresponding to different initial values are similar. To put this another way, we're interested in understanding the structure of the space of all solutions to a fixed k^{th} order homogeneous linear recurrence when the initial conditions range over all k-tuples of complex numbers.

Placing this in a more general setting, we've observed already that the set of all complex sequences forms a vector space over $\mathbb{C}$. In Exercise 2.5 you're asked to prove that the set of solutions to (HL) is a *subspace* of the space of all complex sequences, and this means that every linear combination of solutions is also a solution. We'll call this the **solution space of (HL)** and will denote it by $\mathcal{X}$.

We've seen that any sequence in $\mathcal{X}$ is completely determined by its first k terms. This can be neatly expressed by defining a map π that picks out these terms and writes them as a vector in $\mathbb{C}^k$. Accordingly, we define

$$(2.2) \qquad \pi : \mathcal{X} \to \mathbb{C}^k \text{ by } \pi(\langle s_n \rangle) = (s_{k-1}, \ldots, s_1, s_0)^T.$$

Because every sequence in $\mathcal{X}$ has a unique string of k initial elements, this map is a well-defined function. The fact that any choice of values $s_0, \ldots, s_{k-1}$ can be extended to a infinite sequence that is in $\mathcal{X}$ means that π is an **onto function**. The companion fact that every initial value problem has a *unique* solution translates to the statement that π is a **one-to-one function** (that is, if $\pi(x) = \pi(y)$ for $x, y \in \mathcal{X}$, then $x = y$). Since π is one-to-one and onto, it is called a **bijection** or a **bijective function**, and π has a (two-sided) inverse. This inverse is the map that assigns to any vector $\boldsymbol{\alpha} = (\alpha_{k-1}, \ldots, \alpha_0)^T \in \mathbb{C}^k$ the unique solution in $\mathcal{X}$ that begins with the initial values

$$s_0 = \alpha_0 , \ s_1 = \alpha_1 , \ \ldots , \ s_{k-1} = \alpha_{k-1} .$$

For any two solutions $\langle x_n \rangle, \langle y_n \rangle \in \mathcal{X}$ and complex scalars α and β, we have[1]

$$\alpha * \pi(\langle x_n \rangle) + \beta * \pi(\langle y_n \rangle) = \alpha * \begin{pmatrix} x_{k-1} \\ \vdots \\ x_1 \\ x_0 \end{pmatrix} + \beta * \begin{pmatrix} y_{k-1} \\ \vdots \\ y_1 \\ y_0 \end{pmatrix}$$

$$= \begin{pmatrix} \alpha * x_{k-1} + \beta * y_{k-1} \\ \vdots \\ \alpha * x_1 + \beta * y_1 \\ \alpha * x_0 + \beta * y_0 \end{pmatrix}$$

$$= \pi(\alpha * \langle x_n \rangle + \beta * \langle y_n \rangle),$$

which shows that $\pi : \mathcal{X} \to \mathbb{C}^k$ is a linear transformation. Since we've already shown that it is a bijection, π is an invertible transformation that is often called a **vector space isomorphism**. From linear algebra we know

---

[1] Good notation often leads to seductively simple formulas that can obscure some important details, so it is good to form the habit of making a mental note of multiple uses of the same symbol. In this regard, observe that the symbol "+" has three different meanings in the equation displayed above: first as addition of vectors in $\mathbb{C}^k$, second as addition of complex numbers, and finally as addition of sequences. Check that you can also find the three different uses of the symbol "*".

that the inverse of an invertible linear transformation is also linear, and we obtain the following theorem.

Theorem 2.1.1. *Let $\mathcal{X}$ be the solution space of a kth order homogeneous linear recurrence. Then the map π defined in (2.2) is an isomorphism between the vector spaces $\mathcal{X}$ and $\mathbb{C}^k$. In particular, $\mathcal{X}$ is a k-dimensional complex vector space.*

Since any element of a vector space is a (unique) linear combination of basis vectors, Theorem 2.1.1 can be interpreted as giving a method for constructing solutions $\langle s_n \rangle \in \mathcal{X}$ using the k sequences generated by the basic initial conditions $\mathbf{e}_1, \mathbf{e}_2, \ldots, \mathbf{e}_n$, where $\mathbf{e}_i$ is the ith standard basis vector in $\mathbb{C}^k$, the column vector with 1 in the ith position and zeros elsewhere. The preimage $\pi^{-1}(\mathbf{e}_i) = \langle s_n \rangle$ is the solution whose only non-zero initial condition is $s_{k-i} = 1$. Since the set $\{\mathbf{e}_1, \ldots, \mathbf{e}_k\}$ is a basis for $\mathbb{C}^k$ and π^{-1} is an isomorphism, the set $\{\pi^{-1}(\mathbf{e}_1), \ldots, \pi^{-1}(\mathbf{e}_k)\}$ is a basis for $\mathcal{X}$. Moreover, for given initial conditions

$$ s_0 = \alpha_0, \quad s_1 = \alpha_1, \quad \ldots, \quad s_{k-1} = \alpha_{k-1}, $$

the corresponding solution sequence is

$$ \alpha_0 \pi^{-1}(\mathbf{e}_k) + \alpha_1 \pi^{-1}(\mathbf{e}_2) + \cdots + \alpha_{k-1} \pi^{-1}(\mathbf{e}_1). $$

We illustrate this with two examples, the Fibonacci sequence and the recurrence in (2.1) above.

For the Fibonacci recurrence, the basic sequences are the two sequences

$$ \pi^{-1}(\mathbf{e}_1) = \langle 0, 1, 1, 2, 3, 5, 8, 13, 21, \ldots \rangle $$

and

$$ \pi^{-1}(\mathbf{e}_2) = \langle 1, 0, 1, 1, 2, 3, 5, 8, 13, \ldots \rangle. $$

If we want to find the sequence generated by the recurrence

$$ s_{n+2} = s_{n+1} + s_n, \quad s_0 = -1, s_1 = 3, $$

we would add 3 times $\pi^{-1}(\mathbf{e}_1)$ to -1 times the sequence $\pi^{-1}(\mathbf{e}_2)$, and this gives $-1, 3, 2, 5, 7, 12, \ldots$.

The Fibonacci numbers enjoy many special properties that are not necessarily shared by all recurrences. For instance, the sequence $\pi^{-1}(\mathbf{e}_1)$ is simply the shift of the sequence $\pi^{-1}(\mathbf{e}_2)$ one term to the left. It follows that any sequence $\langle s_n \rangle$ satisfying the Fibonacci recurrence has general term

$$ s_n = a f_n + b f_{n-1}, $$

for $a = s_1, \quad b = s_0$.

For recurrence (2.1), the basic sequences are

$$ \pi^{-1}(\mathbf{e}_1) = \langle 0, 1, 3, 7, 15, 31, 63, 127, 255, 511, \ldots \rangle $$

and

$$ \pi^{-1}(\mathbf{e}_2) = \langle 1, 0, -2, -6, -14, -30, -62, -126, -254, -510, \ldots \rangle. $$

Therefore, the ninth term of the solution to (2.1) with $s_0 = 4$, $s_1 = 11$ is

$$s_8 = 4(-254) + 11(255) = 1789.$$

To summarize: Knowledge of the basic sequences of a homogeneous linear recurrence allows us to express the nth term of any solution as a linear combination of simpler quantities, namely, the nth terms of the basic sequences. Because this procedure uses k basic sequences to find the one sequence of interest, it isn't very useful unless formulas for the basic sequences can be easily obtained. We'd like a basis for $\mathcal{X}$ consisting of sequences that are guaranteed to have an easy-to-find and simple formula for their n^{th} terms. Obtaining such a basis is the goal of the rest of this chapter.

2.2   The Matrix Form

Since each term of a kth order recurrence is determined by the k preceding terms, it is useful to think of it as a function on k-tuples of consecutive terms. Because this function is *linear*, it can be represented by a matrix. For example, the recurrence in (2.1) can be written as

$$(2.3) \qquad \begin{pmatrix} s_{n+2} \\ s_{n+1} \end{pmatrix} = \begin{bmatrix} 3 & -2 \\ 1 & 0 \end{bmatrix} \begin{pmatrix} s_{n+1} \\ s_n \end{pmatrix}, \quad \begin{pmatrix} s_1 \\ s_0 \end{pmatrix} = \begin{pmatrix} 3 \\ 2 \end{pmatrix},$$

while the usual Fibonacci sequence is encoded in the matrix equation

$$(2.4) \qquad \begin{pmatrix} f_{n+2} \\ f_{n+1} \end{pmatrix} = \begin{bmatrix} 1 & 1 \\ 1 & 0 \end{bmatrix} \begin{pmatrix} f_{n+1} \\ f_n \end{pmatrix}, \quad \begin{pmatrix} f_1 \\ f_0 \end{pmatrix} = \begin{pmatrix} 1 \\ 0 \end{pmatrix}.$$

In each case the matrix takes us from one pair of consecutive terms to the next.

In general, we express (HL) in matrix form by introducing the vectors

$$(2.5) \qquad S_n = \begin{pmatrix} s_{n+k-1} \\ \vdots \\ s_{n+1} \\ s_n \end{pmatrix} \quad \text{for all } n \geq 0,$$

and finding a matrix A such that $S_{n+1} = A S_n$. Since

$$s_{n+k} = c_1 s_{n+k-1} + c_2 s_{n+k-2} + \cdots + c_k s_n,$$

this is accomplished using the matrix

$$(2.6) \qquad A = \begin{bmatrix} c_1 & c_2 & \cdots & c_{k-1} & c_k \\ 1 & 0 & \cdots & 0 & 0 \\ 0 & 1 & & \cdots & 0 \\ \vdots & & \ddots & & \vdots \\ 0 & 0 & \cdots & 1 & 0 \end{bmatrix},$$

which when applied to S_n shifts the components down one entry and puts the next term, s_{n+k}, in the first component. The matrix A is called the **companion matrix of the recurrence**. It is also referred to as the **companion matrix of the polynomial** $x^k - c_1 x^{k-1} - \cdots - c_k$.

When the matrix form is used in an initial value problem, we have

$$S_1 = AS_0 , \quad S_2 = AS_1 = A^2 S_0 , \quad S_3 = AS_2 = A^3 S_0 , \text{ and so on,}$$

and this inductively gives the **matrix form**

(2.7) $S_n = A^n S_0 .$

The matrix form of the usual Fibonacci recurrence is

$$\begin{pmatrix} f_{n+1} \\ f_n \end{pmatrix} = \begin{bmatrix} 1 & 1 \\ 1 & 0 \end{bmatrix}^n \begin{pmatrix} 1 \\ 0 \end{pmatrix},$$

and the matrix form of the recurrence in (2.1) with initial conditions $s_0 = 2$ and $s_1 = 3$ is given by

(2.8) $S_n = \begin{bmatrix} 3 & -2 \\ 1 & 0 \end{bmatrix}^n \begin{pmatrix} 3 \\ 2 \end{pmatrix}.$

For the formula in (2.7) to be useful, we must be able to quickly compute powers of the companion matrix. Matrix powers can be easily computed for diagonal matrices, and the next easiest are matrices that are **diagonalizable**, namely, square matrices that have a basis of eigenvectors. When A is not diagonalizable, a modification involving Jordan matrices can be used. (Refer to Appendix C.)

Returning to the second example above, we set

$$A = \begin{bmatrix} 3 & -2 \\ 1 & 0 \end{bmatrix},$$

which has characteristic polynomial

$$ch_A(x) = \det\left(\begin{bmatrix} 3-x & -2 \\ 1 & -x \end{bmatrix}\right) = x^2 - 3x + 2 = (x-2)(x-1)$$

and *distinct* eigenvalues, $\lambda_1 = 2$ and $\lambda_2 = 1$. This means that A is diagonalizable. To find the eigenvectors, solve $(A - 2I)\mathbf{v}_1 = 0$ and $(A - 1I)\mathbf{v}_2 = 0$ to obtain

$$\mathbf{v}_1 = \begin{pmatrix} 2 \\ 1 \end{pmatrix} \text{ and } \mathbf{v}_2 = \begin{pmatrix} 1 \\ 1 \end{pmatrix}.$$

Therefore, for

$$D = \begin{bmatrix} 2 & 0 \\ 0 & 1 \end{bmatrix} \text{ and } P = \begin{bmatrix} 2 & 1 \\ 1 & 1 \end{bmatrix},$$

powers can be computed using

$$A^n = (PDP^{-1})(PDP^{-1})\ldots(PDP^{-1}) = P D^n P^{-1}$$

$$= \begin{bmatrix} 2 & 1 \\ 1 & 1 \end{bmatrix} \begin{bmatrix} 2 & 0 \\ 0 & 1 \end{bmatrix}^n \begin{bmatrix} 2 & 1 \\ 1 & 1 \end{bmatrix}^{-1}$$

$$= \begin{bmatrix} 2 & 1 \\ 1 & 1 \end{bmatrix} \begin{bmatrix} 2^n & 0 \\ 0 & 1 \end{bmatrix} \begin{bmatrix} 1 & -1 \\ -1 & 2 \end{bmatrix}$$

$$= \begin{bmatrix} 2^{n+1} - 1 & -2^{n+1} + 2 \\ 2^n - 1 & -2^n + 2 \end{bmatrix}.$$

Equation (2.8) therefore becomes

$$S_n = \begin{bmatrix} 3 & -2 \\ 1 & 0 \end{bmatrix}^n \begin{pmatrix} 3 \\ 2 \end{pmatrix} = \begin{bmatrix} 2^{n+1} - 1 & -2^{n+1} + 2 \\ 2^n - 1 & -2^n + 2 \end{bmatrix} \begin{pmatrix} 3 \\ 2 \end{pmatrix} = \begin{pmatrix} 2^{n+1} + 1 \\ 2^n + 1 \end{pmatrix},$$

from which we derive $s_n = 2^n + 1$ for all $n \geq 0$. A change of initial conditions is equivalent to changing the vector $S_0 = (3,2)^T$, and the same procedure would yield the general term of the sequence generated from those new initial conditions. In Exercise 2.10 you apply this method to the Fibonacci sequence.

Our general procedure for the case in which A is diagonalizable can be summarized as follows:

Let $s_{n+k} = c_1 s_{n+k-1} + c_2 s_{n+k-2} + \cdots + c_k s_n$ and $S_0 = \alpha$. Suppose the companion matrix A of this recurrence is diagonalizable with eigenvalues $\lambda_1, \ldots, \lambda_k$.

1. Find a basis for $\mathbb{C}^k$ consisting of eigenvectors of A.

2. Use the basis elements as columns to form the matrix P.

3. Form the diagonal matrix D with the eigenvalues of A on the diagonal, written in the order corresponding to the columns of P.

4. Compute $PD^n P^{-1} = A^n$ and the vector $A^n \alpha$.

Then the solution is the sequence of first components of the vectors $A^n \alpha$.

2.3 A Simpler Basis for the Solution Space

In the last section the eigenvalues of the companion matrix played a key role in solving a recurrence. In this section we explore more properties of these eigenvalues with the objective of obtaining a helpful basis for the solution space. We will refer to the polynomial

$$ch(x) = x^k - c_1 x^{k-1} - \cdots - c_{k-1} x - c_k$$

as the **characteristic polynomial of the recurrence** (HL), and its roots (the eigenvalues of the companion matrix A) will be called the **eigenvalues of the recurrence**.

We can obtain an eigenvector associated with an eigenvalue λ of the recurrence by first reminding ourselves that premultiplying a vector $\mathbf{v} = (v_1, \ldots, v_k)^T$ by the companion matrix A shifts its first $k - 1$ components down one position and then inserts $c_1 v_1 + c_2 v_2 + \cdots + c_k v_k$ into the first position. Since λ is an eigenvalue, $c_1 \lambda^{k-1} + c_2 \lambda^{k-1} + \cdots + c_k = \lambda^k$, and we obtain

$$A(\lambda^{k-1}, \lambda^{k-2}, \ldots, \lambda, 1)^T = (\lambda^k, \lambda^{k-1}, \ldots, \lambda)^T = \lambda(\lambda^{k-1}, \lambda^{k-2}, \ldots, \lambda, 1)^T.$$

This means that

$$\mathbf{v}_\lambda = (\lambda^{k-1}, \lambda^{k-2}, \ldots, \lambda, 1)^T$$

is an eigenvector associated with λ.

Theorem 2.3.1. *Consider the homogeneous recurrence*

$$s_n = c_1 s_{n-1} - \cdots - c_{k-1} s_{n-k+1} - c_k s_{n-k}.$$

(a) *If λ is an eigenvalue of the recurrence, then $\pi^{-1}(\mathbf{v}_\lambda) = \langle \lambda^n \rangle$, where π is the vector space isomorphism in (2.2). Consequently, $\langle \lambda^n \rangle$ is in $\mathcal{X}$.*

(b) *If the recurrence has k distinct eigenvalues $\lambda_1, \ldots, \lambda_k$, then the k sequences $\langle \lambda_1^n \rangle, \ldots, \langle \lambda_k^n \rangle$ form a basis for $\mathcal{X}$, and every solution $\langle s_n \rangle$ has the form*

$$(2.9) \qquad s_n = a_1 \lambda_1^n + a_2 \lambda_2^n + \cdots + a_k \lambda_k^n$$

for some constants $a_1, \ldots, a_k \in \mathbb{C}$.

Proof. In part (a), for all $n \geq 0$ we define $s_n = \lambda^n$ and the associated vectors S_n as in (2.5). Then $S_n = \lambda^n \mathbf{v}_\lambda$ and

$$A S_n = \lambda^n A \mathbf{v}_\lambda = \lambda^{n+1} \mathbf{v}_\lambda = S_{n+1},$$

implying $\langle s_n \rangle \in \mathcal{X}$ with initial vector $S_0 = \mathbf{v}_\lambda$. When the recurrence has k distinct eigenvalues, the eigenvectors $\mathbf{v}_{\lambda_i}$ form a basis for $\mathbb{C}^k$, and the k sequences $\langle \lambda_i^n \rangle = \pi^{-1}(\mathbf{v}_{\lambda_i})$ therefore form a basis for $\mathcal{X}$. Since every element of a vector space can be uniquely written as a linear combination of basis vectors, this implies that (2.9) does hold. $\square$

2.3.1   Distinct eigenvalues

For a specific vector $(s_0, s_1, \ldots, s_{k-1}) = \boldsymbol{\alpha}$ of initial conditions, how do we solve the initial value problem, that is, how do we find the a_i in (2.9)? Using the inherent linear algebra and the isomorphism π, this becomes the equivalent problem of solving

$$a_1 \mathbf{v}_{\lambda_1} + \cdots + a_k \mathbf{v}_{\lambda_k} = \boldsymbol{\alpha}$$

for the coefficients $a_1, \ldots, a_k$. Translating this to a matrix equation, we want to find (the unique) $a_1, \ldots, a_k \in \mathbb{C}$ such that

$$(2.10) \qquad \begin{bmatrix} 1 & 1 & 1 & \cdots & 1 \\ \lambda_1 & \lambda_2 & \lambda_3 & \cdots & \lambda_k \\ \vdots & \vdots & \vdots & \ddots & \vdots \\ \lambda_1^{k-1} & \lambda_2^{k-1} & \lambda_3^{k-1} & \cdots & \lambda_k^{k-1} \end{bmatrix} \begin{pmatrix} a_1 \\ a_2 \\ \vdots \\ a_k \end{pmatrix} = \boldsymbol{\alpha}.$$

The coefficient matrix in (2.10) is called the **Vandermonde matrix** associated with $\lambda_1, \ldots, \lambda_k$. It is named after Alexandre Vandermonde (1735–1796), the developer of the modern theory of determinants, and has applications in almost every area of mathematics. (Consult [82] for an interesting survey of some of its many applications.) Because each vector of initial conditions specifies an element of $\mathcal{X}$, the system of equations in (2.10) has a unique solution for each $\boldsymbol{\alpha}$, and the Vandermonde matrix must always be invertible.

Let's look at an example. For $c_1^2 \neq -4c_2$, consider the recurrence

$$s_{n+2} = c_1 s_{n+1} + c_2 s_n, \quad s_0 = \alpha_1, \ s_1 = \alpha_2,$$

whose characteristic polynomial $ch(x) = x^2 - c_1 x - c_2$ has distinct eigenvalues, since its discriminant $D = c_1^2 + 4c_2$ is non-zero. (In Exercise 2.18 you answer the question of what happens when D is zero.) The eigenvalues of the recurrence are

$$\lambda_1 = (c_1 + \sqrt{D})/2 \quad \text{and} \quad \lambda_2 = (c_1 - \sqrt{D})/2,$$

with associated Vandermonde matrix

$$V = \begin{bmatrix} 1 & 1 \\ \lambda_1 & \lambda_2 \end{bmatrix}.$$

The determinant of this matrix is $\lambda_2 - \lambda_1 = -\sqrt{D}$, which is non-zero. Its inverse is

$$V^{-1} = \frac{1}{\sqrt{D}} \begin{bmatrix} -\lambda_2 & 1 \\ \lambda_1 & -1 \end{bmatrix},$$

and

$$\begin{pmatrix} a_1 \\ a_2 \end{pmatrix} = V^{-1} \begin{pmatrix} \alpha_1 \\ \alpha_2 \end{pmatrix} = \frac{1}{\sqrt{D}} \begin{pmatrix} -\lambda_2 s_0 + s_1 \\ \lambda_1 s_0 - s_1 \end{pmatrix}.$$

In general, the inverse of a Vandermonde matrix can be computed fairly easily. For this, define the auxiliary polynomials

$$P_i(x) = \prod_{\substack{j=1 \\ j \neq i}}^{k} (x - \lambda_j) = b_{i,1} + b_{i,2}x + \cdots + b_{i,k}x^{k-1}$$

for all $i = 1, \ldots, k$. Since the λ_i are distinct, then $P_i(\lambda_i)$ is non-zero for all i, and for $j \neq i$, $P_i(\lambda_j) = 0$.

Theorem 2.3.2. *Let B be the $k \times k$ matrix whose ith row records the coefficients of the polynomial $P_i(x)$ written according to increasing powers of x. Then V^{-1} equals DB, where D is the diagonal matrix that has the (non-zero) diagonal entries $1/P_1(\lambda_1), \ldots, 1/P_k(\lambda_k)$.*

Proof. This proof is inspired by a mathematical note of F. D. Parker [124]. The diagonal entries in the matrix product BV have the form

$$(b_{i,1}, b_{i,2}, \ldots, b_{i,k})(1, \lambda_i, \ldots, \lambda_i^{k-1})^T = P_i(\lambda_i) \neq 0,$$

and the off-diagonal entries are

$$(b_{i,1}, b_{i,2}, \ldots, b_{i,k})(1, \lambda_j, \ldots, \lambda_j^{k-1})^T = P_i(\lambda_j) = 0.$$

From this we see that DBV is the identity and $V^{-1} = DB$, as claimed. $\square$

Let's use this result to find an explicit closed form for the general term of

$$s_{n+2} = s_{n+1} + 2s_n, \qquad s_0 = 1, \ s_1 = 5.$$

Since the characteristic polynomial $ch(x) = x^2 - x - 2 = (x-2)(x+1)$ has roots $\lambda_1 = 2, \lambda_2 = -1$, then $P_1(x) = 1 + x$, $P_2(x) = -2 + x$ give

$$B = \begin{bmatrix} 1 & 1 \\ -2 & 1 \end{bmatrix},$$

and from $P_1(2) = 3, P_2(-1) = -3$ we have

$$D = \frac{1}{3}\begin{bmatrix} 1 & 0 \\ 0 & -1 \end{bmatrix} \quad \text{and} \quad V^{-1} = \frac{1}{3}\begin{bmatrix} 1 & 1 \\ 2 & -1 \end{bmatrix}.$$

Therefore,

$$\begin{pmatrix} a_1 \\ a_2 \end{pmatrix} = V^{-1}\begin{pmatrix} 1 \\ 5 \end{pmatrix} = \frac{1}{3}\begin{bmatrix} 1 & 1 \\ 2 & -1 \end{bmatrix}\begin{pmatrix} 1 \\ 5 \end{pmatrix} = \begin{pmatrix} 2 \\ -1 \end{pmatrix},$$

and from (2.9),

$$s_n = 2\lambda_1^n - \lambda_2^n = 2^{n+1} + (-1)^{n+1}.$$

2.3.2 Repeated eigenvalues

In the previous section we showed that the vectors $\mathbf{v}_{\lambda_1}, \ldots, \mathbf{v}_{\lambda_k}$ form a basis for $\mathbb{C}^k$ when the eigenvalues are distinct. What happens when the recurrence has some eigenvalues that are repeated roots of the characteristic polynomial? The vectors $\mathbf{v}_{\lambda_i}$ still form a linearly independent set, but the set is no longer a basis, since it has fewer than k elements. Although we will do considerably more work to identify enough additional vectors to form a basis, the actual result (given in Theorem 2.3.6 below) is only slightly more complicated.

When λ is an eigenvalue of multiplicity m, there exists a polynomial $f(x)$ with complex coefficients such that

$$ch(x) = (x - \lambda)^m f(x), \quad \text{where } f(\lambda) \neq 0.$$

Denoting the differentiation operator on the space of polynomials by D, this implies that the value of $D^j(ch(x))$ at $x = \lambda$ equals zero for all $0 \leq j < m$ and is non-zero for $j = m$. Using the fact that

$$D^j(x^i) = \begin{cases} \dfrac{i!}{(i-j)!} x^{i-j} & \text{if } j \leq i, \\ 0 & \text{if } j > i, \end{cases}$$

in Exercise 2.13 you prove the polynomial identity

(2.11) $$D^j(x^i) = x D^j(x^{i-1}) + j D^{j-1}(x^{i-1}),$$

where D^0 is the identity operator and the second summand on the right side should be interpreted as zero when $j = 0$.

We next construct what are called the cyclic subspaces of $\mathbb{C}^k$. This study is motivated by another concept in linear algebra, the **Rational Canonical Form**, a topic that is more advanced than our review in Appendix C. Because our description is explicit, what follows can be viewed as an illustration of Rational Canonical Form. You can consult [78, Sections 7.1–7.2] if you're interested in more information.

The **cyclic subspace corresponding to the eigenvalue** λ is defined to be

(2.12) $$X_\lambda = \text{Span}\{\mathbf{v}_\lambda^{(0)}, \ldots, \mathbf{v}_\lambda^{(m-1)}\},$$

and the next result implies that the companion matrix A maps X_λ into X_λ, which means that A is an *operator* on the subspace X_λ. This will allow us to apply the theory of linear algebra to the vector space X_λ.

Proposition 2.3.3. *If λ is an eigenvalue of multiplicity m, then*

(2.13) $$A\mathbf{v}_\lambda^{(j)} = \lambda \mathbf{v}_\lambda^{(j)} + j \mathbf{v}_\lambda^{(j-1)} \quad \text{for all } 0 \leq j < m,$$

where $v_\lambda^{(-1)} = 0$, $v_\lambda^{(0)} = v_\lambda$, and for all $j \geq 1$, $v_\lambda^{(j)}$ denotes the element of $\mathbb{C}^k$ obtained by applying D^j to each component of $(x^{k-1}, \ldots, x, 1)^T$ and then evaluating each component at $x = \lambda$.

Proof. Setting $\mathbf{x} = (x^{k-1}, \ldots, x, 1)^T$, we will prove

$$(2.14) \qquad AD^j(\mathbf{x}) = xD^j(\mathbf{x}) + jD^{j-1}(\mathbf{x}) \text{ for all } j < m,$$

which yields (2.13). From the linearity of the operator D we obtain

$$AD^j(\mathbf{x}) = D^j(A\mathbf{x}).$$

Recall that for any vector $\mathbf{v}$, the last $k-1$ components of $A\mathbf{v}$ are obtained by shifting down the first $k-1$ components of $\mathbf{v}$. Also, for any $i \neq 1$, the ith component of $AD^j(\mathbf{x})$ is $D^j(x^{k-i+1})$, which from (2.11) equals $xD^j(x^{k-i}) + jD^{j-1}(x^{k-i})$. This proves the equality of the last $k-1$ components in the vector equation (2.14).

Now a comparison of the first components. The first component of $A\mathbf{x}$ is

$$c_1 x^{k-1} + \cdots + c_{k-1}x + c_k = x^k - ch(x),$$

which implies that the first component of $D^j(A\mathbf{x})$ is

$$D^j(x^k - ch(x)) = D^j(x^k) - D^j(ch(x))$$
$$= xD^j(x^{k-1}) + jD^{j-1}(x^{k-1}) - D^j(ch(x)),$$

again from (2.11). Equality of the first components in (2.14) is obtained from this and the fact that $D^j(ch(x))$ has value 0 at $x = \lambda$. $\qquad \square$

Theorem 2.3.4. *If λ is an eigenvalue of multiplicity m, then $(x - \lambda)^m$ is the minimal polynomial of the operator A on X_λ, and $S = \{v_\lambda^{(0)}, \ldots, v_\lambda^{(m-1)}\}$ is a basis for X_λ.*

Proof. From the definition of X_λ in (2.12), $\dim(X_\lambda) \leq m$ and S is a generating set for X_λ. It suffices to prove $\dim(X_\lambda) \geq m$. We do this by showing that the degree of $\min(x)$, the minimal polynomial of A restricted to the subspace X_λ, equals m. In fact, we prove that $\min(x) = (x - \lambda)^m$. To prove this we will show that for all $0 \leq j < m$,

$$(2.15) \qquad (A - \lambda I)^j v_\lambda^{(j)} \neq \mathbf{0} \text{ and } (A - \lambda I)^m v_\lambda^{(j)} = \mathbf{0}.$$

Fix $j < m$. Then (2.13) can be rewritten in the form

$$(A - \lambda I)v_\lambda^{(j)} = jv_\lambda^{(j-1)},$$

and repeated use of this gives

$$
\begin{aligned}
(A - \lambda I)^j \mathbf{v}_\lambda^{(j)} &= (A - \lambda I)^{j-1}((A - \lambda I)\mathbf{v}_\lambda^{(j)}) \\
&= j(A - \lambda I)^{j-1}\mathbf{v}_\lambda^{(j-1)} \\
&= j(A - \lambda I)^{j-2}((A - \lambda I)\mathbf{v}_\lambda^{(j-1)}) \\
&= j(j - 1) \cdot (A - \lambda I)^{j-2}\mathbf{v}_\lambda^{(j-2)} \\
&\ \ \vdots \\
&= j!\ \mathbf{v}_\lambda^{(0)}\, ,
\end{aligned}
$$

which means that

(2.16) $\qquad\qquad (A - \lambda I)^j \mathbf{v}_\lambda^{(j)} = j!\ \mathbf{v}_\lambda^{(0)}$ for all $j < m$.

Since $\mathbf{v}_\lambda^{(0)}$ is an eigenvector, it is non-zero, and therefore $(A - \lambda I)^j$ is not the zero transformation on X_λ. This proves that the minimal polynomial is not $(x - \lambda)^j$ for any $j < m$. On the other hand, from (2.16) we also obtain for all $j = 0, 1, \ldots, m - 1$,

$$
\begin{aligned}
(A - \lambda I)^m \mathbf{v}_\lambda^{(j)} &= j! \cdot (A - \lambda I)^{m-j}\mathbf{v}_\lambda^{(0)} \\
&= j! \cdot (A - \lambda I)^{m-j-1}((A - \lambda I)\mathbf{v}_\lambda^{(0)}) \\
&= (A - \lambda I)^{m-j-1}\mathbf{0} = \mathbf{0}\, ,
\end{aligned}
$$

since $\mathbf{v}_\lambda^{(0)} = \mathbf{v}_\lambda$ is an eigenvector corresponding to λ. Therefore, $(A - \lambda I)^m$ is the zero transformation on the basis $\{\mathbf{v}_\lambda^{(0)}, \ldots, \mathbf{v}_\lambda^{(m-1)}\}$, and by linearity, $(A - \lambda I)^m$ must be zero on all of X_λ. This proves $\min(x) = (x - \lambda)^m$. □

Corollary 2.3.5. *Let* $\lambda_1, \ldots, \lambda_t$ *be the different eigenvalues of the recurrence, with corresponding multiplicities* $m_1, \ldots, m_t$. *Then*

(2.17) $\qquad \{v_{\lambda_1}^{(0)}, \ldots, v_{\lambda_1}^{(m_1-1)}, \ldots, v_{\lambda_t}^{(0)}, \ldots, v_{\lambda_t}^{(m_t-1)}\}$

is a basis for $\mathbb{C}^k$.

Proof. For each i, let $X_i = X_{\lambda_i}$ and let $\min_i(x)$ denote the minimal polynomial of A restricted to X_i. For $i \neq j$, $X_i \cap X_j$ is an A-invariant subspace whose minimal polynomial divides each of $\min_i(x), \min_j(x)$ and so also divides their greatest common denominator. But $\lambda_i \neq \lambda_j$ gives $\gcd((x - \lambda_i)^m, (x - \lambda_j)^m) = 1$, implying $\min(x) = 1$ and $X_i \cap X_j = \{0\}$. Therefore,

$$
\begin{aligned}
\mathrm{Span}(X_1 \cup X_2 \cup \cdots \cup X_t) &= \mathrm{Span}(X_1) \cup \cdots \cup \mathrm{Span}(X_t) \\
&= X_1 \cup X_2 \cup \cdots \cup X_t\, ,
\end{aligned}
$$

giving

$$\dim(X_1 \cup X_2 \cup \cdots \cup X_t) = m_1 + \cdots + m_t,$$

which does equal $k = \deg(ch)$ since the characteristic polynomial is

$$ch(x) = (x - \lambda_1)^{m_1} \cdots (x - \lambda_t)^{m_t}.$$

$\square$

This basis is the simpler one we promised earlier, and using this basis in Theorem 2.1.1 gives the following result.

Theorem 2.3.6. *Let* $\lambda_1, \ldots, \lambda_t \in \mathbb{C}$ *be distinct. Then* $\langle s_n \rangle$ *is a solution to the recurrence (HL) with* $ch(x) = (x - \lambda_1)^{m_1} \cdots (x - \lambda_t)^{m_t}$ *iff the* n^{th} *term of* $\langle s_n \rangle$ *has the form*

(2.18) $$s_n = a_1(n)\lambda_1^n + \cdots + a_t(n)\lambda_t^n,$$

where each $a_i(x)$ *is a polynomial whose degree is less than* m_i.

Proof. In Exercise 2.15 you check that each sequence whose n^{th} term has the form given in (2.18) satisfies (HL). If $\{\mathbf{b}_1, \ldots, \mathbf{b}_k\}$ is the basis in (2.17), the fact that π is an isomorphism from $\mathcal{X}$ to $\mathbb{C}^k$ implies that every solution to (HL) has the form

(2.19) $$\langle s_n \rangle = \beta_1 \pi^{-1}(\mathbf{b}_1) + \cdots + \beta_k \pi^{-1}(\mathbf{b}_k),$$

for some complex constants $\beta_1, \ldots, \beta_k$. Each $\mathbf{b}_j$ equals $\mathbf{v}_\lambda^{(i)}$ for some eigenvalue λ, and i is less than the multiplicity of λ. The nth term of $\pi^{-1}(\mathbf{v}_\lambda^{(i)})$ is $D^i(x^n)$ evaluated at $x = \lambda$; that is, it equals

$$\begin{cases} \dfrac{n!}{(n-i)!}\lambda^{n-i} & \text{if } j \leq i, \\ 0 & \text{if } j > i, \end{cases}$$

where the coefficient $n!/(n-i)!$ is a polynomial in n whose degree equals i. Therefore, (2.19) yields the result. $\square$

As an illustration of the method consider the following initial value problem:

$$s_{n+4} = s_{n+3} + 3s_{n+2} - 5s_{n+1} + 2s_n \text{ with } s_0 = 1, s_1 = -1, s_2 = 0, s_3 = 1.$$

Then $ch(x) = x^4 - x^3 - 3x^2 + 5x - 2$, and its companion matrix is

$$A = \begin{bmatrix} 1 & 3 & -5 & 2 \\ 1 & 0 & 0 & 0 \\ 0 & 1 & 0 & 0 \\ 0 & 0 & 1 & 0 \end{bmatrix}$$

with eigenvalues $\lambda_1 = 1, \lambda_2 = -2$ and respective multiplicities $m_1 = 3$, $m_2 = 1$. Therefore, there exist polynomials $a_1(x), a_2(x)$ such that any solution $\langle s_n \rangle$ to the recurrence has general term

$$s_n = a_1(n)(1)^n + a_2(n)(-2)^n.$$

Further, $\deg(a_1) < 3$ and $\deg(a_2) < 1$ imply that there exist constants $\beta_i \in \mathbb{C}$ such that

$$a_1(x) = \beta_0 + \beta_1 x + \beta_2 x^2 \text{ and } a_2(x) = \beta_3,$$

which give

$$s_n = (\beta_0 + \beta_1 n + \beta_2 n^2) + \beta_3 (-2)^n.$$

Using the initial values, this reduces to solving a system of four equations in the unknowns $\beta_0, \beta_1, \beta_2, \beta_3$ with augmented matrix

$$\begin{bmatrix} 1 & 0 & 0 & 1 & 1 \\ 1 & 1 & 1 & -2 & -1 \\ 1 & 2 & 4 & 4 & 0 \\ 1 & 3 & 9 & -8 & 1 \end{bmatrix}.$$

Gaussian elimination can be used to obtain the row reduced echelon form

$$\begin{bmatrix} 1 & 0 & 0 & 0 & 8/9 \\ 0 & 1 & 0 & 0 & -8/3 \\ 0 & 0 & 1 & 0 & 1 \\ 0 & 0 & 0 & 1 & 1/9 \end{bmatrix},$$

and from this $a_1(x) = \frac{8}{9} - \frac{8}{3}x + x^2$ and $a_2(x) = \frac{1}{9}$, giving

$$s_n = \frac{8}{9} - \frac{8}{3}n + n^2 + \frac{1}{9}(-2)^n.$$

The $k \times k$ matrix V whose columns consist of the vectors $\mathbf{v}_\lambda^{(i)}$ (written as columns according to the order given in (2.17)) can be called the **generalized Vandermonde matrix** associated with the recurrence. In order to find the polynomials $a_1(x), \ldots, a_t(x)$ corresponding to the initial conditions $(s_0, s_1, \ldots, s_{k-1}) = \alpha$, we solve the matrix equation $V\mathbf{x} = \alpha$. As with the standard Vandermonde matrix, the fact that every initial value problem for (HL) has a unique solution translates to the invertibility of the generalized Vandermonde matrix. The explicit form of V^{-1} is found in Exercise 2.17.

2.4 The Asymptotic Behavior of Solutions

We will now use the matrix form $S_n = A^n S_0$ to estimate the asymptotic size of solutions to (HL) without computing A^n. The basic idea is to write the initial condition S_0 as a linear combination of k linearly independent vectors $C_1, \ldots, C_k$ for which the asymptotic size of each $A^n C_i$ is known.

We first consider the principal case for applications, the case in which A is a diagonalizable matrix. This means that there exists a basis $\{C_1, \ldots, C_k\}$ of eigenvectors for $\mathbb{C}^k$, and we can write S_0 as $S_0 = \sum_{i=1}^{k} a_i C_i$ for some $a_i \in \mathbb{C}$. If all of the dot products $C_j^T C_i = 0$ with $i \neq j$ happen to be zero (in this case the matrix A is said to be **orthogonally diagonalizable** since the basis is an orthogonal set), then

$$C_j^T S_0 = C_j^T \sum_{i=1}^{k} a_i C_i = \sum_{i=1}^{k} a_i C_j^T C_i = a_j \|C_j\|^2,$$

giving $a_j = C_j^T X_0 / \|C_j\|^2$, where $\|C\|$ is the Euclidean length of the vector C. Orthogonality therefore gives an easy expansion of S_0 in terms of the column eigenvectors of A. Although the orthogonality of the eigenvectors can't be guaranteed, we will see that a helpful property that we call biorthogonality always holds when A is diagonalizable.

If λ is an eigenvalue, then each of the pair of homogeneous systems $(A - \lambda I)X = 0$ and $X(A - \lambda I) = 0$ has a non-zero solution. For our purposes, we'll call a non-zero solution to $(A - \lambda I)X = 0$ a **column eigenvector**, while a non-zero solution to the second system will be referred to as a **row eigenvector**. Letting Z be the $k \times k$ matrix whose columns are the column eigenvectors $C_1, \ldots, C_k$ of A, then $AZ = ZD$, where D is the diagonal matrix with diagonal entries $\lambda_1, \ldots, \lambda_k$. From the linear independence of its columns, Z is an invertible matrix and $Z^{-1}A = DZ^{-1}$, which means that the ith row R_i of Z^{-1} is a row eigenvector corresponding to λ_i. From $Z^{-1}Z = I$ we also know that $R_i C_j = 0$ when $i \neq j$, and we call the two sets of vectors $\{R_1, \ldots, R_k\}$ and $\{C_1, \ldots, C_k\}$ a pair of **biorthogonal sets**. (Also note that $R_i C_i = 1$ for all i.) Mimicking the orthogonal case, from $S_0 = a_1 C_1 + \cdots + a_k C_k$ we have

$$R_i S_0 = R_i(a_1 C_1 + \cdots + a_k C_k) = a_i R_i C_i = a_i.$$

Therefore, $S_0 = \sum_{i=1}^{k} (R_i S_0) C_i$, and

$$(2.20) \qquad S_n = A^n S_0 = \sum_{i=1}^{k} (R_i S_0)(A^n C_i),$$

which we record in the following lemma.

Lemma 2.4.1. *Let A be any diagonalizable $k \times k$ matrix. Let Z be the matrix whose columns are linearly independent eigenvectors $C_1, \ldots, C_k$, and let $R_1, \ldots, R_k$ be the rows of Z^{-1}. Then any solution to $S_{n+1} = AS_n$ can be written in the form*

$$S_n = \sum_{i=1}^{k} (R_i S_0) \lambda_i^n C_i.$$

When A has fewer than k linearly independent eigenvectors, then the matrix Z whose columns are the **generalized eigenvectors** (which form a basis for $\mathbb{C}^k$) still satisfies (2.20) where $R_1, \ldots, R_k$ are the rows of the matrix Z^{-1}.

When we order the *different* eigenvalues according to decreasing complex modulus,

(2.21) $$|\lambda_1| \geq |\lambda_2| \geq \cdots \geq |\lambda_t| \quad \text{(where } t \leq k\text{)},$$

λ_1 is called a **dominant eigenvalue of the recurrence.** If $|\lambda_1| > |\lambda_2|$ holds, then $|\lambda_1|$ is called the **strictly dominant eigenvalue.** (Note that strict dominance does not prevent λ_1 from being a multiple eigenvalue.) For the case in which λ_1 is a *simple strictly dominant* eigenvalue, we have

$$S_n = A^n S_0 = \sum_{i=1}^k (R_i S_0)(A^n C_i) = (R_1 S_0)\lambda_1^n C_1 + \sum_{i=2}^k (R_i S_0) A^n C_i \,,$$

and so

$$\frac{S_n}{\lambda_1^n} = (R_1 S_0) C_1 + Y_n \,,$$

where $Y_n = \sum_{i=2}^k (R_i S_0) A^n C_i / \lambda_1^n$. Since each coordinate $|Y_n|_i$ of $|Y_n|$ is bounded above by $|\lambda_2|^n$ times a polynomial in n (refer to Theorem 2.3.6), the limiting value of $|Y_n|_i / \lambda_1^n$ exists and is zero. Therefore,

$$\lim_{n \to \infty} \frac{S_n}{\lambda_1^n} = (R_1 S_0) C_1,$$

where C_1 is a column eigenvector of λ_1 and R_1 is the specific row eigenvector of λ_1 chosen above. For any other row eigenvector R corresponding to λ_1, $(RS_0)/(RC_1) = R_1 S_0$ holds, and we obtain the following theorem.

Theorem 2.4.2. *If λ_1 is a simple strictly dominant eigenvalue of the $k \times k$ matrix A, then*

$$\lim_{n \to \infty} \frac{S_n}{\lambda_1^n} = aC \,,$$

where C is any column eigenvector corresponding to λ_1, R is any row eigenvector corresponding to λ_1, and $a = (RS_0)/(RC)$, a quotient of dot products.

The upshot of this result is that for many reasonable homogeneous recurrences and most initial conditions, the asymptotic behavior of a solution can be considered to be one-dimensional. Full details of a solution would require a somewhat more complicated formula, but as long as the dot product RS_0 is non-zero, the solution converges in a ratio sense to a single vector. When the dot product RS_0 is zero, the theorem says only that the solution grows more slowly than $|\lambda_1|^n$, and it gives no more information about the solution. The other main point is that the value of a can be computed without computing the rest of the coefficients in the full expansion of the solution.

2.5 Exercises

Ex 2.1. Show that every initial value problem has a unique solution.

Ex 2.2. Use induction to verify the closed formulas for the recurrence (2.1) under each of the following initial conditions: $S_0 = (3, 2)^T$; $S_0 = (1, 3)^T$.

Ex 2.3. Suppose that the sequence $\langle s_n \rangle$ satisfies the Fibonacci recurrence and $s_3 = 5, s_6 = -2$. What is s_4? What is s_{10}? What are the initial values s_0 and s_1?

Ex 2.4. Construct an argument that concludes there is a sequence $\langle s_n \rangle$ that satisfies the Fibonacci recurrence for which $s_{20} = 23$ and $s_{1000} = 56$.

Ex 2.5. Show directly that any linear combination of solutions to (HL) is itself a solution.

Ex 2.6. For a fixed polynomial Q in $k \geq 2$ variables consider the set V of all sequences $\langle s_n \rangle$ that are solutions to the kth order recurrence

$$s_{n+k} = Q(s_n, s_{n+1}, \ldots, s_{n+k-1}).$$

If V is a vector space, show that Q must be linear with zero constant term.

We know that the set of all sequences of complex numbers forms a complex vector space under the operations of addition and scaling. Applying the definition of linear independence (refer to Appendix C) to this vector space, we see that sequences $\phi_1(n), \ldots, \phi_k(n)$ are **linearly independent** iff

$$b_1 \phi_1(n) + \cdots + b_k \phi_k(n) = 0 \text{ for all } n \geq 0 \implies b_1 = b_2 = \cdots = b_k = 0.$$

Ex 2.7. Show that every k^{th} order homogeneous linear recurrence has k linearly independent solutions.

Ex 2.8. Let $\phi_1(n), \ldots, \phi_k(n)$ be k linearly independent solutions to a kth order homogeneous linear recurrence. Show that the solution to any initial value problem for this recurrence can be written in the form $\sum_{i=1}^{k} b_i \phi_i(n)$.

Ex 2.9. Let $\phi_1(n), \ldots, \phi_k(n)$ be solutions to a kth order linear recurrence. Show that the following three statements are equivalent:
   (a) $\sum_{i=1}^{k} b_i \phi_i(n)$ is the zero sequence.
   (b) For all $0 \leq n < k$, $\sum_{i=1}^{k} b_i \phi_i(n) = 0$.
   (c) $\sum_{i=1}^{k} b_i \phi_i(n) = 0$ for k consecutive values of n.

Ex 2.10. Use diagonalization of the Fibonacci matrix to obtain Binet's Formula (1.5), the closed form for the Fibonacci numbers.

Ex 2.11. Let $s_{n+2} = c_1 s_{n+1} + c_2 s_n$ be a second–order recurrence with eigenvalues $\lambda_1 = (c_1 + \sqrt{D})/2$ and $\lambda_2 = (c_1 - \sqrt{D})/2$, where $D = c_1^2 + 4c_2$. If $D \neq 0$, show that the sequence whose n^{th} term is

$$\frac{1}{\sqrt{D}}\left(-\alpha_0 \lambda_1 \lambda_2 (\lambda_1^{n-1} - \lambda_2^{n-1}) + \alpha_1 (\lambda_1^n - \lambda_2^n)\right)$$

is the unique solution with initial conditions $s_0 = \alpha_0$, $s_1 = \alpha_1$.

Ex 2.12. Show that the Vandermonde matrix associated with $\lambda_1, \ldots, \lambda_k$ is invertible iff the λ_i are distinct.
Hint: Premultiply the matrix by a row vector where the product for each entry is interpreted as the evaluation of a polynomial at λ_i.

Ex 2.13. Use the product rule for differentiation and induction to substantiate identity (2.11).

Ex 2.14. Solve the initial value problem

$$s_{n+3} = 2s_{n+2} + 5s_{n+1} - 6s_n \,; \quad s_0 = 9, \, s_1 = -18, \, s_2 = 66.$$

Ex 2.15. Check that each sequence whose n^{th} term has the form given in (2.18) satisfies the recurrence (HL).

Ex 2.16. Use Theorem 2.3.6 to find the solution to

$$s_{n+4} = 8s_{n+2} - 16 \,, \, s_0 = -1, \, s_1 = 8, \, s_2 = 4, \, s_3 = 16 \,.$$

Ex 2.17. Let $\lambda_1, \ldots, \lambda_t$ be distinct complex numbers and let $m_1, \ldots, m_t$ be any positive integers that sum to k. Let V be the $k \times k$ **generalized Vandermonde matrix** associated with this data, the matrix V whose columns are the vectors given in (2.17).
(a) Construct an argument that proves V is invertible.
(b) If A is the companion matrix for

$$P(x) = (x - \lambda_1)^{m_1} (x - \lambda_2)^{m_2} \cdots (x - \lambda_t)^{m_t} \,,$$

compute $V^{-1}AV$.
(c) For $1 \leq i \leq t, 1 \leq j \leq m_i$, define $Q(x) = \dfrac{P(x)}{(x - \lambda_i)^j}$, and expand it as a polynomial of degree $k - 1$ in x:

$$Q(x) = b_0 + b_1 x + \cdots + b_{k-j} x^{k-j} + \cdots + b_{k-1} x^{k-1} \,.$$

Show that $\deg(Q) \leq k - j$.
(d) Show that

$$(b_0, b_1, \ldots, b_{k-1}) \cdot v_{\lambda_i}^{(t)} \begin{cases} \neq 0 & \text{if } j = t, \\ = 0 & \text{if } j < t. \end{cases}$$

(e) Compute the precise value of the dot product in (d).

(f) Use the information from the previous parts to construct the inverse of V.

Ex 2.18. Let $s_{n+2} = c_1 s_{n+1} + c_2 s_{n+2}$ be a second–order recurrence such that $D = c_1^2 + 4c_2$ is zero. Using the techniques of Section 2.3.2, show that the sequence with general term

$$s_n = \alpha_0 \left(\frac{c_1}{2}\right)^n + n\left(\alpha_1 - \frac{\alpha_0 c_1}{2}\right)\left(\frac{c_1}{2}\right)^{n-1}$$

is the unique solution to (HL) with $s_0 = \alpha_0$ and $s_1 = \alpha_1$.

Ex 2.19. Write a general procedure for solving (HL) given knowledge of the roots of the characteristic polynomials and their multiplicities.

Ex 2.20. Show that

$$\lim_{n\to\infty} \frac{a(n+1)}{a(n)} = 1$$

holds for any polynomial $a(x) \in \mathbb{C}[x]$.

The next two exercises assume the ordering of eigenvalues given in (2.21).

Ex 2.21. Consider a solution $\langle s_n \rangle$ to the homogeneous linear recurrence $s_{n+k} = c_1 s_{n+k-1} + c_2 s_{n+k-2} + \cdots + c_k s_n$.
  (a) Show that if $|\lambda_1| < 1$, then $\lim_{n\to\infty} s_n = 0$.
  (b) Show that if the coefficients in (HL) are *integers* and $c_k \neq 0$, then $|\lambda_1| \geq 1$.

Ex 2.22. Let $s_{n+2} = c_1 s_{n+1} + c_2 s_n$ be a second–order recurrence with $c_1 \neq 0$ and $c_1^2 + 4c_2 > 0$.
  (a) Show that λ_1 is strictly dominant.
  (b) Prove that for any $\langle s_n \rangle \in X$ the limit $\lim_{n\to\infty} \dfrac{s_n}{\lambda_1^n}$ exists. Moreover, when $s_1 \neq s_0 \lambda_2$, show that this limit must be non-zero.

Ex 2.23. For each $n \geq 1$ let D_n be the determinant of the $n \times n$ tridiagonal matrix

$$\begin{bmatrix} \alpha & 1 & 0 & 0 & \cdots & 0 & 0 & 0 \\ 1 & \alpha & 1 & 0 & \cdots & 0 & 0 & 0 \\ 0 & 1 & \alpha & 1 & \cdots & 0 & 0 & 0 \\ \vdots & & \ddots & \ddots & \ddots & & & \vdots \\ \vdots & & & \ddots & \ddots & \ddots & & \vdots \\ 0 & 0 & 0 & 0 & \cdots & \alpha & 1 & 0 \\ 0 & 0 & 0 & 0 & \cdots & 1 & \alpha & 1 \\ 0 & 0 & 0 & 0 & \cdots & 0 & 1 & \alpha \end{bmatrix},$$

where α is some fixed real number. Show that the sequence $\langle D_n \rangle$ satisfies the second–order recurrence $D_n = \alpha D_{n-1} - D_{n-2}$ with initial conditions

$D_0 = 1$ and $D_1 = \alpha$. Find the values of α for which $\lim_{n \to \infty} |D_n| = \infty$. For what values of α is $D_n = O(n)$? What is the asymptotic size of D_n for other values of α? Find the values of α for which D_n is periodic, and decide what periods are possible.

3

Finite Difference Equations

3.1 Linear Difference Equations

A difference equation is the discrete analog of a differential equation. Although differential equations are typically studied earlier in a mathematical curriculum, there are many respects in which the theory of difference equations is simpler. A **finite difference equation** has the general form

$$(3.1) \qquad s_n = \Phi(s_{n-1}, s_{n-2}, \ldots, s_{n-k}, n),$$

where Φ is a fixed complex-valued function and the integer k is called the **order** of the equation. The **initial value problem** for (3.1) is the problem of finding a sequence $\langle s_n \rangle$ that satisfies (3.1) for a given function Φ and a fixed initial value vector $S_0 = (s_{k-1}, \ldots, s_1, s_0)^T$. An initial value problem has only one solution, because (as in Chapter 2) for any fixed k initial values $s_0, \ldots, s_{k-1}$, the kth term is specified by $s_k = \Phi(s_{k-1}, s_{k-2}, \ldots, s_0, k)$, and all successive values are similarly found. From this we obtain the following result.

Theorem 3.1.1 (Existence and uniqueness theorem). *Every initial value problem for a finite difference equation has a unique solution.*

The essential ingredient in the proof of this result is the assumption that Φ is a *function*, that is, that Φ returns exactly one value for a given input.

The function Φ is called a **linear function** in the variables $s_{n-1}, \ldots, s_{n-k}$ if there exist $k+1$ complex-valued functions $g_1, \ldots, g_k$ and ψ defined on the natural numbers such that for all $n \geq k$,

$$\Phi(s_{n-1}, s_{n-2}, \ldots, s_{n-k}, n) = g_1(n)s_{n-1} + \cdots + g_k(n)s_{n-k} + \psi(n),$$

where ψ is called the **input** or **forcing function**. When Φ is a linear function, then (3.1) is called a **linear difference equation**. For the special case in which the coefficients $g_1, \ldots, g_k$ are constants, (3.1) becomes

$$s_n = c_1 s_{n-1} + \cdots + c_k s_{n-k} + \psi(n) \quad \text{for} \ \ c_1, \ldots, c_k \in \mathbb{C},$$

and it is called a **constant coefficient equation.**

In this chapter we study linear constant coefficient difference equations, and we will simply refer to them as difference equations or recurrences. Every such recurrence can be written in the form

(L) $$s_n - c_1 s_{n-1} - c_2 s_{n-2} - \cdots - c_k s_{n-k} = \psi(n) \ \text{for} \ n \geq k,$$

for some complex-valued function ψ and constants $c_1, \ldots, c_k$, where c_k is non-zero. The *homogeneous* linear constant coefficient equation (HL) studied in Chapter 2 is the special case in which the input function $\psi(x)$ is the zero function.

3.1.1   First–order equations

A **first–order recurrence** has the form

(L1) $$s_n = \lambda s_{n-1} + \psi(n) \qquad \text{for all } n \geq 1 ,$$

where the function ψ is defined for all positive natural numbers and $\lambda \neq 0$. When we consider the associated initial value problem for fixed $s_0 = \alpha_0$, we have

$$s_0 = \alpha_0;$$
$$s_1 = \lambda s_0 + \psi(1) = \lambda \alpha_0 + \psi(1);$$
$$s_2 = \lambda s_1 + \psi(2) = \lambda^2 \alpha_0 + \lambda \psi(1) + \psi(2);$$
$$s_3 = \lambda s_2 + \psi(3) = \lambda^3 \alpha_0 + \lambda^2 \psi(1) + \lambda \psi(2) + \psi(3);$$

$$\vdots$$

and this gives the following form for the solution.

Theorem 3.1.2. *The initial value problem*

(3.2) $$s_n = \lambda s_{n-1} + \psi(n), \quad s_0 = \alpha_0 ,$$

always has the unique solution $\langle s_n \rangle$ *with general term*

(3.3) $$s_n = \alpha_0 \lambda^n + \sum_{i=1}^{n} \psi(i) \lambda^{n-i}$$

(where the sum is defined to be zero when $n = 0$).

Proof. Since we already know that (3.2) has a unique solution, we will show by induction on n that the solution has the form given in (3.3).

For $n = 0$, observe that (3.3) reduces to $s_0 = \alpha_0$, which is the initial condition. For the induction step, assume that (3.3) does satisfy (3.2) for all $n = 0, \ldots, K - 1$ and then use the defining equation (3.2) to find s_K from s_{K-1} and $\psi(K)$:

$$s_K = \lambda s_{K-1} + \psi(K)$$
$$= \lambda \left(\alpha_0 \lambda^{K-1} + \sum_{i=1}^{K-1} \psi(i) \lambda^{K-1-i} \right) + \psi(K)$$
$$= \alpha_0 \lambda^K + \sum_{i=1}^{K} \psi(i) \lambda^{K-i},$$

as required.   □

This proof is remarkably simple when compared to the corresponding theorem for differential equations. To solve a first–order initial value problem in differential equations,

$$D(s(t)) + c\, s(t) = \psi(t), \quad s(0) = \alpha_0,$$

it's necessary to multiply through by an "integrating factor" e^{ct} so that the equation becomes

$$D(s(t))\, e^{ct} + c\, s(t)\, e^{ct} = D\left(e^{ct} s(t) \right) = e^{ct} \psi(t),$$

which has the solution

$$s(t) = e^{-ct} \left(\int_0^t e^{ct} \psi(t)\, dt \right) + e^{-ct} \alpha_0 \,,$$

provided the integral exists. From this we see several ways in which differential equations are more difficult: the existence of a solution is not guaranteed, finding a solution is harder, and some theory is needed to deal with uniqueness.

Let's look at the simplest example of a nonhomogeneous first–order recurrence, when the input $\psi(n) = c$ is a constant,

(3.4) $$s_n = \lambda s_{n-1} + c, \quad s_0 = \alpha_0 \,.$$

From (3.3) we have

$$s_n = \alpha_0 \lambda^n + c(\lambda^{n-1} + \lambda^{n-2} + \cdots + 1),$$

which can be computed to be

(3.5) $$s_n = \begin{cases} \alpha_0 \lambda^n + c \dfrac{1 - \lambda^n}{1 - \lambda} & \text{if } \lambda \neq 1, \\ \alpha_0 + cn & \text{if } \lambda = 1. \end{cases}$$

The difficulty with the form of solution in Equation 3.3 is that it contains a summation, and we would prefer solutions without summations. As the example shows, some summations are easy to replace. In particular, any summation of the form

$$\sum_{i=0}^{n} r^i$$

(which is usually called a geometric series) can easily be replaced by a closed form formula. More generally if $\psi(n)$ can be written as

$$\psi(n) = b_1\gamma_1^n + b_2\gamma_2^n + \cdots + b_l\gamma_l^n,$$

then

$$\sum_{i=1}^{n} \psi(i)\lambda^{n-i} = \sum_{i=0}^{n-1} \psi(n-i)\,\lambda^i$$

$$= \sum_{i=0}^{n-1} \lambda^i \left(\sum_{j=1}^{l} b_j\gamma_j^{n-i}\right)$$

$$= \sum_{j=1}^{l} b_j\gamma_j^n \sum_{i=0}^{n-1} \left(\frac{\lambda}{\gamma_j}\right)^i = \sum_{j=1}^{l} b_j\gamma_j^n \left(\frac{(\frac{\lambda}{\gamma_j})^n - 1}{(\frac{\lambda}{\gamma_j}) - 1}\right)$$

$$= \sum_{j=1}^{l} b_j\gamma_j \left(\frac{\lambda^n - \gamma_j^n}{\lambda - \gamma_j}\right) \qquad (\text{ if } \lambda \neq \gamma_j \text{ for each } j).$$

Of course, a similar formula is possible when some γ_j satisfies $\gamma_j = \lambda$. Further, not much more complicated formulas are possible when $\psi(n)$ has the form

$$p_1(n)\gamma_1^n + p_2(n)\gamma_2^n + \cdots + p_l(n)\gamma_l^n,$$

where each $p_i(n)$ is a polynomial in n. Instead of pursuing these summations in the context of first–order equations, we will move on to k^{th} order equations, and in Section 3.3 we will deal with inputs that have this special form.

3.2   General and Particular Solutions

Because our difference equations are linear, we can conveniently rewrite

$$s_n = c_1 s_{n-1} + c_2 s_{n-2} + \cdots + c_k s_{n-k} + \psi(n)$$

by using

$$L[s_n] = s_n - c_1 s_{n-1} - c_2 s_{n-2} - \cdots - c_k s_{n-k}$$

and writing

$$L[s_n] = \psi(n).$$

When we consider solutions to the homogeneous version of this recurrence

$$L[s_n] = 0,$$

we have

if $L[x_n] = 0$ and $L[y_n] = 0$,
then $L[x_n + y_n] = L[x_n] + L[y_n] = 0$,
and $L[c\,x_n] = c\,L[x_n] = 0$.

Because of these properties we call $L[\ ]$ a **linear operator**. (Refer to Appendix C.)

If we have k linearly independent sequences

$$\langle x_n^{(1)} \rangle, \langle x_n^{(2)} \rangle, \ldots, \langle x_n^{(k)} \rangle$$

such that

$$L[x_n^{(1)}] = \cdots = L[x_n^{(k)}] = 0,$$

then for every choice of constants $\alpha_1, \ldots, \alpha_k$,

$$L\left[\sum_{i=1}^{k} \alpha_i x_n^{(i)}\right] = 0.$$

Since we can choose the α_i's, we can force $L[\sum_{i=1}^{k} \alpha_i x_n^{(i)}]$ to satisfy *any* set of k initial conditions. Specifically, if the initial conditions are $s_0, s_1, \ldots, s_{k-1}$, then

(3.6)
$$\sum_{i=1}^{k} \alpha_i x_0^{(i)} = s_0$$

$$\sum_{i=1}^{k} \alpha_i x_1^{(i)} = s_1$$

$$\vdots$$

$$\sum_{i=1}^{k} \alpha_i x_{k-1}^{(i)} = s_{k-1}$$

and by solving this system of k linear equations in k unknowns we can calculate $\alpha_1, \ldots, \alpha_k$. The assumption that the sequences are linearly independent assures us that this set of linear equations has a unique solution.

Linearity will also allow us to break the the problem of solving

$$L[s_n] = \psi(n)$$

into two parts. We let $s_n = h_n + v_n$ with

$$L[h_n] = 0 \quad \text{and} \quad L[v_n] = \psi(n),$$

and by linearity

$$L[s_n] = L[h_n + v_n] = \psi(n).$$

We are not ready to use the initial conditions. We want to find a v_n that satisfies

$$L[v_n] = \psi(n),$$

but v_n does not need to satisfy the initial conditions. We call such a v_n a **particular solution**. In the next section we will look at some methods for actually finding a particular solution.

As we saw above, if we can find k linearly independent sequences

$$\langle x_n^{(1)} \rangle, \langle x_n^{(2)} \rangle, \cdots, \langle x_n^{(k)} \rangle$$

such that

$$L[x_n^{(1)}] = \cdots = L[x_n^{(k)}] = 0,$$

then for for every choice of α_i's we have

$$L[\sum_{i=1}^{k} \alpha_i x_n^{(i)}] = 0.$$

Since this summation represents a whole family of solutions to the homogeneous difference equation, we call

$$\sum_{i=1}^{k} \alpha_i x_n^{(i)} = g_n$$

the **general solution** to the difference equation and denote this solution by $\langle g_n \rangle$. Then

$$L[g_n + v_n] = L[g_n] + L[v_n] = 0 + \psi(n) = \psi(n),$$

and we can state that every solution of the difference equation can be written as the sum of the *general* solution and a *particular* solution.

Actually, this does not yet solve the initial value problem because g_n still has k unspecified coefficients, $\alpha_1, \alpha_2, \ldots, \alpha_k$. If we choose these coefficients such that $g_n + v_n$ satisfies the initial conditions, we will be done. But this is the same process as solving the system of linear equations (3.6), with the difference that we subtract the initial values of v_n from the original initial conditions. The system of equations to be solved is

$$\sum_{i=1}^{k} \alpha_i x_0^{(i)} = s_0 - v_0$$

$$\sum_{i=1}^{k} \alpha_i x_1^{(i)} = s_1 - v_1$$

$$\vdots$$

$$\sum_{i=1}^{k} \alpha_i x_n^{(i)} = s_{k-1} - v_{k-1}.$$

Letting $\langle h_n \rangle$ be the solution with these computed values for the α_i's gives

$$L[\, h_n + v_n \,] = \psi(n),$$

and $\langle h_n + v_n \rangle$ satisfies the initial conditions.

   We may also view this solution process as

(a) Find any $\langle v_n \rangle$ that satisfies

$$L[\, v_n \,] = \psi(n).$$

(b) Find $\langle h_n \rangle$ such that $\langle h_n + v_n \rangle$ satisfies the initial conditions and

$$L[\, h_n \,] = 0.$$

In this process the second step is simply solving an initial value problem for a homogeneous difference equation.

3.2.1   Finding a particular solution via summation

We have proved (Theorem 3.1.2) that the sequence $\langle v_n \rangle$ defined by

$$v_0 = 0; \quad v_n = \sum_{i=1}^{n} \psi(i)\lambda^{n-i} = (\psi(1), \dots, \psi(n)) \cdot (\lambda^{n-1}, \dots, \lambda, 1)$$

(where the last expression should be read as a summation or dot product) is a particular solution to the first–order equation

$$s_n = \lambda s_{n-1} + \psi(n).$$

The solution can be rewritten as the dot product

$$v_n = (\psi(1), \dots, \psi(n)) \cdot (t_{n-1}, \dots, t_0),$$

where $t_n = \lambda^n$ is the solution to the homogeneous initial value problem

$$s_n = \lambda s_n \quad \text{with} \quad s_0 = 1.$$

We can generalize this formula to general k^{th} order difference equations.

Theorem 3.2.1. *Let $\langle t_n \rangle$ be the solution to the kth order homogeneous initial value problem*

$$s_n = c_1 s_{n-1} + c_2 s_{n-2} + \cdots + c_k s_{n-k} \quad \text{with} \quad (s_{k-1}, \ldots, s_0) = (1, 0, \ldots, 0).$$

Then for any sequence $\psi(n)$, the sequence $\langle v_n \rangle$ given by

$$v_0 = \cdots = v_{k-2} = 0, \quad v_{k-1} = 1, \quad v_n = (\psi(k), \ldots, \psi(n)) \cdot (t_{n-1}, \ldots, t_{k-1})$$

is a particular solution to the difference equation

$$(3.7) \qquad s_n = c_1 s_{n-1} + c_2 s_{n-2} + \cdots + c_k s_{n-k} + \psi(n).$$

Proof. Since $v_0 = \cdots = v_{k-2} = 0$, the value $v_k = \psi(k) t_{k-1} = \psi(k)$ does satisfy (3.7) for $n = k$. For $n > k$, since $t_0 = \cdots = t_{k-2} = 0$ then

$$c_1 v_{n-1} + c_2 v_{n-2} + \cdots + c_k v_{n-k}$$

equals

$$c_1(t_{n-2}\psi(k) + \cdots + t_k\,\psi(n-2) + t_{k-1}\psi(n-1))$$
$$+ c_2(t_{n-3}\psi(k) + \cdots + t_{k-1}\psi(n-2) + t_{k-2}\psi(n-1))$$
$$+ \cdots$$
$$+ c_k(t_{n-1-k}\psi(k) + \cdots + t_1\,\psi(n-2) + t_0\,\psi(n-1)).$$

Adding down the "columns", this becomes

$$\left(\sum_{i=1}^{k} c_i t_{n-1-i}\right)\psi(k) + \cdots + \left(\sum_{i=1}^{k} c_i t_{k-i}\right)\psi(n-1)$$
$$= t_{n-1}\psi(k) + \cdots + t_k\psi(n-1)$$
$$= (t_{n-1}, \ldots, t_k, 1) \cdot (\psi(k), \ldots, \psi(n-1), \psi(n)) - \psi\,i(n)$$
$$= v_n - \psi(n)$$

by definition of v_n. By induction, $\langle v_n \rangle$ is a solution to

$$v_n = c_1 v_{n-1} + \cdots + c_k v_{n-k} + \psi(n).$$

$\square$

Let's find a particular solution to the second order equation

$$(3.8) \qquad s_n = 2s_{n-1} - s_{n-2} + 2^n.$$

As in the theorem, if we let $\langle t_n \rangle$ be the solution to the associated homogeneous initial value problem

$$s_n = 2s_{n-1} - s_{n-2}; \quad (s_1, s_0) = (1, 0),$$

its first few terms are $t_0 = 0, t_1 = 1, t_2 = 2, t_3 = 3, \ldots$, and $t_n = n$ can be proved in general by induction. Then $\psi(n) = 2^n$ gives

$$
\begin{aligned}
v_n &= (\psi(2), \ldots, \psi(n)) \cdot (n-1, \ldots, 1) \\
&= (2^2, \ldots, 2^{n-1}, 2^n) \cdot (n-1, \ldots, 2, 1) \\
&= \sum_{j=2}^{n} 2^j \, (n+1-j)
\end{aligned}
$$

as a particular solution of (3.8). In this case it's relatively easy to get a closed form for $\langle v_n \rangle$ because the sequence satisfies a first–order equation since

$$
\begin{aligned}
v_n &= (2^2, \ldots, 2^n) \cdot (1, \ldots, 1) + (2^2, \ldots, 2^{n-1}, 2^n) \cdot (n-2, \ldots, 1, 0) \\
&= 2^2(1 + \cdots + 2^{n-2}) + v_{n-1} = 2^2(2^{n-1} - 1) + v_{n-1} .
\end{aligned}
$$

Therefore, using $\lambda = 1$ and $\psi(n) = 4(2^{n-1} - 1)$, $\langle v_n \rangle$ is a solution to $v_n = \lambda v_{n-1} + \psi(n)$, and from the result for first–order equations,

$$
\begin{aligned}
v_n &= (\psi(1), \ldots, \psi(n)) \cdot (\lambda^{n-1}, \ldots, \lambda) \\
&= 4 \sum_{i=0}^{n-1} (2^i - 1) = 4(2^n - 1) - 4n = 4(2^n - 1 - n)
\end{aligned}
$$

is a particular solution to (3.8). Notice that determining this closed form relied on being able to compute t_n easily as well as on seeing a linear relationship between consecutive elements of the constructed particular solution $\langle v_n \rangle$.

3.3   A Special Class of Linear Recurrences

We will now generalize the last section and develop an efficient method for obtaining a particular solution for a frequently occurring class of linear recurrences, the ones whose forcing function is $\psi(n) = \lambda^n p(n)$ where λ is a constant and $p(n)$ is a polynomial.

Theorem 3.3.1. *For any polynomial $p(x)$ and any constant λ, consider the linear recurrence*

$$s_n = c_1 s_{n-1} + \cdots + c_k s_{n-k} + \lambda^n p(n).$$

Let $\lambda_1, \ldots, \lambda_t$ be the eigenvalues of the recurrence with respective multiplicities $m_1, \ldots, m_t$ and define δ by

$$\delta = \begin{cases} 0 & \text{if } \lambda \notin \{\lambda_1, \ldots, \lambda_t\}, \\ m_i & \text{if } \lambda = \lambda_i. \end{cases}$$

If $\deg(p) = d$, then there exists a polynomial $q(x)$ with $\deg(q) \leq d$ such that $v_n = \lambda^n n^\delta q(n)$ is a particular solution of the recurrence. Moreover, the polynomial $q(x)$ can be calculated in $O(d^2)$ operations.

Proof. Our proof actually constructs the polynomial q. We may assume that $\lambda \neq 0$, because if $\lambda = 0$, the equation is homogeneous, and $v_n = 0$ satisfies the recurrence. Then $\lambda^n n^\delta q(n)$ is a solution and for all integers $n \geq 0$,

$$\lambda^n n^\delta q(n) - c_1 \lambda^{n-1}(n-1)^\delta q(n-1)$$
$$-c_2 \lambda^{n-2}(n-2)^\delta q(n-2) - \cdots - c_k \lambda^{n-k}(n-k)^\delta q(n-k) = \lambda^n p(n).$$

Dividing both sides by the non-zero λ^{n-k}, the equation becomes a polynomial equation in n, which holds for the (infinite) set of natural numbers, and so in fact, the equation is the polynomial identity

$$\lambda^k x^\delta q(x) - c_1 \lambda^{k-1}(x-1)^\delta q(x-1) - \cdots - c_k(x-k)^\delta q(x-k) = \lambda^k p(x).$$

Recall that $p(x) = p_d x^d + \cdots + p_1 x + p_0$ is a fixed polynomial and our goal is to construct coefficients $q_d, \ldots, q_1, q_0$ such that $q(x) = q_d x^d + \cdots + q_1 x + q_0$ satisfies this identity. Computing a few coefficients, the coefficient of $x^{d+\delta}$ on the left side is

$$\lambda^k q_d - c_1 \lambda^{k-1} q_d - \cdots - c_k q_d,$$

which is $ch(\lambda) q_d$, and the coefficient of $x^{\delta+d-1}$ is

$$\lambda^k q_{d-1} - c_1 \lambda^{k-1}\left((-1)\binom{d+\delta}{1} q_d + q_{d-1}\right)$$
$$- c_2 \lambda^{k-2}\left((-2)\binom{d+\delta}{1} q_d + q_{d-1}\right)$$
$$- \cdots - c_k\left((-k)\binom{d+\delta}{1} q_d + q_{d-1}\right),$$

which we can write as $-t_1\binom{d+\delta}{1}q_d + t_0 q_{d-1}$ using $t_0 = ch(\lambda)$ and $t_1 = \sum_{j=1}^{k} c_j(-j)\lambda^{k-j}$. In general, defining $t_i = \sum_{j=1}^{k} c_j(-j)^i \lambda^{k-j}$ for all $i = 0, 1, \ldots, d$, it can be shown that the coefficient of $x^{\delta+d-j}$ on the left side is

$$-t_j\binom{\delta+d}{j}q_d - \cdots - t_1\binom{\delta+d-(j-1)}{1}q_{d-(j-1)} + t_0 q_{d-j}.$$

We first consider the case in which λ is not an eigenvalue. Then $t_0 = ch(\lambda)$ is non-zero, $\delta = 0$, and the requirement becomes the dot product

$$\left(-t_j\binom{d}{j}, \ldots, -t_1\binom{d-(j-1)}{1}, t_0\right) \cdot (q_d, \ldots, q_{d-(j-1)}, q_{d-j}) = \lambda^k p_j,$$

for all $j = 0, \ldots, d$, a system of $d+1$ equations in the unknowns $q_d, \ldots, q_1, q_0$. The coefficient matrix is a lower triangular matrix in which each diagonal entry is the non-zero t_0, which means that the system has a unique solution. This gives the unique polynomial $q(x)$ whose degree is at most d.

If λ is a simple eigenvalue, then $\delta = 1$ and $t_0 = 0$, which means that the coefficient of $x^{d+\delta}$ on the left side is $q_d t_0 = 0$. Since the right-hand side has degree d, the coefficient of x^{d+1} is 0. These two facts give us the trivial equation $0 = 0$. We have $d+1$ remaining equations for the $d+1$ coefficients of q. These are

$$\left(-t_j\binom{1+d}{j}, \ldots, -t_1\binom{1+d-(j-1)}{1}\right) \cdot (q_d, \ldots, q_{d-(j-1)}) = \lambda^k p_j$$

for all $j = 1, \ldots, d+1$, again a lower triangular system in which each diagonal entry is now t_1. Fortunately,

$$k\,ch(x) - x\,ch'(x) = \sum_{j=1}^{k} c_j(-j)x^{k-j},$$

which means that

$$k\,ch(\lambda) - \lambda\,ch'(\lambda) = t_1.$$

Since λ is a *simple* eigenvalue, then $ch(\lambda) = 0$ but $ch'(\lambda) \neq 0$, and so $t_1 \neq 0$ as required to ensure a unique solution.

What about the remaining cases, in which the multiplicity of λ is $m > 1$? Here it seems that we have $m + d + 1$ equations for the $d + 1$ unknowns $q_d, q_{d-1}, \ldots, q_0$. But $ch(\lambda) = 0$, and for each $j < m$ the value of $D^{(j)}(ch)$ at $x = \lambda$ is zero. This means that the first $\delta = m$ equations are simply the redundant equation $0 = 0$. The remaining $d + 1$ equations will specify the $d + 1$ coefficients of $q(n)$. Notice that t_m can be written as a linear combination of $ch(\lambda), D(ch)(\lambda), \ldots, D^{(m)}(ch)(\lambda)$ with a non-zero coefficient for $D^{(m)}(ch)(\lambda)$ (refer to Exercise 3.6). Since we have assumed that λ has multiplicity m as a root of $ch(\lambda)$, then

$$ch(\lambda) = D(ch)(\lambda) = \cdots = D^{(m-1)}(ch)(\lambda) = 0,$$

and $D^{(m)}(ch)(\lambda) \neq 0$. So $t_m \neq 0$, and the lower triangular system with t_m along the diagonal yields unique values for the coefficients $q_d, \ldots, q_0$.

Now for the operation count. Since k is a constant, each of the $O(d^2)$ entries in the coefficient matrix can be computed with a constant number of multiplications. Further, because the matrix is in triangular form, the associated system of equations can be solved by **back substitution**. That is, q_d can be computed in one division; q_{d-1} is computed using this known value of q_d and one division; in general, each q_{d-j} is computed from the values $q_d, q_{d-1}, \ldots, q_{d-(j-1)}$ using $O(j)$ operations. This means that once the coefficient matrix has been computed in $O(d^2)$ operations, the coefficient set $q_d, \ldots, q_0$ can be found using $O(d^2)$ operations.   □

For the second–order recurrence

$$s_n = 2s_{n-1} - s_{n-2} + 2^n ,$$

$\lambda = 2$ is not an eigenvalue of the recurrence, which means that $\delta = 0$, and $\deg(q) = 0$ follows from $p(n) = 1$. This gives the particular solution $v_n = 2^n c$, where c is a constant. To solve for c,

$$v_n - 2v_{n-1} + v_{n-2} = 2^n ,$$
$$2^n c - 2 \cdot 2^{n-1} c + 2^{n-2} c = 2^{n-2} c = 2^n ,$$

giving $c = 4$. And $v_n = 2^{n+2}$ is a different particular solution from the one we found earlier.

For

$$s_n = 4\, s_{n-1} - 4\, s_{n-2} + 1 \cdot 2^n,$$

the input function is $\psi(n) = 1 \cdot 2^n$; that is, $p(n) = 1$ and $\lambda = 2$. Since 2 is a double root of the characteristic polynomial $\lambda^2 - 4\lambda + 4$, we have $\delta = 2$, and we expect a particular solution v_n to have the form $v_n = \lambda^2 n^\delta q(n) = 2^n n^2 c$ because $p(n)$ is a constant. Substituting this v_n into the equation, we have

$$2^n n^2 c = 4\, (2^{n-1}(n-1)^2 c) - 4\, (2^{n-2}(n-2)^2 c) + 2^n ,$$

and dividing by 2^n gives

$$n^2 c = 2\, (n-1)^2 c - (n-2)^2 c.$$

Equating the coefficients of n^2 gives

$$c = 2\, c - c,$$

which is true but uninformative. Equating coefficients of n gives

$$0 = -4\, c + 4\, c,$$

which again is true but uninformative. Finally, equating the constant terms on each side gives

$$0 = 2c - 4c + 1,$$

which implies that $c = 1/2$, and that the particular solution is $v_n = 2^{n-1}n^2$. As above, this particular solution is equivalent to the polynomial identity

$$n^2 = 2(n-1)^2 - (n-2)^2 + 2,$$

which can be verified by expanding the terms or by checking this 2nd order identity at 3 points. For example, choosing $n = 0$, $n = 1$, and $n = 2$ gives the three valid equations

$$0 = 2 - 4 + 2,$$
$$1 = 0 - 1 + 2,$$
$$4 = 2 - 0 + 2.$$

Appendix A contains more worked problems that use this technique.

3.4 Operator Notation

Operator notation is a convenient way to display some facts about difference equations. Here, we'll use this notation introduced in Section 3.2 to explain the *superposition principle* and to show why it was easy to solve the special class of recurrences in Section 3.3.

Superposition means putting one thing on top of another. In our context, it means combining several sequences to obtain a new sequence. For linear difference equations the **superposition principle** is that

$$L[x_n] = \phi(n) + \psi(n)$$

can be solved by solving the two equations

$$L[y_n] = \phi(n) \quad \text{and} \quad L[z_n] = \psi(n),$$

and then combining these two solutions by letting $x_n = y_n + z_n$ so that

$$L[x_n] = L[y_n + z_n] = L[y_n] + L[z_n] = \phi(n) + \psi(n).$$

Of course, this principle applies to any convenient way in which one can break up the input. Specifically, if

$$L[x_n] = \phi(n),$$

and $\phi(n)$ can be written as

$$\phi(n) = \sum_{i=1}^{r} \alpha_i \psi_i(n),$$

then one can solve r equations:

$$\text{for} \quad i \quad \text{from} \quad 1 \quad \text{to} \quad r \quad \text{solve}$$

$$L[x_n^{(i)}] = \psi_i(n)$$

and sum these to find a solution to the original equation as

$$x_n = \sum_{i=1}^{r} x_n^{(i)}.$$

Operator notation also shows why the special cases of Section 3.3 are easy to deal with. Define the operator S_γ by

$$S_\gamma[\, y_n \,] = y_n - \gamma\, y_{n-1}.$$

Clearly,

$$S_\gamma[\, \gamma^n \,] = \gamma^n - \gamma \cdot \gamma^{n-1} = 0.$$

If one wants to solve the nonhomogeneous difference equation

$$L[x_n] = \gamma^n,$$

then S_γ can be used to reduce the nonhomogeneous equation to a homogeneous equation, because

$$S_\gamma[\, L[x_n]\,] = S_\gamma[\, \gamma^n \,] = 0.$$

Let us define L_γ as the composition of S_γ with L; that is, for all sequences $\langle x_n \rangle$,

$$L_\gamma[\, x_n \,] = S_\gamma[\, L[x_n]\,].$$

The characteristic polynomial of L_γ is simply the product of the characteristic polynomials of S_γ and L. That is,

$$ch_{L_\gamma}(\lambda) = (\lambda - \gamma) \cdot ch_L(\lambda),$$

because the characteristic polynomial of S_γ is $(\lambda - \gamma)$.

Now the reason that this special case is so special becomes clear. The special input functions themselves satisfy linear constant coefficient difference equations. Specifically, letting S_γ^{d+1} be the operator that corresponds to $d + 1$ applications of S_γ, then if $\psi(n) = p(n)\, \gamma^n$, where $p(n)$ is a polynomial of degree d,

$$S_\gamma^{d+1}[\, \psi(n)\,] = S_\gamma^{d+1}[\, p(n)\, \gamma^n \,] = 0.$$

Hence the special nonhomogeneous case

$$L[x_n] = p(n)\, \gamma^n$$

reduces to the higher order homogeneous equation

$$L_{d+1}[x_n] = 0,$$

where L_{d+1} is S_γ composed $d+1$ times with L. Of course, the characteristic polynomial of L_{d+1} is

$$ch_{L_{d+1}}(\lambda) = (\lambda - \gamma)^{d+1} \cdot ch_L(\lambda).$$

This reduction does *not* change the initial values for $\langle x_n \rangle$, but since the order of the recurrence has increased, more initial values are needed. If x_0, ..., x_{k-1} were the original initial conditions, then the extra initial conditions can be computed from these and from some values of $\psi(n)$. Specifically, if

$$L[x_n] = x_n - \sum_{i=1}^{k} c_i \, x_{n-i},$$

then x_k, ..., x_{k+d} can be computed by

$$x_k = \sum_{i=1}^{k} c_i \, x_{k-i} + \psi(k),$$

$$\vdots$$

$$x_{k+d} = \sum_{i=1}^{k} c_i \, x_{k+d-i} + \psi(k+d).$$

Since each new initial condition is a linear combination of k previous conditions and one value of $\psi(n)$, each condition can be computed using $O(k)$ operations, and the set of $d + 1$ conditions can be computed using $O(\, k(d + 1)\,)$ operations.

While considering the special nonhomogeneous case as a homogeneous equation may be psychologically simplifying, it may be computationally more complex. If one uses one of the standard $\Theta(k^3)$ methods to solve a system of k linear algebraic equations, then converting to a $(k + d + 1)^{st}$ order difference equation will lead to solving for $k + d + 1$ coefficients and thus using $\Theta(\, (k + d + 1)^3 \,)$ operations[78]. In contrast, first finding the particular solution, which can be done in $O(d^2)$ operations, and then solving for the k coefficients of a homogeneous solution, which can be done in $\Theta(k^3)$ operations, will take a total of $\Theta(k^3 + d^2 \,)$ operations, which will be fewer than the $\Theta(\, (k + d + 1)^3 \,)$ operations of the increased order homogeneous method.

3.5   The Shift Operator on the Space of Sequences

In contrast with the homogeneous recurrences in Chapter 2, when ψ is not the zero function, the set of solutions to (L) is not a vector space. By looking at things slightly differently, the problem of solving these nonhomogeneous recurrences *can* be put into the context of linear algebra. For the moment we consider the infinite-dimensional complex space $\mathcal{S}$ of *doubly infinite* sequences

$$\ldots, s_{-2}, \ s_{-1}, \ s_0, \ s_1, \ s_2, \ldots ,$$

and define the **shift operator** σ on $\mathcal{S}$ to be the function that shifts a sequence $s \in \mathcal{S}$ one step to the right,

$$\sigma(\langle s_n \rangle) = \langle s_{n-1} \rangle .$$

This operator is defined on the vector space $\mathcal{S}$. We'd like a right-shift operator on $\mathcal{S}^+$, the space of singly infinite sequences, but how is it defined? When we shift a *singly* infinite sequence one step to the right, what fills the vacated 0th term? In Exercise 3.1 you show that there's only one assignment that results in a *linear* operator: the 0th term must equal 0. Accordingly, the linear **shift operator** σ on $\mathcal{S}^+$ is defined as

$$(3.9) \qquad \sigma(\langle s_0, s_1, s_2, s_3, \ldots \rangle) = \langle 0, s_0, s_1, s_2, s_3, \ldots \rangle .$$

(Here we're using the same symbol for the shift operator on both $\mathcal{S}$ and $\mathcal{S}^+$.) In Chapter 2 we used powers of the differentiation operator(repeated composition of the operator D), which are also linear. In general, integer powers of a linear operator are linear, and in fact, any polynomial in a linear operator is a linear operator. For example, the operator σ^k shifts a sequence k steps to the right, and applying the operator $2 - 3\sigma + 4\sigma^3$ to any sequence $\langle s_n \rangle$ results in a sequence whose n^{th} term (for $n \geq 3$) is $2s_n - 3s_{n-1} + 4s_{n-3}$. Since the left side of (L) is the nth term of the sequence $s - c_1\sigma(s) - c_2\sigma^2(s) - \cdots - c_k\sigma^k(s)$, it is precisely $L[s]$, where L is the linear operator $L = I - c_1\sigma - c_2\sigma^2 - \cdots - c_k\sigma^k$. This representation allows us to see another analogy between difference equations and differential equations. The general k^{th} order linear constant coefficient differential equation has the form

$$(3.10) \qquad x(t) - c_1 D(x(t)) - c_2 D^2(x(t)) - \cdots - c_k D^k(x(t)) = \beta(t) ,$$

where D is the differentiation operator on the vector space of infinitely differentiable functions on $\mathbb{R}$ or $\mathbb{C}$ (or some other convenient domain of definition). The left side of (3.10) can be written as $L_D(x)$ for

$$L_D = I - c_1 D - c_2 D^2 - \cdots - c_k D^k ,$$

and (3.10) becomes the functional equation $L_D(x) = \beta$. Likewise, the difference equation (L) can be written as $L[s] = \psi$, where

$$\psi = \langle 0, 0, \ldots, 0, \psi(k), \psi(k+1), \ldots \rangle$$

is an element of the vector space S^+. Then $s \in S^+$ is a solution to (L) iff the sequence $L[s] - \psi$ is zero from the kth term onwards. We can now use linear algebra, because

$$V_k = \{ s \in S^+ : s_n = 0 \text{ for all } n \geq k \}$$

is a k-dimensional subspace of S^+ (refer to Exercise 3.3) and

(3.11) $s \in S^+$ is a solution to (L) $\iff$ $L[s] - \psi \in V_k$.

Theorem 3.5.1. *For any choice of constants $c_1, c_2, \ldots, c_k \in \mathbb{C}$, the operator $L = I - c_1\sigma - c_2\sigma^2 - \cdots - c_k\sigma^k$ is invertible on S^+, and the set of preimages $L^{-1}(V_k)$ is a k-dimensional subspace of S^+.*

Proof. To prove the invertibility of L on S^+ we show that any choice of sequence $y \in S^+$ has a unique preimage $s \in S^+$. For this, define the complex-valued function ψ on $\mathbb{N}$ by $\psi(n) = y_n$, the nth term of the given sequence y, and also $\alpha = (y_{k-1}, \ldots, y_0)$. Then $L[s] = y$ encodes the initial value problem

$$s_n - c_1 s_{n-1} - c_2 s_{n-2} - \cdots - c_k s_{n-k} = \psi(n) \, ; \, S_0^T = \alpha,$$

which we know always has a unique solution $s \in S^+$. This proves that there exists a unique $s \in S^+$ with $L[s] = y$. Since the restriction of L^{-1} to the k-dimensional subspace V_k must be invertible, $L^{-1}(V_k)$ is a k-dimensional subspace of S^+. $\square$

Applying this result to (3.11), we obtain

(3.12) $s \in S^+$ is a solution to (L) $\iff$ $s \in L^{-1}(\psi) + L^{-1}(V_k)$.

Geometrically, this means that the set of solutions to (L) forms a k-dimensional **hyperplane**, a translate of a k-dimensional subspace by a single vector. Algebraically, this means that, as we discussed in Section 3.2, we can find the general solution to (L) by first finding a particular solution to $L[s] = \psi$ and then adding the general solution to the homogeneous equation $L[s] = 0$. Once the general solution is known, any initial value problem can be solved by plugging in the initial values. The technique in Theorem 3.3.1 can be used to find a particular solution for equations with forcing functions of the form $\psi(n) = \lambda^n p(n)$. However, there's no general algorithm that gives a particular solution in closed form, and in fact, some classes of linear recurrences don't even have a closed-form solution [47, 136].

We apply this theory to solve the initial value problem

$$s_0 = 0 \, , \, s_1 = 1 \, , \, s_n = 4s_{n-1} - 4s_{n-2} + 3^n(n-1) \, ,$$

where $\psi(n) = \lambda^n p(n)$ for $\lambda = 3$ and $p(x) = x - 1$, a polynomial of degree $d = 1$. Using Theorem 3.3.1, since $\lambda_1 = 2$ is the only root of $ch(x) = (x-2)^2$ and $\lambda \neq \lambda_1$, then $\delta = 0$, and there is a particular solution of the form

$$v_n = 3^n q(n)$$

for some polynomial $q(x)$ with $\deg(q) \le \deg(p) = 1$. A particular solution $\langle v_n \rangle$ can be constructed using the method in the theorem. For this, $ch(3) = 1$ and $ch'(3) = 2$, which gives

$$t_0 = ch(3) = 1 \quad \text{and} \quad t_1 = 2ch(3) - 3ch'(3) = -4.$$

Since $p_1 = 1$, $p_0 = -1$, the coefficients of $q(x)$ satisfy the associated system of equations

$$\begin{bmatrix} 1 & 0 \\ 4 & 1 \end{bmatrix} \begin{pmatrix} q_1 \\ q_0 \end{pmatrix} = \lambda^2 \begin{pmatrix} 1 \\ -1 \end{pmatrix} = \begin{pmatrix} 9 \\ -9 \end{pmatrix},$$

giving $q_1 = 9$ and $q_0 = -4q_1 - 9 = -45$, which means that $v_n = 3^n(9n - 45)$ is a particular solution to the recurrence. Although $\langle v_n \rangle$ doesn't satisfy the initial conditions, every solution to the recurrence has the form $s_n = h_n + v_n$, where $h_n = (\alpha n + \beta)2^n$ is the general solution to the homogeneous recurrence. From the initial conditions,

$$0 = s_0 = (\alpha \cdot 0 + \beta)2^0 + (9 \cdot 0 - 45)3^0,$$
$$1 = s_1 = (\alpha \cdot 1 + \beta)2^1 + (9 \cdot 1 - 45)3^1,$$

which together yield $\beta = 45$, $\alpha = \frac{19}{2}$, and $s_n = (\frac{19}{2}n + 45)2^n + 3^n(9n - 45)$.

3.6   Formal Power Series

We've already used polynomials in σ, and now we want to go one step further and discuss formal power series in the linear operator σ. The adjective "formal" is used because we aren't concerned with convergence of the power series but rather treat it as a purely formal object. In this chapter we concentrate on the theory of formal power series, and this will form the foundation for the next chapter on generating functions. The interested reader should consult Ivan Niven's paper [121] for a good introduction to the application of formal power series to number theory and combinatorics.

Definition 3.6.1. A **formal power series** in the variable x is an infinite sum

$$\gamma = \sum_{i \ge 0} a_i x^i$$

with coefficients a_i in $\mathbb{C}$. Two formal power series are equal iff all corresponding coefficients are equal.

Note that a polynomial in x is a power series whose coefficients are zero after some point. Because of this, a formal power series can be thought of as an "infinite" polynomial. This is exactly the point of view taken for the operations of addition and multiplication of formal power series: For

$\gamma_1 = \sum_{i \geq 0} a_i x^i$ and $\gamma_2 = \sum_{i \geq 0} b_i x^i$, their sum is

$$\gamma_1 + \gamma_2 = \sum_{i \geq 0} (a_i + b_i) x^i .$$

For multiplication of power series, we also mimic what happens with polynomials. The product $\gamma_1 \cdot \gamma_2$ has the constant term $a_0 b_0$; the coefficient of x in the product is

> The product of power series defined here is called a **convolution**. Exercises 3.20 to 3.25 form a series of exercises on convolution.

$a_0 b_1 + a_1 b_0$; the coefficient of x^2 is $a_0 b_2 + a_1 b_1 + a_2 b_0$, and so on. This leads to the following general formula for the product of two power series:

$$(3.13) \qquad \gamma_1 \cdot \gamma_2 = \left(\sum_{i \geq 0} a_i x^i \right) \left(\sum_{j \geq 0} b_j x^j \right) = \sum_{n \geq 0} \left(\sum_{i+j=n} a_i b_j \right) x^n .$$

An important point to note is that since there are only $n + 1$ pairs of indices i, j that sum to n, each coefficient on the right side is a *finite* sum of products $a_i b_j$ in $\mathbb{C}$, and as such is a complex number. This means that the product $\gamma_1 \gamma_2$ is a formal power series. Using these definitions for addition and multiplication, formal power series behave as polynomials in many respects, and Theorem 3.6.1 below summarizes the algebraic properties of the **integral domain** of formal power series. To get there we will take a detour through a proof technique called the finiteness argument, which explains a good deal about the way formal power series behave algebraically. Given any formal power series $\gamma = \sum a_i x^i$, we define its **partial sums** just as in calculus: for any nonnegative integer d, the dth partial sum is the polynomial $\gamma_d = \sum_{i=0}^{d} a_i x^i$. As an example, let us use partial sums to investigate the computation of the coefficients in γ^2. From the definition of multiplication, the 0th coefficient of γ^2 is a_0^2; the first coefficient is $2a_0 a_1$; the second is $a_1^2 + 2a_0 a_2$; and so on. Compare this to the sequence of squares of the partial sums:

$$\gamma_0^2 = a_0^2 ,$$
$$\gamma_1^2 = a_0^2 + 2a_0 a_1 x + a_1^2 x^2 ,$$
$$\gamma_2^2 = a_0^2 + 2a_0 a_1 x + (a_1^2 + 2a_0 a_2) x^2 + 2a_1 a_2 x^3 + a_2^2 x^4 ,$$
$$\cdots .$$

What happens is that the 0th coefficient of γ^2 is the same as that of γ_d^2 for all d; the first coefficient of γ^2 is the same as the first coefficient of γ_d^2 for $d \geq 1$; the second coefficient of γ^2 is the same as that of γ_d^2 for $d \geq 2$. (We will prove that this pattern continues.) This is what we refer to as the **Finiteness Argument**. It allows us to reduce any question about finitely many coefficients of a formal power series to a question about the coefficients of a well-chosen partial sum.

For example, the statement

$$\text{"The } 17^{\text{th}} \text{ coefficient of } \gamma^2 \text{ is positive."}$$

can be checked by testing the 17th coefficient of any γ_d^2 with $d \geq 17$, and the statement can *not* be checked by looking at the 17th coefficient of γ_{16}^2. In a similar way, consider a statement such as

$$\text{"All the coefficients of } \gamma^2 \text{ are positive."}$$

If we can show that for every $d \geq 0$ the first d coefficients of γ_d^2 are positive, then we could prove this statement as well.

The same ideas apply to more complicated expressions in more than one formal power series. For example, for power series γ, δ, ϵ, the nth coefficient of $3\delta - 4\gamma\epsilon^4 + \epsilon$ coincides with the nth coefficient of the polynomial $3\delta_d - 4\gamma_d\epsilon_d^4 + \epsilon_d$ for any $d \geq n$. Writing this more succinctly, for $\psi(x, y, z) = 3y - 4xz^4 + z$, the nth coefficient of $\psi(\gamma, \delta, \epsilon)$ equals the nth coefficient of $\psi(\gamma_d, \delta_d, \epsilon_d)$ for all $d \geq n$.

The Finiteness Argument

Let $\Psi(x_1, \ldots, x_t)$ be a polynomial in the variables $x_1, \ldots, x_t$.
Let $\gamma_1, \ldots, \gamma_t$ be formal power series in x, and for each i let $(\gamma_i)_d$ denote the dth partial sum of γ_i.
Then for each n, the nth coefficient of $\Psi(\gamma_1, \ldots, \gamma_t)$ coincides with the nth coefficient of the polynomial $\Psi((\gamma_1)_d, \ldots, (\gamma_t)_d)$ for all $d \geq n$.

The idea behind the proof of the Finiteness Argument is the following. If we first consider the most basic polynomials in two variables,

$$\Psi_1(x_1, x_2) = x_1 + x_2 \quad \text{and} \quad \Psi_2(x_1, x_2) = x_1 x_2,$$

and two power series γ_1, γ_2, the conclusion of the Finiteness Argument for each of $\Psi_1(\gamma_1, \gamma_2), \Psi_2(\gamma_1, \gamma_2)$ follows from the respective definitions of addition and multiplication for power series. Since the arithmetic operations inherent in the polynomial Ψ form a *finite* sequence of additions and multiplications of pairs of formal power series, the result follows by induction on the number of operations used to build up to $\Psi(\gamma_1, \ldots, \gamma_t)$.

The Finiteness Argument is a useful way to see that algebraic properties of polynomials transfer to the same properties for formal power series.

Theorem 3.6.1. *The set of formal power series in x (with coefficients in $\mathbb{R}$ or $\mathbb{C}$) forms an* **integral domain** *under the operations of addition and multiplication. This means that the following properties hold:*

  (a) The operations of addition and multiplication satisfy the associative and commutative laws, and also satisfy the usual distributive law of multiplication over addition.

  (b) The constant polynomial 0 is the additive identity.

(c) Every power series has an additive inverse, obtained by negating all of its coefficients.

(d) The constant polynomial 1 is the multiplicative identity.

(e) Whenever the product $\gamma_1 \cdot \gamma_2$ equals the zero power series, at least one of the factors γ_i must equal zero.

Proof. We'll prove that the distributive law holds and then just say that all other properties can be proved in the same manner. Let γ, δ, and ϵ be three arbitrary formal power series in x. The distributive law says that $\gamma(\delta + \epsilon) = \gamma\delta + \gamma\epsilon$. This is an infinite string of statements asserting an equality between the nth coefficient of $\gamma(\delta + \epsilon)$ and the nth coefficient of $\gamma\delta + \gamma\epsilon$. Fix any $n \geq 0$. Since polynomial multiplication distributes over polynomial addition, for every $d \geq n$ we have the partial sum arithmetic $\gamma_d(\delta_d + \epsilon_d) = \gamma_d\delta_d + \gamma_d\epsilon_d$. Therefore, the Finiteness Argument implies that $\gamma(\delta + \epsilon) = \gamma\delta + \gamma\epsilon$, proving that the distributive law does hold for power series. $\qquad\square$

We use the modifier *finiteness* because the argument reduces a question about the infinitely many coefficients in a formal power series to the analogous question for its partial sums that are polynomials or finite power series. There's a more sophisticated finiteness argument that involves the substitution of polynomials into power series. For example, if we substitute the polynomial x^2 into the power series

$$(1 + 3x + 5x^2 + \cdots) = \sum_{i \geq 0}(2i + 1)x^i,$$

we would expect to get the power series

$$(1 + 3x^2 + 5x^4 + \cdots) = \sum_{i \geq 1}(2i + 1)x^{2i}.$$

Substituting the polynomial $x + x^2$ into the same power series should yield

$$1 + 3(x + x^2) + 5(x + x^2)^2 + 7(x + x^2)^3 + \cdots$$
$$= 1 + 3x + 8x^2 + 17x^3 + 35x^4 + \cdots.$$

This type of substitution works because the coefficient of the x^n term in the new power series is a *finite* sum of monomials.

More precisely, consider any power series $\gamma = \sum_{i \geq 0} a_i x^i$ and a polynomial $P(x)$ with *zero* constant term. Because the constant term of $P(x)$ is zero, the exponent of every term in the power $P(x)^i$ is at least i. Thus, in the expansion of

$$\gamma(P(x)) = a_0 + a_1 P(x) + a_2 P(x)^2 + \cdots,$$

no summand $a_i P(x)^i$ with $i > n$ can contribute to the x^n term. This shows two things: that the x^n term in $\gamma(P(x))$ is a finite sum of monomials (proving that $\gamma(P(x))$ is a well-defined formal power series) and that the nth term of $\gamma(P(x))$ is the same as the nth term of the polynomial $\gamma_d(P(x))$ for each $d \geq n$. This allows the following extension of the Finiteness Argument:

The Extended Finiteness Argument

Let $\Psi(x_1, \ldots, x_t)$ be a polynomial in the variables $x_1, \ldots, x_t$.
Let $\gamma_1, \ldots, \gamma_t$ be formal power series in x.
Let $P_1(x), \ldots, P_t(x)$ be polynomials with zero constant term.
Then

$$\Psi(\gamma_1(P_1), \gamma_2(P_2), \ldots, \gamma_t(P_t))$$

is a well-defined power series whose n^{th} term coincides with the nth term of the polynomial

$$\Psi((\gamma_1)_d(P_1), (\gamma_2)_d(P_2), \ldots, (\gamma_t)_d(P_t)), \quad \text{for each } d \geq n.$$

In Theorem 3.6.1 the set of formal powers series was shown to be an integral domain, so called because it has the same algebraic properties as the integers. Unlike what happens in $\mathbb{Z}$, the next result shows that there are many power series that have multiplicative inverses.

Theorem 3.6.2. *Let $P(x)$ be a polynomial in x with zero constant term. Then $\Gamma = \sum_{i \geq 0} P(x)^i$ is a well-defined formal power series and Γ is the multiplicative inverse of $1 - P(x)$ in the integral domain of formal power series. We denote this inverse by $1/(1 - P(x))$.*

Proof. We have already proved that Γ is well-defined. To show that Γ is invertible, set $\Psi(x_1, x_2) = x_1 x_2$ and

$$\gamma_1 = 1 - x; \quad \gamma_2 = \sum_{i \geq 0} x^i; \quad P_1(x) = P_2(x) = P(x).$$

Then for each $d \geq 1$, the dth partial sums of γ_1 and γ_2 are $(\gamma_1)_d = 1 - x$ and $(\gamma_2)_d = 1 + x + \cdots + x^d$. Polynomial multiplication gives

$$(\gamma_1)_d \cdot (\gamma_2)_d = 1 - x^{d+1}$$

and

$$\Psi((\gamma_1)_d(P), (\gamma_2)_d(P)) = 1 - P^{d+1}.$$

Since the 0th and d^{th} terms of $\Psi((\gamma_1)_d(P), (\gamma_2)_d(P))$ are therefore 1 and 0 respectively, from the Extended Finiteness Argument we have

$$1 = \Psi(\gamma_1(P), \gamma_2(P)) = (1 - P(x)) \sum_{i \geq 0} P(x)^i,$$

as required. $\square$

Using $P(x) = \lambda x$ in the last result we have the following theorem.

Theorem 3.6.3 (The Geometric Power Series). *For non-zero $\lambda \in \mathbb{C}$, the inverse of $1 - \lambda x$ is*

$$(3.14) \qquad \frac{1}{1 - \lambda x} = \sum_{n \geq 0} \lambda^n x^n .$$

When λ is restricted to be a real number and x satisfies $|\lambda x| < 1$, Theorem 3.6.3 becomes the formula for the geometric series from calculus. In calculus it's a statement about a function defined within its radius of convergence $|x| < |\lambda|^{-1}$, and (3.14) has no meaning outside its radius of convergence. For us, $\sum_{n \geq 0} \lambda^n x^n$ is a purely formal object that algebraically equals $1/(1 - \lambda x)$.

3.6.1 Formal differentiation

We next use a process of formal differentiation to develop a formal power series for the rational function $(1 - \lambda x)^{-m}$, which is the inverse of $(1 - \lambda x)^m$. They will be used in the next chapter on generating functions.

The **formal derivative** of a power series $\gamma = \sum_{i \geq 0} a_i x^i$ is defined to be the formal power series

$$(3.15) \qquad \mathcal{D}(\gamma) = \sum_{i \geq 1} i a_i x^{i-1}.$$

A direct verification shows that the usual sum rule and the product rule also hold for formal derivatives. Using the definition of convolution,

$$\left(\frac{1}{1-x} \right)^2 = \left(\sum_{k \geq 0} x^k \right)^2 = \sum_{n \geq 0} \left(\sum_{i+j=n} 1^2 x^n \right)$$

$$= \sum_{n \geq 0} (n+1) x^n = \mathcal{D}\left(\sum_{n \geq 0} x^n \right) = \mathcal{D}\left(\frac{1}{1-x} \right),$$

and induction gives (refer to Exercise 3.35)

$$(3.16) \qquad \mathcal{D}^j \left(\sum_{n \geq 0} x^n \right) = \frac{j!}{(1-x)^{j+1}},$$

where $0! = 1$ and $\mathcal{D}^0$ is the identity operator. Letting $\gamma = \sum_{n \geq 0} x^n$, we have

$$\frac{(m-1)!}{(1-x)^m} = \mathcal{D}^{m-1}(\gamma) = (m-1)! \sum_{n \geq 0} \binom{n+m-1}{n} x^n ,$$

and

$$(3.17) \qquad \frac{1}{(1 - \lambda x)^m} = \sum_{n \geq 0} \binom{n + m - 1}{n} (\lambda x)^n \,,$$

the power series for $(1 - \lambda x)^{-m}$ when m is a positive integer. This can be called a generalized Binomial Theorem. The name makes sense because if we define the **generalized binomial coefficient** $\binom{r}{n}$ for any rational r and natural number n by

$$\binom{r}{n} = \frac{r(r - 1) \cdots (r - n + 1)}{n!} \,,$$

then for a negative integer $r = -m$ we have

$$\binom{-m}{n} = (-1)^n \binom{m + n - 1}{n} \,,$$

and (3.17) becomes

$$(1 + x)^{-m} = \sum_{n \geq 0} \binom{-m}{n} (-1)^n (-x)^n = \sum_{n \geq 0} \binom{-m}{n} x^n \,.$$

In [121], the **Generalized Binomial Theorem**

$$(3.18) \qquad (1 + x)^r = \sum_{n \geq 0} \binom{r}{n} x^n$$

is proved for all rational r.

As often happens in mathematics, a good idea like formal power series can be used in a wide variety of areas. Formal power series with real coefficients are widely used in combinatorics (for example, refer to [153, 154]), and formal power series with complex coefficients have applications in physics and statistics. For these applications as well as for technical reasons, the definition of convolution for power series with complex coefficients is defined slightly differently. For other applications, such as the theory of automata and formal languages (refer to [143]), formal power series are further generalized to allow the coefficients to lie in a ring or semi–ring rather than in a field.

3.6.2 An application of formal power series

We now return to the question of inverting the operator $L = I - c_1 \sigma - c_2 \sigma^2 - \cdots - c_k \sigma^k$, where σ is the shift operator $\sigma(\langle s_n \rangle) = \langle 0, s_0, s_1, \ldots \rangle$ on $\mathcal{S}^+$, the space of infinite sequences of complex numbers. In order to make sense of this we need to define the formal power series $\Gamma = \sum a_i \sigma^i$ for any choice of $a_i \in \mathbb{C}$ in such a way that it is a linear operator on $\mathcal{S}^+$. For $s \in \mathcal{S}^+$, it would be natural to interpret $\Gamma(s)$ as an infinite formal sum of the sequences

$$a_0 s + a_1 \sigma(s) + a_2 \sigma^2(s) + \cdots .$$

Writing these summand sequences as rows in an infinite array,

$$
\begin{array}{ccccc}
a_0 s_0 & , & a_0 s_1 & , & a_0 s_2 & , & a_0 s_3 & , & \cdots \\
0 & , & a_1 s_0 & , & a_1 s_1 & , & a_1 s_2 & , & \cdots \\
0 & , & 0 & , & a_2 s_0 & , & a_2 s_1 & , & \cdots & , \\
\vdots & & \vdots & & \vdots & & \vdots & & \cdots
\end{array}
$$

each column has only finitely many non-zero terms, and so the sum of each column is a complex number. This is the motivation for defining $\Gamma(s)$ to be the sequence whose n^{th} term is

$$a_0 s_n + a_1 s_{n-1} + \cdots + a_n s_0 = (a_0, a_1, \ldots, a_n) \cdot (s_n, s_{n-1}, \ldots, s_0).$$

In Exercise 3.34 you prove that the function Γ defined in this way is a linear operator on $\mathcal{S}^+$.

Next suppose the operators $\Gamma_1 = \sum a_i \sigma^i$ and $\Gamma_2 = \sum b_i \sigma^i$ are equal on $\mathcal{S}^+$. This means that $\Gamma_1(s) = \Gamma_2(s)$ for every $s \in \mathcal{S}^+$, which is equivalent to requiring that $\Gamma_1 - \Gamma_2$ be the zero operator on $\mathcal{S}^+$. In particular this must be true for the sequence s^* whose only non-zero term is its nth term, with $s_n^* = 1$. Since the nth term of $(\Gamma_1 - \Gamma_2)(s^*)$ is $a_n - b_n = 0$, and this argument hold for each n, Γ_1 and Γ_2 must be equal as power series. What we've just shown is that algebraic identities involving formal power series are also valid identities for the linear operators they represent. For example, if $\Gamma_1 \Gamma_2 = 1$ holds in the ring of formal power series, then $\Gamma_1 \Gamma_2 = I$ holds for the linear operators, and Γ_2 is the multiplicative inverse of Γ_1.

As an example, let's use the shift operator to solve the first–order recurrence

$$(3.19) \qquad\qquad s_{n+1} = \lambda s_n + n^2 \text{ with } s_0 = \alpha_0 .$$

We first shift indices to get $s_n - \lambda s_{n-1} = (n-1)^2$ and then rewrite the recurrence as $L[s] = \psi$, where $\psi = \langle \alpha_0, 0, 1, 4, 9, \ldots \rangle$ and $L = I - \lambda \sigma$. Applying Theorem 3.6.2 with $P(x) = \lambda x$ yields

$$L^{-1} = I + \lambda \sigma + \lambda^2 \sigma^2 + \lambda^3 \sigma^3 + \cdots = \sum_{i \geq 0} \lambda^i \sigma^i ,$$

and the solution is $s = L^{-1}(\psi)$, whose n^{th} term is

$$s_n = (\alpha_0, 0, 1, 4, \ldots, (n-1)^2) \cdot (\lambda^n, \lambda^{n-1}, \ldots, \lambda, 1).$$

Therefore, the solution to (3.19) is

$$s_n = \alpha_0 \lambda^n + \sum_{i=0}^{n-2} (n-1-i)^2 \lambda^i .$$

(Compare this with Theorem 3.1.2 and our analysis in Section 3.2.1.)

In this example, finding the inverse of L was relatively easy. For higher order recurrences the situation is more difficult. For instance, in the second–order case we must consider the linear operator

$$L = I - c_1 \sigma - c_2 \sigma^2 ,$$

and Theorem 3.6.2 gives

$$L^{-1} = \sum_{i \geq 0} (c_1 \sigma + c_2 \sigma^2)^i = I + (c_1 \sigma + c_2 \sigma^2) + (c_1 \sigma + c_2 \sigma^2)^2 + \cdots .$$

Setting $L^{-1} = \sum_{i \geq 0} a_i \sigma^i$ we obtain

$$a_0 = 1 ,$$
$$a_1 = c_1 ,$$
$$a_2 = c_1^2 + c_2$$
$$a_3 = c_1^3 + 2c_1 c_2 ,$$
$$a_4 = c_1^4 + 3c_1^2 c_2 + c_2^2 ,$$
$$a_5 = c_1^5 + 4c_1^3 c_2 + 4c_1 c_2^2 ,$$
$$\vdots$$

The pattern among the a_i seems to be $a_0 = 1$, $a_1 = c_1$, and $a_{i+2} = c_1 a_{i+1} + c_2 a_i$. The techniques in the next chapter will show that the coefficients of L^{-1} *do* always satisfy the original recurrence. (Refer to Exercise 4.8.)

3.7   Exercises

Ex 3.1. Let $\gamma : \mathcal{S}^+ \to \mathcal{S}^+$ be defined by

$$\gamma(s_0, s_1, s_2, \dots) = \langle a, s_0, s_1, s_2 \dots \rangle \text{ for some } a \in \mathbb{C}.$$

Show that γ is linear iff $a = 0$.

Ex 3.2. Consider the (right) shift operator on $\mathcal{S}$, $\sigma(\langle s_n \rangle) = \langle 0, s_0, s_1, \dots \rangle$

(a) Verify that σ is a linear transformation on $\mathcal{S}$.

(b) Prove that the left shift operator is an inverse for σ, and therefore σ is an invertible operator on $\mathcal{S}$.

Ex 3.3. Show that $\mathcal{S}^+$, the space of singly infinite sequences, is infinite-dimensional by finding n linearly independent vectors for every $n \geq 1$. Verify that for each k, $V_k = \{s \in \mathcal{S}^+ : s_n = 0 \text{ for all } n \geq k\}$ is a k-dimensional subspace of $\mathcal{S}^+$.

Ex 3.4. Show that the Fibonacci sequence $\langle f_n \rangle$ and the shifted sequence $\langle f_{n-1} \rangle$ are linearly independent. This means that every solution to the Fibonacci recurrence $s_n = s_{n-1} + s_{n-2}$ can be written as $\alpha f_{n-1} + \beta f_n$, where the scalars α and β depend on the initial conditions. In particular, find a formula for the **Lucas numbers** (which satisfy the Fibonacci recurrence with $L_0 = 2$, $L_1 = 1$) in the form $L_n = \alpha f_{n-1} + \beta f_n$. Are there integer sequence solutions to $s_n = s_{n-1} + s_{n-2}$ for which α and β are not integers?

Ex 3.5. For this exercise, define a **full solution** to a kth order homogeneous difference equation to be any solution that is *not* a solution to any lower order difference equation.

(a) Show that if s is a full solution to a kth order homogeneous difference equation, then the sequences $s, \sigma(s), \ldots, \sigma^{k-1}(s)$ are linearly independent.

(b) Show that $\langle 2^n \rangle$ is not a full solution to $s_n = 4s_{n-1} - 4s_{n-2}$, but $\langle n2^n \rangle$ is a full solution to the recurrence.

(c) Show that every solution to $s_n = 4s_{n-1} - 4s_{n-2}$ can be written as

$$s_n = a_1 n 2^n + a_2(n-1)2^{n-1},$$

for some constants a_1, a_2.

(d) Show that if $\langle s_n \rangle$ is a full solution to a kth order homogeneous difference equation, then every solution $\langle x_n \rangle$ to the difference equation can be written as

$$x_n = \sum_{i=0}^{k-1} a_i s_{n-i},$$

where the a_i's are constants.

Ex 3.6. For the polynomial $ch(x) = x^k - c_1 x^{k-1} - \cdots - c_k$ and any $i = 1, \ldots, k$ show that $t_i = \sum_{j=1}^{k} c_j(-j)^i \lambda^{k-j}$ can be written as a linear combination of $ch(x), D(ch)(x), \ldots, D^{(j)}(ch)(x)$.

Ex 3.7. Solve

$$s_n = -3s_{n-1} - 2s_{n-2} + (-1)^n \text{ for } n \geq 2, \text{ with } s_0 = 2,\, s_1 = -3.$$

Ex 3.8. For this problem, consider the recurrence

$$s_n = 5s_{n-1} - 6s_{n-2} + 2^n n.$$

(a) Find the general solution to the homogeneous equation.
(b) Show that $v_n = -2^n(n^2 + 7n)$ is a particular solution to the recurrence.
(c) Solve the initial value problem with $s_0 = 5$, $s_1 = 4$.
(d) Solve the initial value problem with $s_0 = 4$, $s_1 = 5$.

Ex 3.9. For this problem, consider the recurrence

$$s_n = 5s_{n-1} - 6s_{n-2} + (-2)^n .$$

(a) Find a particular solution to the recurrence.
(b) Solve the initial value problem with $s_0 = 0$, $s_1 = -7/5$.
(c) Solve the initial value problem with $s_0 = 1$, $s_1 = 2$.

Ex 3.10. Find the general solution to the recurrence

$$s_n = 5s_{n-1} - 6s_{n-2}.$$

Ex 3.11. Find a particular solution to the recurrence

$$s_n = 5s_{n-1} - 6s_{n-2} + n\,3^n .$$

Ex 3.12. Solve the initial value problem

$$s_n = 5s_{n-1} - 6s_{n-2} + n\,3^n$$

with $s_0 = 1$ and $s_1 = 2$.

Ex 3.13. For any fixed α_0, α_1, solve the initial value problem

$$s_n = 6s_{n-1} - 9s_{n-2} + 2^n n ,$$
$$s_0 = \alpha_0 , \quad s_1 = \alpha_1 .$$

Ex 3.14. For $\lambda \neq 3$, find a particular solution to

$$s_n = 6s_{n-1} - 9s_{n-2} + \lambda^n n .$$

Ex 3.15. Solve the initial value problem

$$s_0 = 1, \quad s_1 = 2, \quad s_n = 2s_{n-1} + 4s_{n-2} - 8s_{n-3} + \psi(n) ,$$

for each of the following input functions:

$$\psi(n) = (-1)^n n ; \quad \psi(n) = n^2 ; \quad \psi(n) = (-2)^n n ; \quad \psi(n) = 2^n .$$

Ex 3.16. Use the Finiteness Argument to show that the ring of formal power series has no zero divisors; that is, if γ_1, γ_2 are two power series whose product is the zero polynomial, then at least one of γ_1, γ_2 is the zero polynomial.

Ex 3.17. Let γ_1 and γ_2 be two formal power series in x. Show that $\gamma_1(\gamma_2(x))$ makes sense as a formal power series in x, provided the constant term of γ_2 is zero. Check that the substitution of $\gamma_2 = 1 + x$ (which has non-zero constant term) doesn't work.

Ex 3.18. Let $\Psi(\gamma_1, \ldots, \gamma_t)$ be a polynomial in finitely many power series $\gamma_1, \ldots, \gamma_t$ and let P be a polynomial in x with zero constant term. Use the Extended Finiteness Argument to show that first expanding Ψ and then substituting P for x results in the same power series as first substituting P for x and then expanding ψ.

Ex 3.19 (The Quadratic Formula for Power Series). If $\alpha = \sum_{i \geq 1} a_i x^i$, where each a_i is real, show that there exists a unique $\beta = \sum_{i \geq 1} b_i x^i$ such that β satisfies $(1 + \beta)^2 = 1 + \alpha$ (and we can write $1 + \beta = \sqrt{1 + \alpha}$). Use this to prove that whenever $\gamma_1 = \sum_{i \geq 0} a_i x^i$ and $\gamma_2 = \sum_{i \geq 0} b_i x^i$ are power series with real coefficients and $a_0^2 - 4b_0 \geq 0$, then the equation

$$x^2 + \gamma_1 x + \gamma_0 = 0$$

has the two solutions in power series given by

$$\frac{-\gamma_1 \pm \sqrt{\gamma_1^2 - 4\gamma_2}}{2}.$$

The process of **convolution** appeared when we defined multiplication of power series, and it occurs in many other contexts in mathematics. The next sequence of exercises (Exercises 3.20 to 3.25) explains convolution and gives some examples. The definitions of a group and a ring are needed to work most of these exercises.

Ex 3.20. Show that the product of two formal power series can be defined as

$$\left(\sum_{i \geq 0} a_i x^i\right)\left(\sum_{j \geq 0} b_j x^j\right) = \sum_{n \geq 0}\left(\sum_{l \geq 0} a_l b_{n-l}\right) x^n,$$

where b_{n-l} is considered to be zero for negative subscripts.

Ex 3.21. Let G be an additive abelian group and let $\phi, \psi : G \to \mathbb{R}$ be functions that are zero for all but finitely many $g \in G$. Define the convolution $\phi * \psi$ by

$$(\phi * \psi)(g) = \sum_{h \in G} \phi(h)\psi(g - h).$$

Show that

    (a) $\phi * \psi$ is well-defined (meaning that the sum above is finite);

(b) $(\phi * \psi)(g) = 0$ for all but finitely many $g \in G$;

(c) ϕ commutes with ψ; that is, $\phi * \psi = \psi * \phi$.

Ex 3.22. Let G be as in Exercise 3.21 and consider the set R of all functions $G \to \mathbb{R}$ that are zero for all but finitely many elements of G.

(a) Show that R is a commutative ring under the operations of function addition and convolution.

(b) Show that the function that is 1 at the additive identity of G and is zero elsewhere is a multiplicative identity for this ring.

(c) Either show that R is guaranteed to have no zero divisors or find a group G for which the associated R does have zero divisors.

Ex 3.23. As usual, $\mathbb{R}[x]$ denotes the set of polynomials in the variable x with real coefficients. For any polynomial in $\mathbb{R}[x]$, define a corresponding function on $\mathbb{N}$ whose value at $i \in \mathbb{N}$ is the ith coefficient of the polynomial. Use this to show that $\mathbb{R}[x]$ is the ring defined in Exercise 3.22 with $G = \mathbb{N}$.

Ex 3.24. Consider the set $R = \{\phi : \mathbb{Z} \to \mathbb{R} \mid \phi(n) = 0 \text{ for all } n < 0\}$. Define the convolution of two such functions by

$$(\phi * \psi)(n) = \sum_{i \in \mathbb{N}} \phi(i)\psi(n - i).$$

Show that R forms a commutative ring under the operations of usual function addition and convolution. Identify an identity element for convolution. Note: This ring R is the ring of formal power series with real coefficients.

Ex 3.25. Let $f, g : \mathbb{R} \to \mathbb{R}$ be two integrable functions. Define the convolution of f and g by

$$(f * g)(t) = \int_{-\infty}^{+\infty} f(u)g(t - u)du.$$

Without worrying about the convergence of the integral or the integrability of the resulting function, explain how this is a continuous analog of the convolutions defined above. Show that this serves as multiplication and that the set of integrable functions is a ring. If you know enough, worry about the convergence of the integrals. If you know what the Dirac Delta function is, show that it is an identity for convolution.

We have designed the next string of exercises (Exercises 3.26 to 3.34) to give another perspective on the finiteness arguments used to verify the algebraic properties of power series. These can be seriously attempted only by a student who has had a course in the foundations of real analysis. Other students are encouraged to read through these exercises.

Ex 3.26. Let $\mathbb{R}[x]$ be the set of polynomials in x with real coefficients. Given a non-zero polynomial in $\mathbb{R}[x]$, define its **order** $\text{ord}(f)$ to be the degree of the lowest term with a non-zero coefficient. Thus, the order of a polynomial is 0 unless its constant term is zero; if the constant term is zero, the order is 1 unless there is no linear term; and so on. Show that the order of f is $m \geq 1$ iff $x = 0$ is a root of f with multiplicity m.

Ex 3.27. Pick your favorite constant $c > 1$ and define a function $|\cdot| :$ $\mathbb{R}[x] \to \mathbb{N}$ by $|0| = 0$ and $|f| = c^{-\text{ord}(f)}$ for non-zero f. Show that this **absolute value** satisfies the following familiar properties:
  (a) $|f| \geq 0$ and $|f| = 0$ iff $f = 0$ (Positive definiteness).
  (b) $|f + g| \leq |f| + |g|$ (The triangle inequality).
  (c) $|f - g| = |g - f|$ (Symmetry).
  (d) $|fg| = |f||g|$ (Multiplicativeness).

Ex 3.28. Review the construction of the real numbers $\mathbb{R}$ from the rational numbers $\mathbb{Q}$ using **Cauchy sequences**, paying careful attention to the properties of the absolute value that are used to make this construction work. Now define a Cauchy sequence of polynomials in $\mathbb{R}[x]$ to be a sequence $\langle f_n \rangle$ of polynomials such that for every $\epsilon > 0$, there is some M such that $|f_n - f_m| < \epsilon$ whenever $m, n \geq M$ (where $|\cdot|$ is defined in the previous exercise). Next, define two Cauchy sequences $\langle f_n \rangle$ and $\langle g_n \rangle$ of polynomials to be equivalent if for every $\epsilon > 0$ there is an M such that $|f_n - g_m| < \epsilon$ for $m, n \geq M$. Show that this is an equivalence relation on the set of Cauchy sequences of polynomials. To do this, you do not need to have any idea of what a Cauchy sequence looks like; the only thing you need to use are the properties of the absolute value.

Ex 3.29. Show that every Cauchy sequence of polynomials is equivalent to a unique sequence $\langle g_d \rangle$ of polynomials such that $\deg g_d = d$ and $g_d = g_{d-1} + a_d x^d$ for all $d \geq 1$, in other words, a sequence such that g_d is the partial sum of a formal power series in x. Conclude that equivalence classes of Cauchy sequences of polynomials are the same objects as formal power series.

Ex 3.30. Mimicking the construction of $\mathbb{R}$ from $\mathbb{Q}$, show how to define arithmetic operations on equivalence classes of Cauchy sequences of polynomials and prove that these are well-defined (independent of which equivalent sequences are used.) Show that these operations obey the same algebraic laws as they do when applied to polynomials. (The last exercise says that these are really operations on formal power series, but be sure to work this exercise using only properties of $|\cdot|$.)

Ex 3.31. Show that the arithmetic operations defined in the last exercise are the same ones we defined for formal power series.

Ex 3.32. This is a generalization of Exercise 3.27 from polynomials to power series. Define an absolute value on $\mathcal{S}^+$ as follows. Given a non-zero sequence $\langle x_n \rangle$, define the order of $\langle x_n \rangle$ (which we denote by $\text{ord}(x_n)$) to be the first n for which $x_n \neq 0$. Choosing your favorite $c > 1$ again, let

$$\|\langle x_n \rangle\| = c^{-\operatorname{ord}(x_n)}.$$

Show that this absolute value is positive definite, symmetric, and satisfies the triangle inequality.

Ex 3.33. A space with an absolute value such as the ones above is called a **metric space**. It is called a **complete metric space** if every Cauchy sequence of elements from the space has a limit in the space. Show that $\mathcal{S}^+$ is complete by showing that if we are given a sequence $\langle x^i \rangle$ of sequences $\langle x_n^i \rangle$ that is Cauchy in the sense that for every $\epsilon > 0$ there is an I with the property that $\|x^i - x^j\| < \epsilon$ for $i, j \geq I$, then there is a sequence y such that $\lim_{i \to \infty} x_i = y$. (This limit involves *another* epsilon–I definition, this time with $\|\cdot\|$.)

Ex 3.34. (This exercise requires the preceding exercises.)
(a) Show that if $\lim_{n \to \infty} f_n = \Gamma$, where the f_n are polynomials in σ and Γ is a power series in σ, then for any sequence $s \in \mathcal{S}^+$, we have that $\lim_{n \to \infty} f_n(s)$ exists and is equal to the sequence $\Gamma(s)$ as defined in Section 3.6.2.
(b) Show that any formal power series in σ is a linear operator on $\mathcal{S}^+$.

Ex 3.35. Use induction to show that for all $j \geq 0$,

$$\mathcal{D}^j \left(\sum_{n \geq 0} x^n \right) = \frac{j!}{(1-x)^{j+1}},$$

where $0! = 1$ and $\mathcal{D}^0$ is the identity operator.

Ex 3.36. Let D^j denote the ordinary j^{th} derivative with respect to x. For any formal power series $S(x) = \sum_{k \geq j} s_k x^k$ show that for all $j \geq 1$,

$$\mathcal{D}^j \left(S(x) \right) = \sum_{k \geq j} s_k D^j(x^k) = j! \sum_{k \geq 0} \binom{k+j}{j} s_{k+j} x^k.$$

Ex 3.37. For rational numbers r, s, use the fact that $(1+x)^{r+s} = (1+x)^r(1+x)^s$ to show that the binomial coefficients satisfy

$$\binom{r+s}{k} = \sum_{i+j=k} \binom{r}{i}\binom{s}{j} \quad \text{for all } k \geq 0.$$

Also use $(1 - x^2)^r = (1-x)^r(1+x)^r$ to get

$$\binom{r}{n/2} = (-1)^{n/2} \sum_{k=0}^{n} \binom{r}{k}\binom{r}{n-k} \quad \text{for even } n.$$

Ex 3.38. For $L = 1 - 2\sigma + \sigma^2$, show that $L^{-1} = \sum_{i \geq 0}(i+1)\sigma^i$. Show that the inverse of $1 - \sigma - \sigma^2$ equals $\sum_{i \geq 0} f_{i+1}\sigma^i$, where f_i is the ith Fibonacci number. From this calculation derive a representation of the Fibonacci numbers as sums of binomial coefficients.

Ex 3.39. Show that $\mathcal{D}(\gamma) = 0$ implies that the power series γ is a constant, and use this to show that $\mathcal{D}(\gamma_1) = \mathcal{D}(\gamma_2)$ implies that $\gamma_1 = \gamma_2 + c$ for some constant c. Further show that the general solution to $\mathcal{D}(\gamma) = \sum_{n \geq 0} a_n x^n$ is $\gamma = c + \sum_{n \geq 1} \frac{a_{n-1}}{n} x^n$.

Ex 3.40 (Exponential power series). Show that $\mathcal{D}(\gamma) = \gamma$ has the solution $\gamma = \sum_{n \geq 0} \frac{1}{n!} x^n$, the Taylor series for e^x. Use this fact to show that $\gamma = \sum_{n \geq 0} \frac{1}{n!}(-x)^n$ is the multiplicative inverse of $\gamma = \sum_{n \geq 0} \frac{1}{n!} x^n$. Hint: Multiply by e^{-x}.

Ex 3.41.   (a) Use the product rule for formal differentiation to show that for any formal power series γ, $e^{-x}\mathcal{D}(\gamma) - \gamma e^{-x} = \mathcal{D}(e^{-x}\gamma)$.

 (b) Solve the formal equation $\mathcal{D}(\gamma) - \gamma = \sum_{n \geq 0} \frac{1}{n!} x^n$ subject to the condition that $\gamma = 1 + \sum_{n \geq 1} a_n x^n$.
  Hint: Multiply both sides by e^{-x}.

4

Generating Functions

This chapter is devoted to the study of generating functions, which we use for solving difference equations. After some motivation for the term "generating function" is given in the first section, the remainder of the chapter concentrates on using generating functions to solve recurrences. We include examples from combinatorics and number theory.

> The term **generating function** is said to have been coined by Pierre-Simon Laplace (1749–1827), but it formed an integral part of Abraham de Moivre's [48] 1730 paper on linear recurrences.

4.1 Counting Strings with Some Restrictions

To introduce generating functions we begin with a combinatorial problem whose solution uses generating *polynomials*. For any $n \geq 0$, an ordered list of n symbols from some alphabet is called an **n-string**. The only 0-string is the string with no elements, which is usually called the **nullstring**.

For the alphabet $\mathbf{A} = \{x, y, z\}$, there are three 1-strings, the symbols in the alphabet $\mathbf{A}$. One way to obtain a listing of all 2-strings from $\mathbf{A}$ is to use the formal algebraic sum obtained from the product $(x + y + z) \cdot (x + y + z)$

according to the distributive law for multiplication,

$$(x + y + z) \cdot (x + y + z) = (x + y + z)x + (x + y + z)y + (x + y + z)z$$
$$(4.1) \qquad\qquad\qquad = xx + yx + zx + xy + yy + zy + xz + yz + zz.$$

Since each 2-string appears exactly once as a summand on the right side of (4.1), the polynomial $(x + y + z) \cdot (x + y + z)$ is said to generate the 2-strings. Using the commutative law and the laws of exponents, (4.1) can be compressed to

$$(4.2) \qquad (x + y + z)^2 = 1x^2 + 2xy + 2xz + 2yz + 1y^2 + 1z^2 .$$

Although the compressed polynomial in (4.2) is no longer a complete listing of 2-strings, it does encode some important information. For instance, the term $2xy$ indicates that there are two 2-strings with one x and one y; $1z^2$ says that there is one 2-string that uses only the symbol z. In general, for all $n \geq 0$, the polynomial $P_n(x, y, z) = (x + y + z)^n$ generates all n-strings formed using the alphabet $\mathbf{A} = \{x, y, z\}$, and the coefficient of $x^i y^j z^k$ in its compressed representation counts the number of n-strings that have i x's, j y's, and k z's.

When $x = 1, y = 1, z = 1$ is substituted into $P_n(x, y, z)$, each summand is 1, and accordingly, $P_n(1, 1, 1) = 3^n$, the total number of n-strings. The value of $P_n(x, y, z)$ for other choices of the variables x, y, z can be used to count the number of strings with certain characteristics. For instance, suppose we want to calculate the four quantities $s_{00}^{(n)}, s_{01}^{(n)}, s_{10}^{(n)}, s_{11}^{(n)}$, where each $s_{ij}^{(n)}$ equals the number of n-strings in which the number of y's is congruent to i (mod 2) and the number of z's is congruent to j (mod 2). For example, for $n = 2$, from (4.1) we obtain

$$s_{00}^{(2)} = |\{xx, yy, zz\}| = 3; \qquad\qquad s_{01}^{(2)} = |\{zx, xz\}| = 2;$$
$$s_{10}^{(2)} = |\{yx, xy\}| = 2; \qquad\qquad s_{11}^{(2)} = |\{zy, yz\}| = 2.$$

Since each n-string is counted once in the numbers $s_{00}^{(n)}, s_{01}^{(n)}, s_{10}^{(n)}, s_{11}^{(n)}$, then

$$(4.3) \qquad 3^n = (1 + 1 + 1)^n = P_n(1, 1, 1) = s_{00}^{(n)} + s_{01}^{(n)} + s_{10}^{(n)} + s_{11}^{(n)}$$

holds for all $n \geq 0$. When we substitute $x = 1, y = -1, z = 1$ into $P_n(x, y, z)$, each string with an odd number of y's is counted as negative. Since strings with an odd number of y's are counted in $s_{10}^{(n)}$ and $s_{11}^{(n)}$, then

$$(4.4) \qquad 1 = (1 - 1 + 1)^n = P_n(1, -1, 1) = s_{00}^{(n)} + s_{01}^{(n)} - s_{10}^{(n)} - s_{11}^{(n)}.$$

Similarly,

$$(4.5) \qquad 1 = (1 + 1 - 1)^n = P_n(1, 1, -1) = s_{00}^{(n)} - s_{01}^{(n)} + s_{10}^{(n)} - s_{11}^{(n)}$$

and

$$(4.6) \quad (-1)^n = (1 - 1 - 1)^n = P_n(1, -1, -1) = s_{00}^{(n)} - s_{01}^{(n)} - s_{10}^{(n)} + s_{11}^{(n)}.$$

The equations in (4.3) through (4.6) form a nonsingular system of four equations in four unknowns, that can be solved to obtain all of $s_{00}^{(n)}$, $s_{01}^{(n)}$, $s_{10}^{(n)}$, $s_{11}^{(n)}$.

For example, consider the question of finding a formula for $s_{00}^{(n)}$, calculating the number of n-strings that have an even number of y's and an even number of z's. Actually identifying all n-strings with this property would be a daunting task for large n. On the other hand, summing the four equations (4.3) through (4.6) results in

$$3^n + 2 + (-1)^n = 4s_{00}^{(n)},$$

and we have the exact formula

$$(4.8) \qquad s_{00}^{(n)} = \frac{3^n + 2 + (-1)^n}{4},$$

obtained without listing all permissible strings. Checking this formula for $n = 2$, we see that xx, yy, zz is a complete listing of permissible strings and $(3^2 + 2 + (-1)^2)/4$ does equal 3. The values of $P_n(1, y, z)$ for $y, z \in \{\pm 1\}$ allowed us to do this computation.

We close this section with an indication of how recurrence techniques from earlier chapters can also be used to calculate $s_{00}^{(n)}$. To get a recurrence that relates $s_{00}^{(n)}$ to some $s_{jk}^{(i)}$ for earlier $i < n$, we decompose each n-string into two substrings: its first character and the remaining string of length $n - 1$. Any permissible n-string has one of three forms: Its first character is x and the rest is a permissible $(n - 1)$-string; its first character is y and the remaining $(n-1)$-string has an odd number of y's and an even number of z's; its first character is z and the remaining $(n - 1)$-string has an even number of y's and an odd number of z's. Among the n-strings counted in $s_{00}^{(n)}$ there are $s_{00}^{(n-1)}$ strings of the first type, $s_{10}^{(n-1)}$ of the second type, and $s_{01}^{(n-1)}$ of the third type. Therefore,

$$(4.9) \qquad s_{00}^{(n)} = s_{00}^{(n-1)} + s_{10}^{(n-1)} + s_{01}^{(n-1)}.$$

Similar arguments give

$$(4.10) \qquad s_{01}^{(n)} = s_{01}^{(n-1)} + s_{11}^{(n-1)} + s_{00}^{(n-1)};$$

$$(4.11) \qquad s_{10}^{(n)} = s_{10}^{(n-1)} + s_{00}^{(n-1)} + s_{11}^{(n-1)};$$

$$(4.12) \qquad s_{11}^{(n)} = s_{11}^{(n-1)} + s_{01}^{(n-1)} + s_{10}^{(n-1)}.$$

This approach to the problem is completed in Exercise 4.2.

4.2   An Overview of the Generating Function Technique

We've just shown how the polynomial $P_n(x, y, z) = (x+y+z)^n$ can be used to generate all n-strings from the alphabet $\mathbf{A} = \{x, y, z\}$. The remainder of this chapter will consider generating functions in *one* variable. For a more comprehensive discussion of generating functions and their applications the reader is referred to [169]. Also, in [152] Richard Stanley gives a survey of generating functions as used in combinatorics.

We first consider the simplest case, *polynomial* generating functions in one variable. For this, recall that a polynomial in the variable x with complex coefficients can be written in the form $p(x) = s_0 + s_1 x + \cdots + s_r x^r$ for some complex constants $s_0, s_1, \ldots, s_r$. Complete knowledge of the polynomial is equivalent to knowledge of its coefficients, the sequence $s_0, s_1, \ldots, s_r$. In this way $p(x)$ can be considered to be the generating polynomial for the finite sequence $\langle s_0, s_1, \ldots, s_r \rangle$, and conversely the sequence $\langle s_0, s_1, \ldots, s_r \rangle$ is the sequence generated by the polynomial $p(x)$. For example, the **Binomial Theorem** says that

$$(1 + x)^n = \sum_{k=0}^{n} \binom{n}{k} x^k, \text{ for any positive integer } n$$

where the **binomial coefficient** $\binom{n}{k} = \dfrac{n!}{k!(n-k)!}$ is the number of combinations of n things taken k at a time, and the polynomial $p_n(x) = (1+x)^n$ records the sequence $\binom{n}{0}, \binom{n}{1}, \ldots, \binom{n}{n}$ (and so each $p_n(x)$ can be viewed as the generating function for the nth row of Pascal's triangle). Because the solution of an initial value problem is an infinite sequence, its generating function is an infinite series rather than a polynomial unless the solution sequence is eventually zero.

The Generating Function of a Sequence

The **generating function** *for an infinite sequence* $\langle s_n \rangle$ *is the formal power series*

$$S(x) = \sum_{n \geq 0} s_n x^n.$$

When the sequence is finite, say $\langle s_0, s_1, \cdots, s_r \rangle$, *its generating function is the polynomial*

$$s_0 + s_1 x + \cdots + s_r x^r.$$

As in Chapter 3, a generating function is a power series, an algebraic object that can be formally manipulated using algebraic operations. The

sum and product of the generating functions $S(x) = \sum_{n\geq0} s_n x^n$ and $T(x) = \sum_{n\geq0} t_n x^n$ are respectively
(4.13)

$$S(x) + T(x) = \sum_{n\geq0} (s_n + t_n)x^n \text{ and } S(x) \cdot T(x) = \sum_{n\geq0} \left(\sum_{j=0}^{n} s_j t_{n-j} \right) x^n .$$

Remember that a formal power series is never evaluated at a specific value of $x = a$ unless it can be shown that the series is more than purely formal by proving it converges on an open disk containing a.

In this chapter we develop the generating function technique for solving recurrences, and the following examples demonstrate the method.

Example 4.2.1. (**The general first–order homogeneous recurrence**)
For $\lambda \in \mathbb{C}$, the generating function $S(x)$ for solutions to $s_{n+1} = \lambda s_n$ satisfies

$$S(x) = s_0 + x \sum_{n\geq0} s_{n+1} x^n = s_0 + x \sum_{n\geq0} \lambda s_n x^n = s_0 + \lambda x S(x),$$

which can be solved for $S(x)$ to get

(4.14)
$$S(x) = \frac{s_0}{1 - \lambda x}.$$

By Theorem 3.6.3,

$$S(x) = \frac{s_0}{1 - \lambda x} = \sum_{n\geq0} s_0 \lambda^n x^n ,$$

and equality of formal power series gives $s_n = s_0 \lambda^n$, as expected. It is worthwhile to take a minute to summarize the steps in this solution. We first used the recurrence to find a rational form for the generating function and then "expanded" this rational function into another power series. The solution sequence is the sequence of coefficients of this power series.

Example 4.2.2. (**The Fibonacci sequence**) The Fibonacci generating function is the formal power series $F(x) = \sum_{n\geq0} f_n x^n$, which satisfies

$$F(x) = f_0 + f_1 x + \sum_{n\geq2} f_n x^n = x + \sum_{n\geq2} (f_{n-1} + f_{n-2})x^n.$$

At this point a creative re-indexing can be performed to get

$$F(x) = x + \sum_{n\geq1} f_n x^{n+1} + \sum_{n\geq0} f_n x^{n+2} = x + x \sum_{n\geq0} f_n x^n + x^2 \sum_{n\geq0} f_n x^n ,$$

which can be rewritten as

$$F(x) = x + xF(x) + x^2 F(x).$$

Collecting the terms containing $F(x)$ and solving, we obtain

(4.15) $$F(x) = \frac{x}{1 - x - x^2},$$

which is more complicated than the same stage (4.14) of the last example, and we use the technique of **partial fractions** from calculus to decompose $F(x)$ into the sum of two generating functions, each of which is a geometric power series. (The general technique of partial fractions is developed in Section 4.3.)

In order to decompose the rational function on the right side of (4.15) into partial fractions, we find the roots of $1 - x - x^2 = 0$, which are

$$\lambda_1^{-1} = \frac{-1 + \sqrt{5}}{2} \quad \text{and} \quad \lambda_2^{-1} = \frac{-1 - \sqrt{5}}{2},$$

and from $\lambda_1^{-1} \lambda_2^{-1} = -1$,

$$1 - x - x^2 = -(\lambda_1^{-1} - x)(\lambda_2^{-1} - x) = (1 - \lambda_1 x)(1 - \lambda_2 x).$$

By partial fractions there exist constants A, B such that

(4.16) $$F(x) = \frac{x}{1 - x - x^2} = \frac{A}{1 - \lambda_1 x} + \frac{B}{1 - \lambda_2 x}.$$

Before finding the constants A and B, we note that Theorem 3.6.3 allows us to write the right side of (4.16) as the formal power series

$$F(x) = \sum_{n \geq 0} \left(A\lambda_1^n + B\lambda_2^n \right) x^n,$$

and the uniqueness of formal power series implies

(4.18) $$f_n = A\lambda_1{}^n + B\lambda_2{}^n, \quad \text{for all } n \geq 0.$$

To find A and B return to (4.16). Since

$$(1 - \lambda_1 x)(1 - \lambda_2 x) = 1 - x - x^2,$$

we have

$$\frac{x}{1 - x - x^2} = \frac{A(1 - \lambda_2 x) + B(1 - \lambda_1 x)}{1 - x - x^2}.$$

Because these two rational expressions for $F(x)$ have the same denominator, the numerators must also be equal, giving the polynomial equation

$$x = A(1 - \lambda_2 x) + B(1 - \lambda_1 x).$$

As polynomials this implies

$$0 = A + B \quad \text{and} \quad 1 = -A\lambda_2 - B\lambda_1,$$

and (4.18) becomes **Binet's Formula**

$$f_n = \frac{1}{\sqrt{5}}\left(\left(\frac{1+\sqrt{5}}{2}\right)^n - \left(\frac{1-\sqrt{5}}{2}\right)^n\right).$$

Example 4.2.3. Consider the recurrence

(4.21) $s_0 = 2, \ s_1 = 3, \ s_{n+2} = 6s_{n+1} - 9s_n$.

Its generating function is

$$S(x) = \sum_{n\geq 0} s_n x^n = s_0 + s_1 x + \sum_{n\geq 0} s_{n+2} x^{n+2}$$

$$= 2 + 3x + \sum_{n\geq 0} (6s_{n+1} - 9s_n)x^{n+2}$$

$$= 2 + 3x + 6x \sum_{n\geq 0} s_{n+1}x^{n+1} - 9x^2 \sum_{n\geq 0} s_n x^n$$

$$= 2 + 3x + 6x \sum_{n\geq 1} s_n x^n - 9x^2 \sum_{n\geq 0} s_n x^n$$

$$= 2 + 3x + 6x(S(x) - 2) - 9x^2 S(x),$$

and

(4.22) $(1 - 6x + 9x^2)S(x) = 2 - 9x$.

Therefore,

$$S(x) = \frac{2 - 9x}{1 - 6x + 9x^2} = \frac{2 - 9x}{(1 - 3x)^2} = \frac{A}{1 - 3x} + \frac{B}{(1 - 3x)^2},$$

and we can solve to get $A = 3$, $B = -1$. Using this and (3.17),

$$S(x) = \frac{3}{1 - 3x} - \frac{1}{(1 - 3x)^2} = 3\sum_{n\geq 0} 3^n x^n - \sum_{n\geq 0} (n+1)(3x)^n,$$

and

$$s_n = 3^{n+1} - (n+1)3^n = (2-n)3^n.$$

As noted earlier, the first step in these three examples was to represent the generating function as a rational function. In each case its denominator bears a strong resemblance to the characteristic polynomial of the recurrence, namely, it is the **reciprocal polynomial**

$$ch^R(x) = 1 - c_1 x - \cdots - c_k x^k$$

of the characteristic polynomial

$$ch(x) = x^k - c_1 x^{k-1} - c_2 x^{k-2} - \cdots - c_k.$$

Before proving the general result, we consider one more example, a **non-homogeneous** equation

(4.24) $s_0 = 1,\ s_1 = 4,\ s_{n+2} = 6s_{n+1} - 9s_n + 4(n-1)$.

As in the last example,

$$S(x) = 1 + 4x + \sum_{n \geq 0} s_{n+2} x^{n+2}$$

$$= 1 + 4x + 6x \sum_{n \geq 1} s_n x^n - 9x^2 \sum_{n \geq 0} s_n x^n + 4x^3 \sum_{n \geq 0} n x^{n-1} - 4x^2 \sum_{n \geq 0} x^n.$$

At this point we want the rational representation of $\sum_{n \geq 0} n x^{n-1}$, which is a special case of the following formula proved in Chapter 3:

(4.25) $$\frac{1}{(1 - \lambda x)^m} = \sum_{n \geq 0} \binom{n + m - 1}{n} (\lambda x)^n \text{ for any integer } m \geq 1.$$

From this,

$$S(x) = 1 + 4x + 6x(S(x) - 1) - 9x^2 S(x) + \frac{4x^3}{(1-x)^2} - \frac{4x^2}{1-x},$$

which gives

(4.26) $$(1 - 6x + 9x^2)S(x) = \frac{(1 - 3x)(1 - x - 2x^2)}{(1-x)^2}$$

and

$$S(x) = \frac{1 - x - 2x^2}{(1-x)^2(1-3x)}.$$

(Here the denominator is not the reciprocal polynomial, because the input function contributes a factor of $(1-x)^2$ and allows the cancellation of $1 - 3x$.) The solution is completed as above, with the partial fraction decomposition

$$S(x) = \frac{1}{1 - 3x} + \frac{1}{(1-x)^2} - \frac{1}{1-x}$$

yielding the solution

$$s_n = 3^n + (n+1) - 1 = 3^n + n.$$

It is interesting to note that because 3 is not a double root of the denominator of the rational function in this example, the solution sequence is actually a solution to a *first–order* recurrence with $ch(x) = x - 3$. Explicitly, we see that $\langle s_n \rangle$ is a solution to $s_{n+1} = 3s_n + 1 - 2n$, since

$$s_{n+1} - 3s_n = (3^{n+1} + n + 1) - 3(3^n + n) = 1 - 2n.$$

4.2.1 Rational representation

We're now ready to determine the rational form that encodes an initial value problem for

(L) $s_{n+k} = c_1 s_{n+k-1} + c_2 s_{n+k-2} + \cdots + c_k s_n + \psi(n)$

where $\psi(n)$ is the input sequence. Let $S(x) = \sum_{n \geq 0} s_n x^n$ and $F(x) = \sum_{n \geq 0} \psi(n) x^n$ be the generating functions of $\langle s_n \rangle$ and $\langle \psi(n) \rangle$ respectively. Then

$$S(x)\, ch^R(x) = S(x)(1 - c_1 x - \cdots - c_k x^k)$$

$$= \sum_{n \geq 0} s_n x^n - c_1 \sum_{n \geq 0} s_n x^{n+1} - \cdots - c_k \sum_{n \geq 0} s_n x^{n+k}$$

$$= \sum_{n \geq 0} s_n x^n - c_1 \sum_{n \geq 1} s_{n-1} x^n - \cdots - c_k \sum_{n \geq k} s_{n-k} x^n$$

$$= \sum_{n \geq k} \left(s_n - c_1 s_{n-1} - \cdots - c_k s_{n-k} \right) x^n + \sum_{n=0}^{k-1} s_n x^n$$

$$- c_1 \sum_{n=1}^{k-1} s_{n-1} x^n - \cdots - c_{k-2} \sum_{n=k-2}^{k-1} s_{n-k+2} x^n - c_{k-1} s_0 x^{k-1} .$$

The first summand is

$$\sum_{n \geq 0} \left(s_{n+k} - c_1 s_{k+n-1} - \cdots - c_k s_n \right) x^{n+k} = x^k \sum_{n \geq 0} \psi(n) x^n = x^k F(x) ,$$

which depends only on $F(x)$ and k. The remainder of the expression for $S(x) ch^R(x)$ contains the information about the initial conditions, and is the polynomial

$$
\begin{array}{lllll}
s_0 & + s_1 x & + s_2 x^2 & \cdots & + s_{k-1} x^{k-1} \\
 & - c_1 s_0 x & - c_1 s_1 x^2 & \cdots & - c_1 s_{k-2} x^{k-1} \\
 & & - c_2 s_0 x^2 & \cdots & - c_2 s_{k-3} x^{k-1} \\
 & & & \cdots & \\
 & & & & - c_{k-1} s_0 x^{k-1}.
\end{array}
$$

Writing this polynomial as

$$d(x) = d_0 + d_1 x^{k-1} + \cdots + d_{k-1} x^{k-1}$$

gives $d_i = s_i - c_1 s_{i-1} - \cdots - c_i s_0$ and in matrix-vector form

(4.27)

$$(d_{k-1}, \ldots, d_0)^T = M_0 S_0 = \begin{bmatrix} 1 & -c_1 & -c_2 & \cdots & -c_{k-1} \\ 0 & 1 & -c_1 & \cdots & -c_{k-2} \\ 0 & 0 & 1 & \cdots & -c_{k-3} \\ \vdots & \vdots & \vdots & & \vdots \\ 0 & 0 & 0 & \cdots & 1 \end{bmatrix} \begin{pmatrix} s_k \\ s_{k-1} \\ s_{k-2} \\ \vdots \\ s_0 \end{pmatrix},$$

where S_0 is the column vector of initial conditions and M_0 is the above $k \times k$ matrix, which depends only on the coefficients of the difference equation (L). This gives the following rational representation of the generating function for (L).

The Rational Representation

If $F(x)$ is the generating function of $\langle \psi(n) \rangle$ then the generating function of the solution to (L) with initial values S_0 is

(4.28)
$$S(x) = \frac{d(x) + x^k F(x)}{ch^R(x)},$$

where $d(x)$ is the polynomial with coefficients $(d_{k-1}, \ldots, d_0)^T = M_0 S_0$.

From this we see that the rational representation is a **rational function** (a ratio of polynomials) iff the generating function for the input function can be written as a rational function.

Returning to the example in (4.24),

$$F(x) = 4 \sum_{n \geq 0} (n-1)x^n = 4x \sum_{n \geq 0} nx^{n-1} - 4 \sum_{n \geq 0} x^n$$

$$= \frac{4x}{(1-x)^2} - \frac{4}{1-x} = \frac{4(2x-1)}{(1-x)^2}$$

and $M_0 S_0 = (-2, 1)^T$ give

$$ch^R(x)S(x) = 1 - 2x + \frac{4x^2(2x-1)}{(1-x)^2} = \frac{(1-3x)(1-x-2x^2)}{(1-x)^2}$$

as in (4.26), since $ch^R(x) = (1-3x)^2$.

4.3   A Review of Partial Fractions

If you've taken calculus, you've probably seen the technique of partial fractions, which is used there to decompose a rational function into a sum of rational functions that are simpler to integrate. For recurrences whose generating function $S(x)$ is a ratio of polynomials, the partial fraction decomposition allows us to write $S(x)$ as a sum of formal power series that are in the form given in (4.25). The fact that a partial fraction expansion *exists* is an algebraic result that we prove here.

Assume that p and q are polynomials with $\deg(p) < \deg(q)$ and that the leading coefficient of $q(x)$ equals 1. Such a polynomial is called a **monic polynomial**. From the Fundamental Theorem of Algebra (refer to Appendix B) we know that any polynomial with complex coefficients can be factored into linear polynomials,

$$q(x) = (x - \rho_1)^{m_1} \cdots (x - \rho_t)^{m_t}$$

for different elements $\rho_1, \ldots, \rho_t$ of $\mathbb{C}$. The first step in obtaining a useful partial fraction decomposition will be to show that there exist polynomials $A_1(x), \ldots, A_t(x)$ such that

$$(4.29) \qquad \frac{p(x)}{q(x)} = \frac{A_1(x)}{(x - \rho_1)^{m_1}} + \cdots + \frac{A_t(x)}{(x - \rho_t)^{m_t}}$$

and each $\deg(A_i) < m_i$. We will show not only that such polynomials $A_i(x)$ exist and are unique, but our proof also gives a constructive method for finding them. Once the initial decomposition in (4.29) is found, each $A_i(x)$ is expressed as a polynomial in $x - \rho_i$, say $A_i(x) = \alpha_1(x - \rho_i)^{m-1} + \cdots + \alpha_{m-1}(x - \rho_i) + \alpha_m$ (for $m = m_i$), and then the summand $A_i(x)/(x - \rho_i)^m$ is expanded as

$$\frac{A_i(x)}{(x - \rho_i)^m} = \frac{\alpha_1}{x - \rho_i} + \frac{\alpha_2}{(x - \rho_i)^2} + \ldots + \frac{\alpha_m}{(x - \rho_i)^m} \,,$$

where $\alpha_1, \alpha_2, \ldots, \alpha_m$ are now elements of $\mathbb{C}$. The **partial fraction decomposition** we use is found by replacing each of the summands in (4.29) by this finer decomposition. This is helpful, since each of the summands now has the form given in (4.25), and the formula can be applied to obtain the power series for the generating function of $p(x)/q(x)$.

It remains to prove that (4.29) holds. First we note that when $t = 1$, $p(x)/q(x) = p(x)/(x - \rho_1)^{m_1}$ already has the form in (4.29). Our proof for the case in which q has more than one root uses the **Euclidean Algorithm for polynomials**, which is similar to the Euclidean Algorithm for integers, which we discuss later, in Chapters 7 and 8. This algorithm relies on the fact that every non-zero element of $\mathbb{C}$ has an inverse, and so the **Division Algorithm** can be applied to any pair of non-zero polynomials a, b to give unique polynomials Q_1, R_1 for which

$$a(x) = Q_1(x)b(x) + R_1(x) \text{ with } \deg(R_1) < \deg(b) \,.$$

Once the Division Algorithm has been applied once, the process can be continued by dividing $b(x)$ by $R_1(x)$, and continuing in this way a sequence of polynomials $\langle R_i(x) \rangle$ is obtained. Since $\deg(R_1), \deg(R_2), \ldots$ is a strictly decreasing sequence of natural numbers, the process ends after finitely many steps with $R_{n+1}(x) = 0$. In this way we have constructed two finite sequences of polynomials, $Q_1, \ldots, Q_{n+1}$ and $R_1, \ldots, R_n$, such that

$$a(x) = Q_1(x)b(x) + R_1(x), \quad b(x) = Q_2(x)R_1(x) + R_2(x), \quad \ldots,$$
$$R_{n-2}(x) = Q_n(x)R_{n-1}(x) + R_n(x), \quad R_{n-1}(x) = Q_{n+1}(x)R_n(x) + 0 \,.$$

This sequence of equations is the Euclidean Algorithm for polynomials. From the last equation, we note that R_n divides R_{n-1}. Using this information in the previous equation, R_n also divides R_{n-2}, and continuing back we find that R_n divides both a and b, and so R_n is a common divisor of the original polynomials a and b. (In Exercise 4.7 you show that R_n is the gcd of a and b, since it has the largest degree among all polynomials that are common divisors.) Also, we can use the equations in reverse to obtain polynomials B_1 and B_2 such that

$$B_1(x)b(x) + B_2(x)a(x) = R_n(x).$$

Let us now return to the problem at hand, justification of the existence of unique polynomials $A_1(x), \ldots, A_t(x)$ with $\deg(A_i) < m_i$ such that (4.29) holds. As we have already mentioned, there is nothing to prove when $t = 1$. For $t > 1$, the polynomials $b(x) = (x - \rho_2)^{m_2} \cdots (x - \rho_t)^{m_t}$ and $a(x) = (x - \rho_1)^{m_1}$ have no common factors, and so the last non-zero remainder when the Euclidean Algorithm is applied to a and b is a constant polynomial, $R_n(x) = \alpha \in \mathbb{C}$. Also, the equation

$$B_1(x)b(x) + B_2(x)a(x) = \alpha$$

can be multiplied by $\alpha^{-1}p(x)$ to obtain the two polynomials $C_1(x) = \alpha^{-1}p(x)B_1(x)$ and $C(x) = \alpha^{-1}p(x)B_2(x)$ such that

$$(4.30) \qquad\qquad C_1(x)b(x) + C(x)a(x) = p(x),$$

giving

$$\frac{C_1(x)}{(x - \rho_1)^{m_1}} + \frac{C(x)}{b(x)} = \frac{p(x)}{q(x)}.$$

This process can then be inductively continued using $C(x)/b(x)$ to obtain polynomials $A_2(x), \ldots, A_t(x)$ such that (4.29) holds, provided $\deg(C) < \deg(b)$. For this, we note that by the Division Algorithm there exists a polynomial t such that $A_1 = C_1 - ta$ has small degree, $\deg(A_1) < \deg(a)$. Substituting A_1 and $p_1 = C + tb$ in (4.30), we have

$$A_1(x)b(x) + p_1(x)a(x) = p(x).$$

Then

$$\deg(A_1 b) < \deg(a) + \deg(b) = \deg(q) \quad \text{and} \quad \deg(p) < \deg(q)$$

and

$$\deg(p_1) + \deg(a) = \deg(p_1 a) \le \max\{\deg(p), \deg(A_1 b)\} < \deg(q)$$

combine to give $\deg(p_1) < \deg(b)$, as required.

In our application, the denominator in the rational representation of the generating function $S(x) = p(x)/q(x)$ has a non-zero constant term, which means that all of its roots $\rho_1, \ldots, \rho_t$ are non-zero, and therefore each can be written as $\rho_i = 1/\lambda_i$ for some constant λ_i. The representation in (4.29) becomes

$$(4.31) \qquad \frac{p(x)}{q(x)} = \frac{A_1(x)}{(1 - \lambda_1 x)^{m_1}} + \cdots + \frac{A_t(x)}{(1 - \lambda_t x)^{m_t}},$$

where each $\lambda_i^{m_i}$ has been absorbed into the old A_i.

How do we actually find the polynomials $A_1(x), \ldots, A_t(x)$? Assume for the moment that the denominator q has the property that it can be factored into the form

$$q(x) = (1 - \lambda_1 x) \cdots (1 - \lambda_k x),$$

where the non-zero constants $\lambda_1, \ldots, \lambda_k \in \mathbb{C}$ are different. We look at this case first only because its analysis is somewhat simpler. What we want to do is determine the *constants* $A_1, \ldots, A_k$ such that

$$(4.32) \qquad \frac{p(x)}{q(x)} = \frac{A_1}{1 - \lambda_1 x} + \cdots + \frac{A_k}{1 - \lambda_k x}.$$

For this, set $\lambda = \lambda_1$ and multiply both sides of (4.32) by the polynomial $1 - \lambda x$ to obtain

$$\frac{p(x)(1 - \lambda x)}{q(x)} = A_1 + \frac{A_2(1 - \lambda x)}{1 - \lambda_2 x} + \cdots + \frac{A_k(1 - \lambda x)}{1 - \lambda_k x},$$

where

$$\lim_{x \to 1/\lambda} \frac{A_i(1 - \lambda x)}{1 - \lambda_i x} = 0$$

for all $i = 2, \ldots, k$. This gives

$$\lim_{x \to 1/\lambda} \frac{p(x)(1 - \lambda x)}{q(x)} = A_1,$$

and A_1 has been found.

Our method for computing the polynomials $A_1(x), \ldots, A_t(x)$ when q has repeated roots is only a slight modification of the one we've just given for simple roots. Setting $\lambda = \lambda_1$, $A(x) = A_1(x)$, $m = m_1$ and multiplying (4.31) by $(1 - \lambda x)^m$, we have

$$(4.33) \qquad \frac{p(x)(1 - \lambda x)^m}{q(x)} = A(x) + (-\lambda)^m (x - a)^m T(x),$$

where

$$a = \frac{1}{\lambda} \quad \text{and} \quad T(x) = \frac{A_2(x)}{(1 - \lambda_2 x)^{m_2}} + \cdots + \frac{A_t(x)}{(1 - \lambda_t x)^{m_t}}.$$

Since $\lambda \notin \{\lambda_2, \ldots, \lambda_t\}$, the function $T(x)$ is defined for all complex numbers x in the disk

$$|x - a| < R \text{ with radius } R = \min\left\{ \left|\frac{1}{\lambda_i} - a\right| : i = 2, \ldots, t \right\} > 0.$$

The function $f(x) = \dfrac{p(x)(1 - \lambda x)^m}{q(x)}$ therefore has a Taylor series (refer to Appendix B)

$$f(x) = \sum_{i \geq 0} b_i (x - a)^i, \text{ for } b_i = \frac{D^i(f)(a)}{i!},$$

which converges on $|x - a| < R$. (Here D is the differentiation operator and $D^i(f)(a)$ is the evaluation of the ith derivative of f at $x = a$.) Because $T(x)$ also has a Taylor series

$$T(x) = \sum_{i \geq 0} t_i (x - a)^i \text{ on } |x - a| < R,$$

from (4.33) we have

$$\sum_{i \geq 0} b_i (x - a)^i = f(x) = A(x) + \sum_{i \geq 0} (-\lambda)^m t_i (x - a)^{i+m}.$$

The uniqueness of the Taylor series implies that

$$A(x) = \sum_{i=0}^{m-1} b_i (x - a)^i = \sum_{i=0}^{m-1} \frac{D^i(f)(a)}{i!} (x - a)^i,$$

the Taylor polynomial of degree $m - 1$. This result is summarized in the following.

The Partial Fraction Decomposition

If $q(x) = \prod_{i=1}^{t}(1 - \lambda_i x)^{m_i}$ (for distinct non-zero $\lambda_1, \ldots, \lambda_t$), then the partial fraction decomposition of any rational function $p(x)/q(x)$ with $\deg(p) < q(x)$ has the form

$$\frac{p(x)}{q(x)} = \sum_{j=1}^{m_1} \frac{\alpha_{1j}}{(1 - \lambda_1 x)^j} + \cdots + \sum_{j=1}^{m_t} \frac{\alpha_{tj}}{(q - \lambda_t x)^j},$$

where for each $i = 1, \ldots, t$, $\sum_{j=1}^{m_i} \alpha_{ij}(1 - \lambda_i x)^{m_i - j}$ is the Taylor polynomial about $a_i = 1/\lambda_i$ of degree $m_i - 1$ for $f_i(x) = \dfrac{p(x)(1 - \lambda_i x)^{m_i}}{q(x)}$.

There are other ways to determine the constants α_{ij}. For instance, using the notation above we again suppose q has the factorization $q(x) = a(x)b(x)$, where

$$a(x) = (1 - \lambda_1 x)^m, \quad b(x) = (1 - \lambda_2 x)^{m_2} \cdots (1 - \lambda_t x)^{m_t},$$

and $\lambda_1, \ldots, \lambda_t$ are distinct and non-zero. Setting $q_1(x) = q(x)/(1 - \lambda_1 x)$, from the form of the partial fraction decomposition we know that there exists $\alpha_{1m} \in \mathbb{C}$ and a polynomial p_1 such that

$$\frac{p(x)}{q(x)} = \frac{\alpha_{1m}}{(1 - \lambda_1 x)^m} + \frac{p_1(x)}{q_1(x)}.$$

Multiplying this by $(1 - \lambda_1 x)^m$, we obtain

$$\alpha_{1m} = \frac{p(x)}{b(x)} - \frac{p_1(x)(1 - \lambda_1 x)}{b(x)},$$

and since $b(1/\lambda_1)$ is non-zero, then

$$\alpha_{1m} = \lim_{x \to 1/\lambda_1} \left(\frac{p(x)}{b(x)} - \frac{p_1(x)(1 - \lambda_1 x)}{b(x)} \right) = \frac{p(1/\lambda_1)}{b(1/\lambda_1)}.$$

This determines α_{1m}. Also, $p(1/\lambda_1) - \alpha_{1m}b(1/\lambda_1) = 0$, implying that $x = 1/\lambda_1$ is a root of $P(x) = p(x) - \alpha_{1m}b(x)$, which means that $p_1(x) = P(x)/(1 - \lambda_1 x)$ is a polynomial for which $\deg(p_1) < \max(\deg(p), \deg(b))$, and so $\deg(p_1) < \deg(q_1)$. So, the process can be continued with $p_1(x)/q_1(x)$, and all the constants α_{ij} can be found in this way.

We end this section with two examples. First we find the partial fraction decomposition of $1/(1 - x)(1 - 2x)^2$ by calculating the required Taylor polynomials for $\rho = 1, 2$. For $\rho = 1$, we want the Taylor polynomial of degree 0 for $f_1(x) = 1/(1 - 2x)^2$, which is $f_1(1) = 1$, while for $\rho = 2$ the Taylor polynomial of degree 1 is required for $f_2(x) = 1/(1 - x)$, which is

$$f_2(1/2) + f_2'(1/2)(x - 1/2) = 2 + 4(x - 1/2) = 2 - 2(1 - 2x).$$

This gives the partial fraction decomposition

$$\frac{p(x)}{q(x)} = \frac{1}{1 - x} + \frac{2}{(1 - 2x)^2} - \frac{2}{1 - 2x}.$$

Now let's use the second method for determining $\alpha_{11}, \alpha_{22}, \alpha_{21}$ such that

$$\frac{p(x)}{q(x)} = \frac{\alpha_{11}}{1 - x} + \frac{\alpha_{22}}{(1 - 2x)^2} + \frac{\alpha_{21}}{1 - 2x}.$$

From $p(x) = 1$ and $b(x) = (1 - 2x)^2$, $\alpha_{11} = p(1)/b(1) = 1$. Also,

$$p_1(x) = \frac{1 - 1(1 - 2x)^2}{1 - x} = 4x$$

which means that we want to find α_{22}, α_{21} such that

$$\frac{4x}{(1-2x)^2} = \frac{p(x)}{q(x)} = \frac{\alpha_{22}}{(1-2x)^2} + \frac{\alpha_{21}}{1-2x},$$

where now $p(x) = 4x$ and $q(x) = (1-2x)^2$. For this, the new $b(x)$ equals 1, and so $\alpha_{22} = p(1/2)/b(1/2) = 2$ and

$$p_1(x) = \frac{4x-2}{1-2x} = -2 = \alpha_{21},$$

as found above.

For another example, consider $p(x)/q(x)$ for $p(x) = -4 - 13x - 2x^2 - 7x^3 + 8x^4$ and $q(x) = 1 + x - 5x^2 - x^3 + 8x^4 - 4x^5$. To find the factorization of $q(x)$, we note that $q(1) = 0$ and divide $q(x)$ by $1 - x$ to get

$$q(x) = (1-x)(4x^4 - 4x^3 - 3x^2 + 2x + 1).$$

The quotient is also divisible by $1 - x$, and in fact,

$$q(x) = (1-x)^3(4x^2 + 4x + 1) = (1-x)^3(1+2x)^2.$$

We know that
$$\frac{p(x)}{q(x)} = \frac{A_1(x)}{(1-x)^3} + \frac{A_2(x)}{(1+2x)^2},$$

where $A_1(x)$ is the Taylor polynomial of degree 2 about $x = 1$ for

$$f_1(x) = \frac{p(x)(1-x)^3}{q(x)} = \frac{p(x)}{(1+2x)^2},$$

and $A_2(x)$ is the Taylor polynomial of degree 1 for $f_2(x) = \dfrac{p(x)(1+2x)^2}{q(x)}$ about $x = -\frac{1}{2}$. Calculation gives

$$A_1(x) = -2 + 2(x-1) + (x-1)^2 = -2 - 2(1-x) + (x-1)^2,$$
$$A_2(x) = 1 - 4(x + \frac{1}{2}) = 1 - 2(1+2x),$$

and the decomposition is

$$(4.34) \quad \frac{p(x)}{q(x)} = -\frac{2}{(1-x)^3} - \frac{2}{(1-x)^2} + \frac{1}{1-x} + \frac{1}{(1+2x)^2} - \frac{2}{(1+2x)}.$$

4.4   Examples of the Generating Function Technique

The generating function technique we've just described allows us to solve the linear difference equation $L(s) = \psi$ when the generating function $F(x)$

for the input function is a rational function. When ψ is a polynomial, from the method in the last sections we can write $F(x)$ as a sum of rational functions whose power series have the form given in (4.25). As an example, let us apply the generating function technique to solve

$$s_n = -s_{n-1} + 5s_{n-2} + s_{n-3} - 8s_{n-4} + 4s_{n-5},$$
$$S_0 = (9, -43, -13, -9, -4)^T,$$

which has characteristic polynomial $ch(x) = x^5 + x^4 - 5x^3 - x^2 + 8x - 4$, giving

$$\begin{bmatrix} 1 & 1 & -5 & -1 & 8 \\ 0 & 1 & 1 & -5 & -1 \\ 0 & 0 & 1 & 1 & -5 \\ 0 & 0 & 0 & 1 & 1 \\ 0 & 0 & 0 & 0 & 1 \end{bmatrix} \begin{pmatrix} 9 \\ -43 \\ -13 \\ -9 \\ -4 \end{pmatrix} = \begin{pmatrix} 8 \\ -7 \\ -2 \\ -13 \\ -4 \end{pmatrix}.$$

By (4.28) the generating function of $\langle s_n \rangle$ is

$$S(x) = \frac{-4 - 13x - 2x^2 - 7x^3 + 8x^4}{1 + x - 5x^2 - x^3 + 8x^4 - 4x^5},$$

whose partial fraction decomposition is given in (4.34), and from (4.25),

$$s_n = \left(-2\binom{n+2}{2} - 2\binom{n+1}{1} + 1 \right) + \left(\binom{n+1}{1} - 2 \right)(-2)^n$$
$$= -(n^2 + 5n + 3) + (n - 1)(-2)^n.$$

The rest of this section considers two *nonlinear* recurrences that can be solved using generating functions.

4.4.1 The Catalan numbers

For integer $n \geq 0$ consider a string of $n + 1$ elements from a set with a binary operation. If the operation is not associative, then the result of

$$a_0 a_1 \cdots a_n$$

is ambiguous until parentheses are inserted, and we can ask how many different arrangements of parentheses can be inserted in this string. This number is called the nth **Catalan number** C_n. For instance, when $n = 2$ the possibilities are

$$a_0(a_1 a_2) \text{ and } (a_0 a_1)a_2,$$

giving $C_2 = 2$, and $C_3 = 5$ since the choices for $n = 3$ are

$$(a_0 a_1)(a_2 a_3); \ a_0((a_1 a_2)a_3); \ (a_0(a_1 a_2))a_3; \ a_0(a_1(a_2 a_3)); \ ((a_0 a_1)a_2)a_3.$$

In [26], Catalan considers this sequence, and he points out that Lamé [91] already proved that the sequence satisfies the nonlinear recurrence

$$\text{for all } n \geq 0, \quad C_{n+1} = C_0 C_n + C_1 C_{n-1} + \cdots + C_n C_0,$$

with initial values $C_0 = 1$, $C_1 = 1$. Some authors (among them [18]) claim that this sequence was originally investigated by Euler [60].

To see that the sequence of Catalan numbers does indeed satisfy the recurrence, consider any arrangement of parentheses in an $(n+2)$-string as defining a sequence of multiplications. The final multiplication is a product of a $(k+1)$-string and an $(n+1-k)$-string (for some k, $0 \leq k \leq n$), where the parentheses in the two factors can be arranged in C_k and C_{n-k} ways, respectively. This means that for each $k \geq 0$ there is a total of $C_k C_{n-k}$ different arrangements of parentheses in which the first factor of the final multiplication is a $(k+1)$-string. Summing over allowable $k = 0, 1, \ldots, n$ we obtain $C_{n+1} = C_0 C_n + C_1 C_{n-1} + \cdots + C_n C_0$, as required.

Let's use generating functions to solve this recurrence. Applying the definition of multiplication of power series as in (4.13), the generating function $C(x)$ for the Catalan numbers satisfies

$$(C(x))^2 = \sum_{n \geq 0} (C_0 C_n + C_1 C_{n-1} + \cdots + C_n C_0) x^n = \sum_{n \geq 0} C_{n+1} x^n,$$

which gives

$$1 + xC(x)^2 = 1 + x \sum_{n \geq 0} C_{n+1} x^n = C_0 + \sum_{n \geq 1} C_n x^n = C(x).$$

Setting $z = C(x)$, then z is a power series solution to the quadratic equation $xz^2 - z + 1 = 0$, and Exercise 3.19 implies that $C(x)$ must be one of

$$C(x) = \frac{1 \pm \sqrt{1 - 4x}}{2x}.$$

By the Generalized Binomial Theorem in (3.18) for $r = 1/2$, the power series of $\sqrt{1 - 4x}$ is $\sum_{n \geq 0} \binom{1/2}{n}(-4x)^n$, and we obtain

$$xC(x) = \frac{1 \pm \sum_{n \geq 0} \binom{1/2}{n}(-4x)^n}{2}.$$

Comparing the first coefficients of these power series, we see that the negative sign must be chosen and

$$C(x) = -\frac{1}{2} \sum_{n \geq 1} \binom{1/2}{n}(-4)^n x^{n-1}$$

$$= -\frac{1}{2} \sum_{k \geq 0} (-1)^{k+1} \binom{1/2}{k+1} 4^{k+1} x^k$$

$$= \sum_{k \geq 0} (-1)^k \binom{1/2}{k+1} 2^{2k+1} x^k.$$

This gives

$$C_k = (-1)^k \binom{1/2}{k+1} 2^{2k+1} = \frac{1}{k+1} \binom{2k}{k}$$

by Exercise 4.24.

4.4.2 Stirling numbers of the second kind

The second combinatorial problem we consider is associated with partitioning a finite set, and our counting argument uses partial fractions. A **partition** of the n-set $S = \{1, 2, \ldots, n\}$ is a set of disjoint subsets (often called **equivalence classes**) whose union is S. For each pair of positive integers k, n with $k \leq n$, the **Stirling number of the second kind** is denoted by $\left\{ {n \atop k} \right\}$ and is defined to be the number of partitions of an n-set in which there are k classes. For instance, considering the partitions of a 4-set into $k = 2$ equivalence classes, there are three in which both equivalence classes have two elements (consider the possibilities for the class containing 1) and four that have a 1-set and a 3-set. This gives $\left\{ {4 \atop 2} \right\} = 7$.

First we obtain a recurrence. For this we note that the number of partitions with k classes in which $\{n\}$ is one of the equivalence classes is $\left\{ {n-1 \atop k-1} \right\}$. In the other partitions, n is in an equivalence class with at least one other element, and when n is removed what remains is a partition of an $(n-1)$-set in which there are k classes. There are $\left\{ {n-1 \atop k} \right\}$ such partitions, and n might have originally been in any of the k classes. Therefore, for all $1 \leq k \leq n$,

$$(4.35) \qquad \left\{ {n \atop k} \right\} = \left\{ {n-1 \atop k-1} \right\} + k \left\{ {n-1 \atop k} \right\}.$$

We extend this to all natural numbers n, k by defining $\left\{ {n \atop k} \right\}$ to be zero when $k > n$ or $k = 0$ (except for $\left\{ {0 \atop 0} \right\} = 1$). For each fixed $k \geq 1$ define the generating function

$$S_k(x) = \sum_{n \geq 0} \left\{ {n \atop k} \right\} x^n = \sum_{n \geq k} \left\{ {n \atop k} \right\} x^n,$$

which from (4.35) becomes

$$S_k(x) = \sum_{n \geq k} \left\{ {n-1 \atop k-1} \right\} x^n + k \sum_{n \geq k} \left\{ {n-1 \atop k} \right\} x^n = x S_{k-1}(x) + kx S_k(x).$$

This gives $S_k(x) = \dfrac{x}{1 - kx} S_{k-1}(x)$, and inductively

$$S_k(x) = \frac{x^2 S_{k-2}(x)}{(1 - kx)(1 - (k-1)x)} = \cdots = \frac{x^k}{(1 - kx) \cdots (1 - x)},$$

since $S_0(x) = 1$. From the theory of partial fractions we know that there exist $\alpha_j \in \mathbb{C}$ such that

$$\frac{1}{(1 - x)(1 - 2x) \cdots (1 - kx)} = \sum_{j=1}^{k} \frac{\alpha_j}{1 - jx},$$

which means that

$$S_k(x) = x^k \sum_{j=1}^{k} \alpha_j \frac{1}{1 - jx} = x^k \sum_{j=1}^{k} \alpha_j \sum_{n \geq 0} (jx)^n$$

$$= \sum_{n \geq 0} \sum_{j=1}^{k} \alpha_j j^n x^{n+k}$$

$$= \sum_{n \geq k} \left(\sum_{j=1}^{k} \alpha_j j^{n-k} \right) x^n.$$

Combining this with $S_k(x) = \sum_{n \geq k} \left\{ {n \atop k} \right\} x^n$ gives

$$\left\{ {n \atop k} \right\} = \sum_{j=1}^{k} \alpha_j j^{n-k}.$$

For each $j = 1, \ldots, k$, α_j can be calculated (refer to Exercise 4.25) to be

$$\alpha_j = \frac{(-1)^{k-j} j^{k-1}}{(j-1)!(k-j)!} \quad \text{(for all } n, k > 0\text{)},$$

and this gives the formula

$$\left\{ {n \atop k} \right\} = \sum_{j=1}^{k} \frac{(-1)^{k-j} j^{n-1}}{(j-1)!(k-j)!}.$$

4.5   Reversion of Generating Functions

In this section we explore another method for recovering the elements of a sequence from its generating function. (Be sure to review the material in Appendix B.) It's called the **Fourier Transform Method** and is a preview of the **Fast Fourier Transform** which we describe in detail in Chapter 9 (refer to Section 9.5). The basic idea is to replace the formal differentiation process by an appropriately selected weighted sum of some values of the generating function. Remember that in order to evaluate the generating function at any complex number $x = \alpha$, the power series must be more than simply a formal series—it must *converge* at $x = \alpha$. The disk of convergence must contain a disk around each complex number at which the series is to be evaluated.

In Appendix B we discussed **primitive n^{th} roots of unity**, complex numbers ω such that $\omega^n = 1$ and $\omega^j \neq 1$ for all $1 \leq j < n$. For example, $\omega_2 = -1$ is the only primitive second root of unity. For general $n \geq 1$,

$$\omega_n = e^{\frac{2\pi i}{n}} = \cos\left(\frac{2\pi}{n}\right) + i \sin\left(\frac{2\pi}{n}\right)$$

is always a primitive n^{th} root of unity and is often called the **principal n^{th} root of unity**. When ω is any n^{th} root of unity, then

$$0 = \omega^n - 1 = (\omega - 1)(\omega^{n-1} + \cdots + \omega + 1)$$

implies

(4.36) $\omega^{n-1} + \cdots + \omega + 1 = 0, \quad \text{provided } \omega \neq 1.$

The next result shows how roots of unity can be used to recover the elements of *finite* sequences.

Theorem 4.5.1. *Let $\langle s_0, \ldots, s_{p-1} \rangle$ be a finite sequence of length $p \geq 2$ with generating polynomial $S(x) = s_0 + s_1 x + \cdots + s_{p-1} x^{p-1}$. If ω is a primitive p^{th} root of unity, then s_j can be computed by the formula*

$$s_j = \frac{1}{p} \sum_{n=0}^{p-1} \omega^{-jn} S(\omega^n), \quad \text{for any } j = 1, \ldots, p-1.$$

Proof. For any fixed $j = 1, 2, \ldots, p-1$ we have

$$\omega^{-j} S(\omega) = s_0 \omega^{-j} + s_1 \omega^{1-j} + \cdots + s_{p-1} \omega^{p-1-j},$$
$$\omega^{-2j} S(\omega^2) = s_0 \omega^{-2j} + s_1 \omega^{2-2j} + \cdots + s_{p-1} \omega^{2(p-1)-2j},$$

$$\vdots$$

$$\omega^{-(p-1)j} S(\omega^{p-1}) = s_0 \omega^{-(p-1)j} + s_1 \omega^{p-1-(p-1)j} + \cdots + s_{p-1} \omega^{(p-1)^2-(p-1)^2 j}.$$

Adding these $p-1$ equations to $S(1) = s_0 + s_1 + \cdots + s_{p-1}$, we obtain

$$\sum_{n=0}^{p-1} \omega^{-jn} S(\omega^n) = s_0 \sum_{n=0}^{p-1} \omega^{(0-j)n} + s_1 \sum_{n=0}^{p-1} \omega^{(1-j)n} + \cdots + s_{p-1} \sum_{n=0}^{p-1} \omega^{(p-1-j)n} ,$$

where the coefficient of s_i in this equation is $C_i = \sum_{n=0}^{p-1} \omega^{(i-j)n}$. When $0 \le i, j < p-1$ and $i \ne j$, the primitivity of ω implies $\omega^{i-j} \ne 1$, and (4.36) gives $C_i = 0$. Therefore,

$$C_i = \begin{cases} 0 & \text{if } i \ne j, \\ p & \text{if } i = j, \end{cases}$$

and

$$\sum_{n=0}^{p-1} \omega^{-jn} S(\omega^n) = p s_j,$$

proving the theorem.   □

This result can be easily extended to **periodic sequences**, infinite sequences formed by juxtaposing copies of a fixed finite sequence,

$$s_0, s_1, \ldots, s_{p-1}, s_0, s_1, \ldots, s_{p-1}, \ldots.$$

(The number p is called the **period** of $\langle s_n \rangle$ when it's the least integer with this property.) Since the first p terms of such sequences are the terms of the finite sequence $\langle s_0, s_1, \ldots, s_{p-1} \rangle$, using Theorem 4.5.1 its terms can be calculated as

$$s_j = \frac{1}{p} \sum_{n=0}^{p-1} \omega^{-jn} S(\omega^n).$$

Since the sequence $\langle s_n \rangle$ has been assumed to be the infinite juxtaposition of these p terms, we've proved the following result.

Corollary 4.5.2. *Let* $\langle s_n \rangle$ *be a periodic sequence with period p and let ω be a primitive p^{th} root of unity. If $S(x)$ is the generating function for the period $s_0, s_1, \ldots, s_{p-1}$, then the terms are*

$$s_{j+kp} = \frac{1}{p} \sum_{n=0}^{p-1} \omega^{-jn} S(\omega^n) \ \text{ for all } k, j.$$

It should be noted that the technique used in these results can be extended to more general sequences, provided we can evaluate the generating function of the sequence in some neighborhood of $x = 0$. For this, we cannot look at the series expansion as only a *formal* power series, but rather require it to satisfy the analytic property of convergence.

Let us first illustrate the technique by using it to recover s_0 and s_1. Suppose $S(x)$ converges on $|x| < R$. Then each of the series

$$S(x) = s_0 + s_1 x + \cdots + s_n x^n + \cdots$$

and

$$S(-x) = s_0 - s_1 x + \cdots + (-1)^n s_n x^n + \cdots$$

converges for every $|x| < R$, as does the sum

(4.37) $$S(x) + S(-x) = 2(s_0 + s_2 x^2 + \cdots + s_{2n} x^{2n} + \cdots) = 2 f_2(x),$$

where here $f_2(x)$ is the function $f_2(x) = \sum_{n \geq 0} s_{2n} x^{2n}$ discussed in Exercise 4.26(b). Therefore, (4.37) becomes

$$\lim_{x \to 0} \frac{S(x) + S(-x)}{2} = s_0,$$

and we've recovered s_0! Also, considering the difference $S(x) - S(-x)$ and Exercise 4.26, we have

$$S(x) - S(-x) = 2x(s_1 + s_3 x^2 + \cdots + s_{2n+1} x^{2n} + \cdots) = 2x f_2(x);$$

(4.38) $$\lim_{x \to 0} \frac{S(x) - S(-x)}{2x} = \lim_{x \to 0} f_2(x) = s_1,$$

the next element of the sequence.

To generalize this to a general reversion formula, notice that (4.38) can be rewritten as

$$\lim_{x \to 0} \frac{S(x) + \omega S(\omega x)}{2x} = s_1$$

when $\omega = -1$, the principal 2nd root of unity. Although this might seem like a complicated way to write (4.38), we will see that this type of identity works in general.

As a warm-up for the general case, let ω be the principal 3rd root of unity. Since

$$|x| = |\omega x| = |\omega^2 x| \quad \text{for all } x,$$

then for x within $|x| < R$ the generating function $S(x)$ converges at all three of x and ωx and $\omega^2 x$. Therefore, $\omega^3 = 1$ implies

$$S(\omega x) = s_0 + s_1 \omega x + s_2 \omega^2 x^2 + s_3 x^3 + s_4 \omega x^4 + s_5 \omega^2 x^5 + \cdots,$$

$$S(\omega^2 x) = s_0 + s_1 \omega^2 x + s_2 \omega x^2 + s_3 x^3 + s_4 \omega^2 x^4 + s_5 \omega x^5 + \cdots,$$

and

$$S(x) + \omega S(\omega x) + \omega^2 S(\omega^2 x)$$
$$= s_0 + s_1 x + s_2 x^2 + s_3 x^3 + s_4 x^4 + s_5 \omega x^5 + \cdots$$
$$+ s_0 \omega + s_1 \omega^2 x + s_2 x^2 + s_3 \omega x^3 + s_4 \omega^2 x^4 + s_5 x^5 + \cdots$$
$$+ s_0 \omega^2 + s_1 \omega x + s_2 x^2 + s_3 \omega^2 x^3 + s_4 \omega x^4 + s_5 x^5 + \cdots$$
$$= 0 s_0 + 0 x s_1 + 3 s_2 x^2 + 0 s_3 + 0 s_4 x^4 + 3 s_5 x^5 + \cdots,$$

the last from (4.36). This can be reworded as

(4.39) $$S(x) + \omega S(\omega x) + \omega^2 S(\omega^2 x) = 3x^2 f_3(x),$$

where again from Exercise 4.26 we obtain

$$\lim_{x \to 0} \frac{S(x) + \omega S(\omega x) + \omega^2 S(\omega^2 x)}{3x^2} = s_2.$$

In Exercise 4.27 you show that the pattern in (4.39) generalizes to

(4.40) $$S(x) + \omega S(\omega x) + \cdots + \omega^{m-1} S(\omega^{m-1} x) = m x^{m-1} f_m(x),$$

and from Exercise 4.26 we therefore obtain

$$\lim_{x \to 0} \frac{S(x) + \omega S(\omega x) + \cdots + \omega^{m-1} S(\omega^{m-1} x)}{m x^{m-1}} = \lim_{x \to 0} f_m(x) = s_{m-1}.$$

We have proved the following Reversion Formula.

The Reversion Formula

If the generating function $S(x)$ for the sequence $\langle s_n \rangle$ converges in some neighborhood of $x = 0$, then for any integer $m \geq 2$,

(4.41) $$s_{m-1} = \lim_{x \to 0} \frac{1}{m \, x^{m-1}} \sum_{n=0}^{m-1} \omega_m^n S(\omega_m^n x),$$

where ω_m is the principal m^{th} root of unity.

Note that we could have allowed ω_m to be any primitive m^{th} root of unity in this Reversion Formula.

As an example, let us use the Reversion Formula to find the first two terms of the Fibonacci sequence. In Exercise 4.26 the radius of convergence for the Fibonacci generating function $F(x) = x/(1 - x - x^2)$ is shown to be $R = (\sqrt{5} - 1)/2$. We can therefore use $m = 1, 2$ in the Reversion Formula to obtain the first two terms of the sequence,

$$f_0 = \lim_{x \to 0} \frac{F(x) + F(-x)}{2} = \lim_{x \to 0} \frac{x^2}{(1 - x - x^2)(1 + x - x^2)} = 0,$$

and

$$f_1 = \lim_{x \to 0} \frac{F(x) - F(-x)}{2x} = \lim_{x \to 0} \frac{1 - x^2}{(1 - x - x^2)(1 + x - x^2)} = 1.$$

Of course, for more complicated generating functions the limits in the Reversion Formula might be difficult to compute. In such situations, approximations to the limit give an estimate for the elements in the sequence.

4.5.1 Using the Fourier Transform

For an approximation, we would like to truncate the power series and make it a polynomial. Of course, we don't know the power series or its approximating polynomial. But we do know a formula for the generating function. We can use this formula to calculate approximate values for the approximating polynomial. The really clever trick here is that the inverse Fourier transform can be used to go from a set of values to the coefficients of a polynomial. With the FFT algorithm this calculation can be done quickly, i.e. the n coefficients of a polynomial can be computed from the n values of the polynomial using $\Theta(n \log n)$ arithmetic operations. (See Section 9.5 for more details on the FFT.)

Consider the polynomial $p(x) = 1 + x$. Evaluating at the 4th roots of unity $1, i, -1, -i$ gives

$$\begin{aligned}
p(1) &= 2, \\
p(i) &= 1 + i, \\
p(-1) &= 0, \\
p(-i) &= 1 - i.
\end{aligned}$$

If we express $p(x)$ as $p(x) = c_0 + c_1x + c_2x^2 + c_3x^3$, then the coefficients can be computed via the inverse Fourier transform as

$$c_0 = \frac{1}{4}\left[p(1) + p(i) + p(-1) + p(-i)\right]$$
$$= 1\,,$$

$$c_1 = \frac{1}{4}\left[p(1) + p(i)\cdot(-i) + p(-1)\cdot(-i)^2 + p(-i)\cdot(-i)^3\right]$$
$$= \frac{1}{4}\left[2 \quad\quad +1-i \quad\quad +0 \quad\quad +1+i\right]$$
$$= 1\,,$$

$$c_2 = \frac{1}{4}\left[p(1) + p(i)\cdot(-1) + p(-1)\cdot(-1)^2 + p(-i)\cdot(-1)^3\right]$$
$$= \frac{1}{4}\left[2 \quad\quad -1-i \quad\quad +0 \quad\quad -1+i\right]$$
$$= 0\,,$$

$$c_3 = \frac{1}{4}\left[p(1) + p(i)\cdot(i) + p(-1)\cdot(i)^2 + p(-i)\cdot(i)^3\right]$$
$$= \frac{1}{4}\left[2 \quad\quad -1+i \quad\quad +0 \quad\quad -1-i\right]$$
$$= 0\,.$$

This evaluation at the 4th roots of unity is the 4-point Fourier transform. Notice that the evaluation points are i^0, i^1, i^2, i^3, the powers of i taken in this order. Back-calculating $p(x)$'s coefficients from these values is the inverse Fourier transform. This inverse calculation can also be viewed as treating the values of $p(x)$ as coefficients of another polynomial and evaluating this other polynomial at the powers of $-i$, and then dividing these values by 4. These results will be the coefficients of $p(x)$.

If we think of the roots of unity as being arranged around the unit circle in the complex plane, the powers of i are found by going counterclockwise around this circle, and the powers of $-i$ are found by going clockwise around the circle. So by calculating in the counterclockwise direction, we compute the Fourier transform, and by calculating in the clockwise direction we compute the inverse Fourier transform. The factor $1/n$ (in this example $1/4$) is needed to normalize the result.

The inverse Fourier transform (except for the normalization) can be calculated by an algorithm for the Fourier transform by making use of complex conjugation. Section 9.5 explains this in more detail.

What does all this have to do with reversion of generating functions? Generally, generating functions are *not* polynomials, but near $x = 0$, one hopes that a generating function may be approximated by a Taylor polynomial. For example, the Fibonacci generating function

$$F(x) = 0 + x + x^2 + 2x^3 + 3x^4 + 5x^5 + 8x^6 + \cdots$$

may be approximated close to $x = 0$ by the polynomial $0 + x$. On the other hand, evaluating $F(x)$ at the roots of unity looks "iffy" because, for example, the series for $F(1)$ does not converge. A better-behaved series can be obtained by *scaling*. For example, replacing x by $.1\,z$ in $F(x)$ gives

$$G(z) = F(.1z) = 0 + .1z + .01z^2 + .002z^3 + .0003z^4 + \ldots.$$

Since the series for $G(z)$ converges for $z = 1$ and $G(z)$ may be "reasonably approximated" by the polynomial

$$0 \quad + \quad .1z \quad + \quad .01z^2 \quad + \quad .002z^3$$

when $|z| = 1$, we expect that if we had the values of $G(z)$ for $z = 1, i, -1, -i$, we would be able to use the inverse Fourier transform to calculate a polynomial of degree at most 3 whose coefficients should be reasonably close to the coefficients of

$$G_4(z) \quad = \quad 0 \quad + \quad .1z \quad + \quad .01z^2 \quad + \quad .002z^3.$$

We can use

$$G(z) = \frac{.1z}{1 - .1z - .01z^2}$$

to find

$$
\begin{aligned}
G(1) &\approx .11236, \\
G(i) &\approx -.0097 + .098i, \\
G(-1) &\approx -.0917, \\
G(-i) &\approx -.0097 - .098i.
\end{aligned}
$$

Then using the inverse Fourier transform we find that

$$G_4(z) \approx .00046 \quad + \quad .1z \quad + \quad .01z^2 \quad + \quad .002z^3,$$

and rescaling to x we find that

$$F_4(z) \approx .00046 \quad + \quad x \quad + \quad x^2 \quad + \quad 2x^3,$$

which is a reasonable approximation to the first four terms in the generating function for the Fibonacci numbers.

The following box gives the general form of this Fourier Transform reversion technique.

The Fourier Reversion Method

Let $S(x)$ be the generating function for the sequence $\langle s_n \rangle$.
Find a scaling factor α such that $G(z) = S(\alpha x)$ and $G(z)$ is reasonably approximated by a polynomial $G_m(z)$ with m coefficients.
Let ω be the principal m^{th} root of unity.
Evaluate $G(z)$ at $1, \quad \omega, \quad \omega^2, \quad \ldots, \quad \omega^{m-1}$.
Use the inverse Fourier transform to calculate the approximate coefficients of $G_m(z)$.
Re-scale to obtain approximations to the first m coefficients of the series for $S(x)$ and hence approximations for the first m terms of $\langle s_n \rangle$.

4.6   Exercises

Ex 4.1. Using the alphabet $\{x, y\}$, find an exact formula for the number of n-strings with an even number of x's.

Ex 4.2. For this exercise, use the notation established in Section 4.1.
   (a) Prove $s_{01}^{(n)} = s_{10}^{(n)}$ for all $n \geq 0$. Denote the common value by t_n.
   (b) Let $s_n = s_{00}^{(n)}$ and $t_n = s_{01}^{(n)} = s_{10}^{(n)}$. Use this and equations (4.3) and (4.9) through (4.12) to obtain

$$t_{n+1} = 2t_n + 3t_{n-1} \; ; \; t_0 = 0, t_1 = 1;$$

$$s_{n+2} = 3s_{n+1} + s_n - 3s_{n-1} \; ; \; s_0 = 1, s_1 = 1, s_2 = 3;$$

$$s_n = s_{n-1} + 2t_{n-1} \; ; \; s_0 = 1, s_1 = 1.$$

   (c) Find exact formulas for s_n and t_n.

Ex 4.3. Using the alphabet $\{x, y, z\}$, let v_n be the number of n-strings in which the number of y's is odd and the number of z's is odd. Find an exact formula for v_n.

Ex 4.4. Use the technique of generating functions to find a formula for the nth term of

$$s_0 = 0, s_1 = 1, s_n = 5s_{n-1} - 6s_{n-2}.$$

Ex 4.5. Let $f(x) = a_k x^k + \cdots + a_1 x + a_0$ be a nonconstant polynomial with complex coefficients and $a_0 \neq 0$. (For instance, the characteristic polynomial of (L) has these properties.) Let $f^R(x) = a_k + a_{k-1}x + \cdots + a_1 x^{k-1} + a_0 x^k$, the reciprocal polynomial of $f(x)$. Show that λ is a non-zero root of $f(x) = 0$ iff $1/\lambda$ is a root of $f^R(x) = 0$.

Ex 4.6. Let $q(x) = (1 - \lambda x)(1 - \bar{\lambda}x)$ for nonreal $\lambda \in \mathbb{C}$.

(a) Show that for any $p(x) \in \mathbb{R}[x]$ with $\deg(p) < 2n$, there exists a unique $\beta(x) \in \mathbb{C}[x]$ with $\deg(\beta) < n$ such that

$$(4.42) \qquad \frac{p(x)}{(q(x))^n} = \frac{\beta(x)}{(1 - \lambda x)^n} + \frac{\overline{\beta(x)}}{(1 - \overline{\lambda} x)^n},$$

where $\overline{\beta(x)}$ denotes the polynomial obtained from $\beta(x)$ by conjugating all coefficients.

(b) For any constant $\beta \in \mathbb{C}$,

$$(4.43) \qquad \frac{\beta}{1 - \lambda x} + \frac{\overline{\beta}}{1 - \overline{\lambda} x} = 2 \sum_{j \geq 0} \Re(\beta \lambda^j) x^j,$$

where $\Re(\beta \lambda^j)$ denotes the real part of the complex number $\beta \lambda^j$.

Ex 4.7. Let $a(x), b(x) \in \mathbb{C}[x]$ be two non-zero polynomials. Show that any common divisor of $a(x)$ and $b(x)$ divides the last non-zero remainder $R_n(x)$ in the Euclidean Algorithm applied to $a(x)$ and $b(x)$. This proves that $\alpha^{-1} R_n(x)$ is the **gcd** of $a(x)$ and $b(x)$, where α is the leading coefficient of $R_n(x)$.

Ex 4.8. As established in Chapter 3, any operator of the form $L = I - c_1 \sigma - \cdots - c_k \sigma^k$ is an invertible operator on $\mathcal{S}^+$ and is also invertible as a formal power series. We claimed that the coefficients of $L^{-1} = ch^R(\sigma)^{-1} = \sum_{n \geq 0} a_n \sigma^n$ satisfy

$$a_n = c_1 a_{n-1} + c_2 a_{n-2} + \cdots + c_k a_{n-k} \text{ for all } n \geq k.$$

Use results from generating functions to show this.

Ex 4.9. If $L(\sigma)$ is the shift operator associated with the initial value problem

$$s_0 = 0, \; s_1 = 1, \; s_n = 4 s_{n-1} - 4 s_{n-2},$$

show that its inverse is the operator

$$L^{-1} = \sum_{n \geq 0} (n + 1) 2^n \sigma^n.$$

Ex 4.10. Verify that the partial fraction expansion of the generating function for

$$s_0 = 0, \; s_1 = 1, \; s_n = 4 s_{n-1} - 4 s_n + 3^n (n - 1)$$

is

$$\frac{19}{2} \frac{1}{(1 - 2x)^2} + \frac{71}{2} \frac{1}{1 - 2x} + \frac{9}{(1 - 3x)^2} - \frac{54}{1 - 3x}.$$

Ex 4.11.   (a) Find the generating function of $s_n = n^2$.

(b) Show that the generating function of the sequence $s_n = 1^2 + 2^2 + \cdots + n^2$ is

$$\frac{2}{(1-x)^4} - \frac{3x}{(1-x)^3} - \frac{2}{(1-x)^2}.$$

Ex 4.12. (a) If $S(x)$ is the generating function for the sequence $\langle s_n \rangle$, show that $S(x)/(1-x)$ is the generating function for the sequence of partial sums $\sum_{j=0}^{n} s_j$.

(b) Use part (a) to prove the Fibonacci identity

$$F_0 + F_1 + \cdots + F_n = F_{n+2} - 1.$$

Ex 4.13 (L'Hôpital's Rule). Let $p(x), q(x) \in \mathbb{C}[x]$ be polynomials.

(a) If $\alpha \in \mathbb{C}$ is a root of $q(x)$, show $q(x) = (x - \alpha)q_1(x)$ for some polynomial $q_1(x)$ and that $q'(\alpha) = q_1(\alpha)$.

(b) Suppose α is a root of $p(x)$ and a *simple* root of $q(x)$. Show that

$$\lim_{x \to \alpha} \frac{p(x)}{q(x)} = \lim_{x \to \alpha} \frac{p'(x)}{q'(x)}.$$

Ex 4.14. Use the generating function method to find an exact formula for the nth term of

$$s_0 = 5, \; s_1 = 13, \; s_n = -4s_{n-1} + 5s_{n-2}.$$

Ex 4.15. Use the generating function method to find a formula for the nth term

$$s_0 = -1, \; s_1 = 2, \; s_2 = 14, \; s_n = s_{n-1} - 4s_{n-2} + 4s_{n-3}.$$

Ex 4.16. Expand $\dfrac{4x^2 - 2x - 1}{(1+x)(1-x-3x^2)}$ into a power series.

Ex 4.17. Find the sequence with generating function $\dfrac{1}{1-2x} + \dfrac{1}{(1-3x)^3}$.

Ex 4.18. Use the generating function technique to show that

$$s_n = \begin{cases} 2^n & \text{if } n \text{ is even}, \\ 2^n + 2 & \text{if } n \equiv 1 \bmod 4, \\ 2^n - 2 & \text{if } n \equiv 3 \bmod 4 \end{cases}$$

is the solution to

$$(s_0, s_1, s_2) = (1, 4, 4), \quad s_{n+3} = 2s_{n+2} - s_{n+1} + 2s_n.$$

Ex 4.19. (a) Use the technique of generating functions to solve

$$s_0 = 1, \; s_1 = 0, \quad s_n = 4s_{n-1} - 4s_{n-2} + 3^n(n-1).$$

(b) Find all solutions to

$$s_n = 4s_{n-1} - 4s_{n-2} + 3^n(n-1).$$

Ex 4.20. Consider the following coupled pair of initial value problems:

$$s_0 = 1,\ s_1 = 0,\ t_0 = 0,\ t_1 = 1,\ s_n = 2t_{n-1} + s_{n-2},\ t_n = -s_{n-1} + t_{n-2}.$$

Let $S(z)$ and $T(z)$ denote the generating functions of $\langle s_n \rangle$ and $\langle t_n \rangle$. Find a system of two equations that relate $S(z)$ and $T(z)$, and use this to solve for the generating functions.

Ex 4.21. We consider a nonhomogeneous recurrence with characteristic polynomial $ch(x)$ and forcing function $\psi(n) = \lambda^n p(n)$ for some polynomial $p(x)$ of degree $m \geq 1$.

(a) Show that there exists a polynomial $Q(x)$ with $\deg(Q) \leq m$ such that the generating function of every initial value problem for this recurrence has the form

$$(4.44) \qquad S(x) = \frac{d(x)(1 - \lambda x)^{m+1} + x^k Q(x)}{ch^R(x)(1 - \lambda x)^{m+1}}$$

for some polynomial $d(x)$ with $\deg(d) < \deg(ch)$.

(b) Show that for any polynomial $d(x)$ with $\deg(d) < \deg(ch)$, there exists an initial value problem whose generating function has the form given in (4.44).

Ex 4.22. Let $s_n = \alpha\lambda_0^n + a_1(n)\lambda_1^n + \cdots + a_{k-1}(n)\lambda_{k-1}^n$ be a solution to the kth order recurrence $s_n = c_1 s_{n-1} + c_2 s_{n-2} + \cdots + c_k s_{n-k}$. If λ_0 is a simple eigenvalue of the recurrence, show that

$$\alpha = \lim_{x \to \lambda_0} \frac{(d_0 x^{k-1} + d_1 x^{k-2} + \cdots + d_{k-1})(x - \lambda_0)}{ch(x)},$$

where $d_i = s_i - c_1 s_{i-1} - \cdots - c_{i-1} s_0$.

Ex 4.23 (Generalized Fibonacci sequence). For $k \geq 2$ define the sequence $\langle f_n \rangle$ by

$$f_0 = f_1 = \cdots = f_{k-2} = 0,\ f_{k-1} = 1 \quad \text{and} \quad f_n = f_{n-1} + f_{n-2} + \cdots + f_{n-k}.$$

(a) Show that for all $x \neq 1$,

$$ch(x) = \frac{x^k(x-2) + 1}{x - 1}.$$

(b) Use the last exercise and part (a) to show that if λ_0 is the largest eigenvalue of the recurrence (later we'll prove that it is the only *positive* eigenvalue), then $f_n = \alpha\lambda_0^n + d_n$, where d_n has no λ_0^n component and

$$\alpha = \frac{\lambda_0 - 1}{\lambda_0[(k+1)\lambda_0 - 2k]}.$$

Ex 4.24. Show that for all positive integers k,

$$\binom{1/2}{k+1} 2^{2k+1} = \frac{(-1)^k}{k+1}\binom{2k}{k},$$

where $\binom{1/2}{n}$ is the **generalized binomial coefficient** given by

$$\binom{1/2}{n} = \frac{1/2(-1/2)\cdots(3/2-n)}{n!}.$$

Ex 4.25. For any $k \geq 1$, show that the partial fraction decomposition of

$$\frac{1}{(1-x)(1-2x)\cdots(1-kx)} = \frac{\alpha_1}{1-x} + \cdots + \frac{\alpha_k}{1-kx}$$

has

$$\alpha_j = \frac{(-1)^{k-j}j^{k-1}}{(j-1)!(k-j)!}.$$

Ex 4.26. (Refer to the information on convergence given in Appendix B.) Let $\gamma(x) = \sum_{n\geq 0} a_n x^n$ be a fixed power series with disk of convergence $|x| \leq R$.

(a) For any strictly increasing infinite sequence $\langle n_i \rangle$ of positive integers, consider the power series $\gamma_1(x) = \sum_{i\geq 0} a_{n_i} x^{n_i - n_0}$. If $\alpha \in \mathbb{C}$ such that $|\alpha| < R$, show the power series $\gamma_1(x)$ also converges at $x = \alpha$.

(b) Show that

$$\lim_{x\to 0} \frac{\gamma(x) + \gamma(-x)}{2} = a_0.$$

(c) For any positive integer m use part (a) to define the complex-valued function f_m on the open disk $|x| < R$ by

$$f_m(x) = \sum_{i\geq 0} a_{(i+1)m-1} x^{im}.$$

Show that

$$\lim_{x\to 0} \frac{f_m(x)}{mx^m} = a_{m-1}.$$

(d) Show that $(\sqrt{5}-1)/2$ is the radius of convergence of the Fibonacci generating function.

Ex 4.27. Verify the pattern in (4.40) and thereby construct an argument to convince yourself that the Reversion Formula holds for general $m \geq 2$.

Ex 4.28. Let $S(x) = \dfrac{2}{1-4x^2}$.

(a) Use the Reversion Formula (4.41) to calculate the first four elements of the sequence with generating function $S(x)$.

(b) Use the Fourier Reversion Method to calculate approximations to the first four elements of this sequence.

(c) Use the partial fraction decomposition of $S(x)$ to verify that you have computed these terms correctly.

Ex 4.29. For any sequence $\langle a_n \rangle$, define its **exponential generating function** by

$$\mathcal{E}(a_n) = \sum_{n \geq 0} \frac{a_n}{n!} x^n .$$

In particular, $\mathcal{E}((-1)^n) = e^{-x}$ and $\mathcal{E}(n!) = 1/(1-x)$. Show that
(a) For any positive integer m, $\mathcal{E}(a_{n+m}) = \mathcal{D}^m(\mathcal{E}(a_n))$.
(b) For any polynomial $p(x)$, $\mathcal{E}(p(n)a_n) = p(x\mathcal{D})(\mathcal{E}(a_n))$.
(c) $\mathcal{E}(a_n)\mathcal{E}(b_n) = \mathcal{E}(\sum_{k=0}^{n} \binom{n}{k} a_k b_{n-k})$.

Ex 4.30. For the recurrence

$$a_{n+1} = (n+1)a_n + (-1)^n \text{ with } a_0 = 1 ,$$

show directly that $\mathcal{E}(a_n) = \mathcal{E}((-1)^n)\mathcal{E}(n!)$, and then use the last problem to show that $a_n = \sum_{i=0}^{n} \frac{n!}{i!}(-1)^n$.

Ex 4.31. A **derangement** of the finite set $\{1, 2, \ldots, n\}$ is a permutation of the set in which every element is moved from its original position. Let d_n equal the number of derangements of the set $\{1, 2, \ldots, n\}$. For instance, $d_1 = 0$, $d_2 = 1$, $d_3 = 2$. Let $d_0 = 1$.

(a) Show that d_n is the number of derangements of $\{1, 2, \ldots, n+1\}$ in which $n+1$ is in the first position and 1 is not in the last place. Use this idea to show that $\langle d_n \rangle$ satisfies the second order recurrence

$$d_{n+1} = n(d_n + d_{n-1}) .$$

(b) Setting $b_{n+1} = d_{n+1} - (n+1)d_n$, show that $b_{n+1} + b_n = 0$ holds for all $n \geq 0$. This shows that $\langle d_n \rangle$ satisfies the first–order recurrence in the last problem.

(c) Find a finite sum that describes the probability that a permutation is a derangement. Show that this probability converges to $1/e$ as $n \to \infty$.

5

Nonnegative Difference Equations

In this chapter we consider **nonnegative difference equations**. Our prototype, the Fibonacci sequence

$$f_n = f_{n-1} + f_{n-2}; \quad f_0 = 0, \, f_1 = 1 \,,$$

is a nonnegative system because all coefficients and all initial values are nonnegative. We've shown earlier (in Chapter 1) that elements of the Fibonacci sequence have the form given in Binet's Formula

$$f_n = \frac{1}{\sqrt{5}}(\lambda_0^n - \lambda_1^n) \,,$$

where $\lambda_0 = (1 + \sqrt{5})/2$ and $\lambda_1 = (1 - \sqrt{5})/2$ are the roots of the characteristic polynomial, $ch(x) = x^2 - x - 1$. Although λ_0 and λ_1 are irrational, f_n is an integer, and in fact,

$$f_n = \text{Round}(\lambda_0^n/\sqrt{5}) \text{ for all } n \geq 0,$$

where $\text{Round}(X)$ is the function that returns the integer nearest to X. This implies the asymptotic size $f_n = \Theta(\lambda_0^n)$, where λ_0 is the positive eigenvalue of the recurrence. We'd like to generalize this example and discover what properties hold for solutions of the generalized problem. For example, for integer $k \geq 3$ the **generalized Fibonacci sequence** is the kth order recurrence

$$f_n^{(k)} = f_{n-1}^{(k)} + f_{n-2}^{(k)} + \cdots + f_{n-k}^{(k)}$$

with initial values

$$f_0^{(k)} = 0, \quad f_1^{(k)} = 1, \quad f_2^{(k)} = 2, \quad f_3^{(k)} = 2^2, \quad \ldots, \quad f_{k-1}^{(k)} = 2^{k-2} \,.$$

We might guess that $f_n^{(k)} = \Theta(\lambda_0^n)$, where $\lambda_0 = \lambda_0(k)$ is some nonnegative number that depends on k. A further generalization is the *homogeneous nonnegative* equation

$$(\text{HNN}) \qquad s_n = c_1 s_{n-1} + c_2 s_{n-2} + \cdots + c_k s_{n-k} \,,$$

where the c_i's are nonnegative and all initial conditions $s_0, \ldots, s_{k-1}$ are nonnegative. (We also assume that at least one of these initial conditions is positive, or else the solution is the zero sequence.) We will investigate whether any additional conditions are needed to ensure the existence of a nonnegative number λ_0 such that $s_n = \Theta(\lambda_0^n)$. We also consider the asymptotic size of solutions to *nonhomogeneous nonnegative* equations

$$(\text{NN}) \qquad s_n = c_1 s_{n-1} + c_2 s_{n-2} + \cdots + c_k s_{n-k} + g(n) \,,$$

where the c_i's are nonnegative, the initial conditions are nonnegative, and the $g(n)$ are functions with nonnegative values. Here we ask when $s_n = \Theta(g(n))$ or $s_n = \Theta(\lambda_0^n)$ for some λ_0. As in previous chapters, much information about the recurrence is encoded in its characteristic polynomial.

5.1   Nonnegative Polynomials

The purpose of this section is to study polynomials of the form

$$(5.1) \quad p(x) = x^k - c_1 x^{k-1} - \cdots - c_{k-1} x - c_k \,, \quad \text{with all } c_i \geq 0 \text{ and } c_k \neq 0.$$

These polynomials are called **nonnegative**, and are the characteristic polynomials of nonnegative linear recurrences, recurrences that have the form (NN). The study of these polynomials takes us through some classical areas of mathematics that are often omitted from an undergraduate education but are currently experiencing at least a small revival because of applications to computing, robotics, and other areas.

5.1.1   The dominant root

From the Fundamental Theorem of Algebra (refer to Appendix B) we know that any polynomial $p(x)$ has $\deg(p)$ complex roots when the roots are counted according to multiplicity. It's difficult to say how many of these roots are real. One of the oldest results in this direction is Descartes' Rule of Signs, which relates the number of positive roots of a polynomial with real coefficients to the number of sign changes among its coefficients. For us, a **sign change** occurs in the polynomial when some coefficient is positive and the next non-zero coefficient is negative, or vice versa. For instance, $x^2 - 2x + 1$ has two sign changes, and $x^{30} + x^{20} - 5x^{10} - 7$ has just one sign change. Descartes' Rule bounds the number of positive roots of a real polynomial by its number of sign changes. Here's the statement of the rule and a proof.

Theorem 5.1.1 (Descartes' Rule of Signs). *The number of positive roots (counted according to multiplicity) of a polynomial with real coefficients is no more than the number of sign changes among its coefficients.*

Proof. We proceed by induction both on the number of sign changes n in the polynomial $p(x) = c_0 x^k + \cdots + c_k$ and on the degree k. When there are no sign changes, $p(x)$ has the same sign at every positive number and therefore has no positive root. Because of this we may assume $n > 0$. The induction hypothesis is that any polynomial with fewer than n sign changes or with exactly n sign changes but whose degree is less than k has at most n positive (real) roots.

If $c_k = 0$, then $p(x) = xP(x)$ for some polynomial $P(x)$ that also has n sign changes and whose degree is $k - 1$. The polynomial $P(x)$ is covered by our induction hypothesis, and its number of positive roots is therefore at most n. Since $p(x) = xP(x)$, every non-zero root of $p(x)$ is a root of $P(x)$. This gives the conclusion for such $p(x)$ and allows us to assume that $c_k \neq 0$.

Without loss of generality we may assume that $p(x)$ has at least one positive root, and set λ_0 to be the least of its positive roots. Consider the derivative of $p(x)$,

$$p'(x) = kc_0 x^{k-1} + (k-1)c_1 x^{k-2} + \cdots + 2c_{k-2} x + c_{k-1} .$$

The number of sign changes in $p'(x)$ either equals the number for $p(x)$ or is one less, depending on whether c_{k-1} and c_k have the same or opposite sign. By the continuity of $p(x)$ and $p'(x)$, between every two different consecutive real roots of $p(x)$ there must be a local extremum and so at least one root of $p'(x)$. Together with Exercise 5.1 this implies that $p(x)$ can have at most one more positive root than $p'(x)$ on the interval $[\lambda_0, \infty)$. The proof is completed by showing that $p'(x)$ has at most $n - 1$ roots in this interval (because then the number of positive roots of $p(x)$ is at most n, since λ_0 was chosen as its least positive root.)

If c_k and c_{k-1} have opposite signs, then $p'(x)$ has $n - 1$ sign changes. Our induction hypothesis therefore limits the number of positive roots of $p'(x)$ to $n - 1$, as claimed. If c_k and c_{k-1} have the same sign, both $p(x)$ and $p'(x)$ have n sign changes, and by induction $p'(x)$ has at most n positive roots. We will show that $p'(x)$ has at least one root in the interval $(0, \lambda_0)$. This follows from a bit of calculus. First considering the case in which $c_k > 0$, the values of both $p(x)$ and $p'(x)$ on any sufficiently small interval containing zero are dominated by their constant terms and therefore are both positive. So $p(x)$ is increasing and positive at $x = 0$ but must decrease to $p(\lambda_0) = 0$, and $p(x)$ has a relative maximum (which is a root of $p'(x)$) on the interval $(0, \lambda_0)$. When $c_k < 0$, this argument can be applied to

$q(x) = -p(x)$ to obtain a relative minimum for $p(x)$. In either case, $p(x)$ has a relative extremum on $(0, \lambda_0)$, and continuity yields a root of $p'(x)$ in the interval $(0, \lambda_0)$ as claimed.   □

In 1637 Descartes stated this rule of signs without proof in his famous book *La Géométrie* [149]. During the next two centuries several others proved and refined the rule. Among these was Carl Friedrich Gauss, [70] who in 1828 provided the additional information that the difference between the number of positive roots and the number of sign changes is always an even number. **Sturm's Theorem** (*Mémoire sur la résolution des équations numériques*, published in 1829) is another classic theorem in this vein. In its simplest version it yields an algorithm for counting the number of real roots of $P(x)$ in any interval. More complicated versions allow the determination of the number of roots of a polynomial $P(x)$ subject to sign constraints on another polynomial $Q(x)$. These theorems fell into relative obscurity until revived by Tarski in 1940 [157] to prove an abstract result in mathematical logic.

Nonnegative polynomials $p(x)$ have exactly one sign change and so can have *at most* one positive root. On the other hand, since $p(0) = -c_k$ is negative and $\lim_{x \to +\infty} p(x) = +\infty$, the Intermediate Value Theorem implies that $p(x)$ has *at least* one positive root. Therefore, we obtain the following corollary.

Corollary 5.1.2. *A nonnegative polynomial has exactly one positive root and it is a simple root.*

We next show that among all roots of a nonnegative polynomial, its sole positive root has the largest complex modulus.

Theorem 5.1.3. *If λ_0 is the positive root of a nonnegative polynomial $p(x)$, then λ_0 is a **dominant root**, in the sense that any other root $\lambda \in \mathbb{C}$ satisfies $|\lambda| \le \lambda_0$.*

Proof. If λ is any root of $p(x)$, then

$$\lambda^k = c_1 \lambda^{k-1} + c_2 \lambda^{k-2} + \cdots + c_k ,$$

and the absolute value inequality yields

$$|\lambda|^k = |\lambda^k| \le |c_1 \lambda^{k-1}| + |c_2 \lambda^{k-2}| + \cdots + |c_k|$$
$$= c_1 |\lambda|^{k-1} + c_2 |\lambda|^{k-2} + \cdots + c_k .$$

We conclude that

$$p(|\lambda|) = |\lambda|^k - c_1 |\lambda|^{k-1} - \cdots - c_k \le 0 ,$$

and from Exercise 5.2 this implies $|\lambda| \le \lambda_0$.   □

Do nonnegative polynomials ever have other roots of modulus λ_0? Not only is it possible to show that the answer to this question is yes, but we can actually find all roots whose modulus is λ_0 without doing much work at all. This is described in the next theorem. We again make use of the mth **roots of unity** (refer to Appendix B), complex numbers ζ such that $\zeta^m = 1$.

Theorem 5.1.4. *Let λ_0 be the positive root of the nonnegative polynomial $p(x) = x^k - c_1 x^{k-1} - \cdots - c_k$. If the* **index of imprimitivity** *g is defined to be the gcd of the set of indices of non-zero coefficients in $p(x)$, then $p(x)$ has exactly g roots of modulus λ_0, and these are the complex numbers of the form $\lambda_0 \zeta$, where ζ is a gth root of unity.*

Before proving this result we consider a few examples. For a polynomial of the form $p_1(x) = x^k - 1$ we have $g = k$, and the roots of the polynomial *are* the gth roots of unity. For example, $x^4 - 1$ has the four roots $\pm 1, \pm i$, which can be written as i^0, i^1, i^2, i^3 since i is a primitive fourth root of unity. Next, let's look at the polynomial $p_2(x) = x^4 - x^2 - 1 = (x^2)^2 - (x^2) - 1$. To find its positive root, we apply the quadratic formula to get

$$x^2 = \frac{1 \pm \sqrt{5}}{2},$$

which leads to its positive root

$$\lambda_0 = \sqrt{\frac{1 + \sqrt{5}}{2}}.$$

Since the only non-zero coefficients are c_2 and c_4, then $g = 2$, the gth roots of unity are ± 1, and the theorem correctly says that $\pm \lambda_0$ are the two roots whose modulus is maximal. For a more interesting example, consider the nonnegative polynomial $p_3(x) = x^9 - 3x^6 - 7$. Even though we don't know the value of λ_0, from the theorem we know that it has exactly three roots of modulus λ_0, since $\gcd\{3, 9\} = 3$.

Proof of Theorem 5.1.4. If ζ is both a second root *and* a third root of unity, ζ is in the set $\{-1, 1\}$ and also in the set $\{1, e^{2\pi/3}, e^{4\pi/3}\}$, which implies that $\zeta = 1$. This observation is generalized in Exercise 5.6 where you show that if g is the gcd of a finite set G of integers, then ζ is a gth root of unity iff ζ is an ith root of unity for every $i \in G$. We apply this here with $G = \{i : c_i > 0\}$, the set of indices of the negative coefficients of $p(x)$. Since λ_0 is the dominant root of $p(x)$, it suffices to prove that ω is a root of $p(x)$ with $|\omega| = \lambda_0$ iff $\zeta = \omega/\lambda_0$ is an ith root of unity for all $i \in G$.

If ζ is a complex number with $\zeta^i = 1$ for all $i \in G$, then

$$\zeta^{-k} \cdot p(\lambda_0 \zeta) = \lambda_0^k - \sum_{i \in G} c_i \lambda_0^{k-i} \zeta^{-i} = \lambda_0^k - \sum_{i \in G} c_i \lambda_0^{k-i} = p(\lambda_0) = 0,$$

and $\omega = \lambda_0 \zeta$ is indeed a root of $p(x)$. On the other hand, suppose $\omega = \zeta \lambda_0$ is a root of $p(x)$ with $|\omega| = \lambda_0$. Then

$$0 = p(\zeta \lambda_0) = \zeta^k \lambda_0^k - c_1 \zeta^{k-1} \lambda_0^{k-1} - \cdots - c_k \,;$$

$$\lambda_0^k = \sum_{i \in G} c_i \lambda_0^{k-i} \zeta^{-i} \,,$$

and $p(\lambda_0) = 0$ implies

$$\lambda_0^k = c_1 \lambda_0^{k-1} + \cdots + c_k = \sum_{i \in G} c_i \lambda_0^{k-i} \,.$$

Subtracting these two equations gives

$$(5.2) \qquad\qquad 0 = \sum_{i \in G} c_i \lambda_0^{k-i} (1 - \zeta^{-i}) \,.$$

Notice that since each ζ^{-i} lies on the unit circle, the real part of each $1 - \zeta^{-i}$ is $1 - \cos(-i\theta)$ (where θ is the argument of ζ) and so must be nonnegative. Taking the real part of the two sides of (5.2), the left side is 0 and the right side contains only nonnegative terms, since all λ_0^{k-i} are positive (remember that $\lambda_0 > 0$). Therefore, the real part $\Re(1 - \zeta^{-i})$ must equal zero for all $i \in G$. Again using the fact that each ζ^{-i} lies on the unit circle, we note that $\Re(1 - \zeta^{-i}) = 0$ iff $\zeta^{-i} = 1$, and we've proved that ζ is an ith root of unity for all $i \in G$. $\qquad\square$

Returning to our examples, we see the fact that the index of imprimitivity for $p_2(x)$ and $p_3(x)$ doesn't equal 1 allows us to write the polynomials as $p_2(x) = (x^2)^2 - (x^2) - 1$ and $p_3(x) = (x^3)^3 - 3(x^3)^2 - 7$, displaying a sort of periodicity. Because of this, we call a nonnegative polynomial **periodic** when its index of imprimitivity does not equal one, otherwise we call it **primitive** (or **aperiodic**). Combining Theorem 5.1.3 with Theorem 5.1.4, we obtain the following corollary.

Corollary 5.1.5. *Any primitive nonnegative polynomial has a unique root of largest modulus, usually referred to as the* **strictly dominant root***.*

5.2   When are integer solutions rounded powers of an eigenvalue?

We've shown that the Fibonacci sequence satisfies

$$(5.3) \qquad\qquad f_n = \text{Round}(\lambda_0^n / \sqrt{5}) \text{ for all } n \geq 0 \,,$$

where $\text{Round}(X)$ returns the integer nearest to X. Is this a special property of the Fibonacci sequence, or are there more general sequences with this feature? The proof of (5.3) did use properties of the Fibonacci sequence. For instance, the absolute value of the nondominant eigenvalue is relatively small when compared to λ_0, and the initial deviations are small. When we speak of **deviation** here we mean the difference between the actual element of the sequence and the approximation. (Occasionally we may use the term deviation when we mean the absolute value of the deviation.) For example,

$$d_n = f_n - \lambda_0^n/\sqrt{5}$$

is the nth Fibonacci deviation. More generally, we want to approximate the elements of an integer recurrence sequence $\langle s_n \rangle$ by numbers of the form $\alpha\lambda_0^n$, where λ_0 is the dominant eigenvalue and α is a constant that depends on the initial values. We'd like the absolute values of the deviations $d_n = s_n - \alpha\lambda_0^n$ to start small and stay small, although they need not be decreasing. (The results in this section are based on Capocelli and Cull [23].)

Let's first consider the special case in which the eigenvalues, $\lambda_0, \ldots, \lambda_{k-1}$, are all simple. From previous work we know that then s_n can be written as

$$s_n = \sum_{i=0}^{k-1} \alpha_i \lambda_i^n$$

for some constants $\alpha_0, \ldots, \alpha_{k-1} \in \mathbb{C}$. For the approximation of s_n by $\alpha_0\lambda_0^n$ the deviations satisfy

$$d_n = s_n - \alpha_0\lambda_0^n = \sum_{i=1}^{k-1} \alpha_i \lambda_i^n,$$

and the familiar absolute value inequality gives

(5.4) $$|d_n| \leq \sum_{i=1}^{k-1} |\alpha_i||\lambda_i|^n.$$

For instance, if each nondominant eigenvalue were to satisfy $|\lambda_i| \leq 1$, then $s_n = \text{Round}(\alpha_0\lambda_0^n)$ holds whenever $\sum_{i=1}^{k-1} |\alpha_i| < 1/2$. This result isn't as good as it might seem, because in order to apply it we have to calculate all $\alpha_1, \ldots, \alpha_{k-1}$! In an effort to obtain a more effective result, let's perform a more thorough analysis of (5.4) from a graphical point of view. The deviations can be plotted as points (n, d_n), which can dance around (perhaps erratically) beneath the envelope formed by the bounding function. If the deviations behave very irregularly, any upper bound on them may be a dramatic overestimation. One possible type of irregularity is a **spiking behavior**, where the deviations might be close to zero for a while, say for $d_0, d_1, \ldots, d_5$, but then d_6 is relatively large in absolute value. Such spiking could occur if $\lambda_1 = (1 - \epsilon)\omega$, where ϵ is a small positive number and

ω is a primitive 6th root of unity. Longer period spiking is also possible. For instance, if ω_i is a primitive i^{th} root of unity with $\lambda_1 = (1 - \epsilon_1)\omega_3$ and $\lambda_2 = (1 - \epsilon_2)\omega_5$, then spiking behavior with period 15 can occur if the period-3 spike augments the period-5 spike to give a large spike of period 15. It's possible to construct sequences with even longer periods, because a number of short periods could join together in such a way to give a long period. The moral here is that one pays for the smoothness of the bound, since the estimating curve might not be very close to the deviations. For general recurrences, the enveloping curve given by the upper bound in (5.4) may well be the best *easy* estimate on the deviations. We will now show an additional restriction on the characteristic polynomial that can give a stronger result that is often quick and easy to check.

Notice that since λ_0 is a *simple* eigenvalue, then any solution has the form

$$s_n = \alpha\lambda_0^n + a_1(n)\lambda_1^n + \cdots + a_{k-1}(n)\lambda_{k-1}^n \text{ for all } n \geq 0,$$

for some constant α and polynomials $a_1(x), \ldots, a_{k-1}(x)$.

Theorem 5.2.1. *Let $s_n = \alpha\lambda_0^n + a_1(n)\lambda_1^n + \cdots + a_{k-1}(n)\lambda_{k-1}^n$ be an integer solution to (HNN) with dominant eigenvalue λ_0. Set $d_n = s_n - \alpha\lambda_0^n$ for all $n \geq 0$ and let $M = \max\{|d_0|, \ldots, |d_{k-1}|\}$, the maximum of the initial deviations. We further assume that $f(x) = (x - 1)ch(x)/(x - \lambda_0)$ is a nonnegative polynomial.*

(a) If $M < \frac{1}{2}$, then $s_n = \text{Round}(\alpha\lambda_0^n)$ for all $n \geq 0$.

(b) If $f(x)$ is primitive, $s_n = \text{Round}(\alpha\lambda_0^n)$ for all sufficiently large n because there is an n_0 with $\max\{|d_{n_0}|, \ldots, |d_{n_0+k-1}|\} < 1/2$.

Proof. There exist nonnegative $b_1, \ldots, b_k$ such that $f(x) = x^k - b_1 x^{k-1} - \cdots - b_k$, where $1 - b_1 - b_2 - \cdots - b_k = 0$, since $\lambda = 1$ is a root of $f(x)$. We rewrite this as $\sum b_i = \sum |b_i| = 1$ and obtain that any solution $\langle y_n \rangle$ of the homogeneous recurrence with characteristic polynomial $f(x)$ must satisfy

(5.5) $|y_n| \leq \max\{|y_0|, |y_1|, \ldots, |y_{k-1}|\}$ for all n.

(Refer to Exercise 5.9.) Since

$$d_n = s_n - \alpha\lambda_0^n = a_1(n)\lambda_1^n + \cdots + a_{k-1}(n)\lambda_{k-1}^n$$

is a solution to the difference equation with characteristic polynomial $f(x)$, (5.5) holds with $y_n = d_n$. Therefore, if the hypothesis of (a) holds, all deviations satisfy $|d_n| \leq M < 1/2$, and $s_n = \text{Round}(\alpha\lambda_0^n)$.

In part (b), the fact that f is primitive means that its only positive root $\lambda = 1$ is *strictly* dominant. Ordering the roots by

$$\lambda_0 = 1 > |\lambda_1| \geq |\lambda_2| \geq \cdots \geq |\lambda_{k-1}|,$$

we have

(5.6) $$|d_n| \leq \sum_{i=1}^{k-1} |a_i(n)||\lambda_i|^n \leq |\lambda_1|^n \sum_{i=1}^{k-1} |a_i(n)|.$$

Since each $a_i(n)$ is a polynomial whose degree is less than the multiplicity of λ_i, then $\deg(a_i) < k$ and we can find positive constants N, C such that each $|a_i(n)|$ is less than Cn^k for all $n \geq N$. Equation (5.6) gives

$$|d_n| \leq |\lambda_1|^n Ckn^k \quad \text{for all } n \geq N.$$

Since $|\lambda_1|^n$ is exponentially decreasing to 0 while n^k is only growing at polynomial rate, the inequality $|d_n| < 1/2$ eventually holds, and part (a) yields $s_n = \text{Round}(\alpha \lambda_0^n)$ for sufficiently large n. $\qquad \square$

Do we have to explicitly find $f(x)$ (and so calculate λ_0) in order to apply this result? Fortunately, there is a relatively easy condition that ensures that f is nonnegative.

Let $\langle H_n \rangle$ be the sequence defined by the first–order recurrence

$$H_0 = 1, \quad H_n = \lambda_0 H_{n-1} - \psi(n),$$

where $\psi(n)$ is the sequence whose first k terms are the coefficients $c_1, \ldots, c_k$ and all later terms are zero. Since λ_0 is a root of $ch(x) = x^k - c_1 x^{k-1} - \cdots - c_k$, then

$$H_k = \lambda_0 H_{k-1} - c_k = \lambda_0(\lambda_0 H_{k-2} - c_{k-1}) - c_k = \cdots = ch(\lambda_0) = 0,$$

and Theorem 3.1.2 gives

$$(5.7) \qquad H_n = \begin{cases} \lambda_0^n - \sum\limits_{i=1}^{n} c_i \lambda_0^{n-i} & \text{for } n < k, \\ 0 & \text{for } n \geq k. \end{cases}$$

Also, it can be checked that

$$(x - \lambda_0)(x^{k-1} + H_1 x^{k-2} + \cdots + H_{k-1}) = ch(x),$$

and the polynomial $f(x)$ is

$$f(x) = (x - 1)\frac{ch(x)}{x - \lambda_0} = (x - 1)(x^{k-1} + H_1 x^{k-2} + \cdots + H_{k-1})$$

$$= x^k - (1 - H_1)x^{k-1} - (H_1 - H_2)x^{k-2} - \cdots - (H_{k-2} - H_{k-1})x - H_{k-1}.$$

The nonnegativity of $f(x)$ is therefore equivalent to

$$(5.8) \qquad 1 \geq H_1 \geq H_2 \geq \cdots \geq H_{k-1} > 0,$$

where the condition $H_{k-1} > 0$ is free because

$$0 = H_k = \lambda_0 H_{k-1} - c_k \quad \text{and} \quad H_{k-1} = c_k/\lambda_0 > 0.$$

It might seem that testing $H_1 = \lambda_0 - c_1 \leq 1$ requires knowledge of λ_0. But recall (refer to Exercise 5.2) that we have the characterization of λ_0 given by

$$\text{for positive } x, \quad ch(x) \geq 0 \iff x \geq \lambda_0,$$

a condition that is practical and quick to check. (In Section 5.4 we discuss Horner's Method, in which polynomial evaluation is performed in at most k multiplications.) When the strict inequality $ch(c_1 + 1) > 0$ holds, the coefficient of x^{k-1} in $f(x)$ must be positive, and so $f(x)$ is definitely primitive. As far as the other conditions in (5.8), from (5.7) we know that for all $n < k - 1$,

$$H_{n+1} = \lambda_0^{n+1} - c_1\lambda_0^n - \sum_{i=2}^{n+1} c_i\lambda_0^{n+1-i} = \lambda_0^n(\lambda_0 - c_1) - \sum_{i=1}^{n} c_{i+1}\lambda_0^{n-i},$$

which gives

$$H_n - H_{n+1} = \lambda_0^n(c_1 + 1 - \lambda_0) + \sum_{i=1}^{n}(c_{i+1} - c_i)\lambda_0^{n-i}.$$

Therefore, $H_n > H_{n+1}$ is implied by the conditions

$$c_1 + 1 > \lambda_0 \quad \text{and} \quad c_{n+1} \geq c_{n+1} \geq \cdots \geq c_2 \geq c_1,$$

which are easy-to-check restrictions on the coefficients of the original characteristic polynomial $ch(x)$. These observations are collected in the following theorem.

Theorem 5.2.2 (The Rounding Theorem). *Let λ_0 be the dominant eigenvalue of a homogeneous nonnegative recurrence (HNN) whose coefficients satisfy*

(5.9) $c_{k-1} \geq \cdots \geq c_1 \quad$ *and* $\quad ch(c_1 + 1) \geq 0.$

Suppose $\langle s_n \rangle$ is a integer solution of (HNN), then:
  (a) If $\max\{|d_0|, \ldots, |d_{k-1}|\} < 1/2$, then $s_n = \text{Round}(\alpha\lambda_0^n)$ for all $n \geq 0$.
  (b) If the strict inequality $ch(c_1+1) > 0$ holds, then $s_n = \text{Round}(\alpha\lambda_0^n)$ for all $n \geq n_0$, where n_0 is such that $\max\{|d_{n_0}|, \ldots, |d_{n_0+k-1}|\} < 1/2$.

5.2.1   Using the Rounding Theorem

As we've seen before, the Fibonacci recurrence can be generalized to a kth order recurrence for any $k \geq 3$ by

$$f_0 = 0, \quad f_1 = 1, \quad f_2 = 2, \quad \ldots, \quad f_{k-1} = 2^{k-2},$$
$$f_n = f_{n-1} + f_{n-2} + \cdots + f_{n-k},$$

which has

$$ch(c_1 + 1) = ch(2) = 2^k - 2^{k-1} - \cdots - 2 - 1 = 2^k - (2^k - 1) = 1 > 0$$

and $c_{i+1} \geq c_i$, since each c_i is 1. Hence the Rounding Theorem ensures the existence of some α and $n_0 \geq 0$ such that $f_n = \text{Round}(\alpha\lambda_0^n)$ for $n \geq n_0$, where the values of α and n_0 depend on the initial conditions. We can work backwards from the initial values and obtain an equivalent set of initial conditions, $f_{-(k-2)} = f_{-(k-3)} = \cdots = f_{-1} = f_0 = 0$ and $f_1 = 1$. (This translation doesn't change the value of either α or n_0 and allows an easier computation of α and the deviations.) From Exercise 4.23,

$$\alpha = \frac{\lambda_0 - 1}{\lambda_0[(k+1)\lambda_0 - 2k]},$$

and the corresponding deviations are

$$d_{-(k-2)} = 0 - \alpha\lambda_0^{-(k-2)},$$
$$d_{-(k-3)} = 0 - \alpha\lambda_0^{-(k-3)},$$
$$\vdots$$
$$d_0 = 0 - \alpha,$$
$$d_1 = 1 - \alpha\lambda_0.$$

Notice that $\max\{|d_{-(k-2)}|, |d_{-(k-3)}|, \ldots, |d_0|, |d_1|\} = \max\{|d_0|, |d_1|\}$ because $ch(1) = -(k-1)$ implies $\lambda_0 > 1$. Also, d_1 is positive, because otherwise d_n is always negative and so has a λ_0^n component. If we can show that both $\alpha < 1/2$ and $1 - \alpha\lambda_0 < 1/2$, then we can take $n_0 = -(k-2)$, and the generalized Fibonacci numbers can be calculated by $f_n = \text{Round}(\alpha\lambda_0^n)$ for all $n \geq -(k-2)$.

To show that $1 - \alpha\lambda_0 < 1/2$, it suffices to have $\dfrac{2(\lambda_0 - 1)}{(k+1)\lambda_0 - 2k} > 1$. From Exercise 5.4, the denominator is positive, and the requirement can be written as $0 > (k-1)(\lambda_0 - 2)$, which is true because $2 > \lambda_0$.

To show that $1/2 > \alpha$, we want

$$\frac{1}{2} > \frac{\lambda_0 - 1}{\lambda_0[(k+1)\lambda_0 - 2k]}$$

which can be rewritten as $2 > (k+1)\lambda_0(2 - \lambda_0)$ (since the denominator is positive). Using $2 - \lambda_0 = \lambda_0^{-k}$, this is equivalent to the inequality $2\lambda_0^{k-1} > k+1$. For $k = 2$, this reduces to $\lambda_0 > 3/2$, which is easy to verify. For $k > 2$, we use $\lambda_0^{k-1} = \lambda_0^{k-2} + \cdots + 1 + \frac{1}{\lambda_0}$ to get $2\lambda_0^{k-1} = 2(\lambda_0^{k-2} + \cdots + \frac{1}{\lambda_0}) > 2(k-1)$ using the fact that $\lambda_0 > 1$. Finally, $2(k-1) \geq k+1$ if $k \geq 3$, and $1/2 > \alpha$ is established. For another example, let's consider

(5.10) $$s_n = 2s_{n-1} + 2s_{n-2} + 3s_{n-3}$$

with $ch(x) = x^3 - 2x^2 - 2x - 3$, a nonnegative polynomial with dominant root $\lambda_0 = 3$. Here $c_2 = 2 \geq c_1$ and $c_1 + 1 = 3 = \lambda_0$. Because $c_1 + 1$ equals the dominant root, eventual roundability is not ensured but depends on the deviations, which in turn are specified by the initial conditions. Next we look at this recurrence under various choices of initial conditions, using the fact that $\alpha = \frac{1}{13}(s_0 + s_1 + s_2)$. (Refer to Exercise 5.10.)

For $s_0 = 1, s_1 = 3, s_2 = 9$, then $\alpha = 1$ and the deviations are:

$$
\begin{aligned}
d_0 &= s_0 - \alpha &&= 1 - 1 = 0\,, \\
d_1 &= s_1 - \alpha\lambda_0 = 3 - 3 = 0\,, \\
d_2 &= s_2 - \alpha\lambda_0^2 = 9 - 9 = 0\,,
\end{aligned}
$$

and the Rounding Theorem implies

$$
s_n = \text{Round}(\alpha\lambda_0^n) = \text{Round}(3^n) = 3^n\,.
$$

(This sequence could have been obtained directly without using the Rounding Theorem.) For the initial conditions $s_0 = 0, s_1 = 0, s_2 = 1$, roundability is not obvious without the theorem. The coefficient is $\alpha = 1/13$ with deviations

$$
\begin{aligned}
d_0 &= 0 - 1/13 = -1/13\,, \\
d_1 &= 0 - 3/13 = -3/13\,, \\
d_2 &= 1 - 9/13 = 4/13\,.
\end{aligned}
$$

Since the absolute values of the initial deviations are bounded by $1/2$, then $s_n = \text{Round}(\frac{1}{13}\lambda_0^n)$ for all $n \geq 0$. Our final choice of initial conditions for the recurrence (5.10) is $s_0 = 0, s_1 = 3, s_2 = 9$. Here, $\alpha = 12/13$ and the deviations are

$$
\begin{aligned}
d_0 &= 0 - \frac{12}{13} &&= -\frac{12}{13}\,, \\
d_1 &= 3 - \frac{12}{13} * 3 &&= \frac{3}{13}\,, \\
d_2 &= 9 - \frac{12}{13} * 3^2 &&= \frac{9}{13}\,.
\end{aligned}
$$

In this case, the initial deviations are not all less than $1/2$ in absolute value and neither immediate rounding nor eventual rounding is promised by the Rounding Theorem since $c_1 + 1 = \lambda_0$. It is easy to calculate that

$$
\begin{aligned}
d_3 &= 24 - \frac{12}{13} * 3^3 &&= -\frac{12}{13}\,, \\
d_4 &= 75 - \frac{12}{13} * 3^4 &&= \frac{3}{13}\,, \\
d_5 &= 225 - \frac{12}{13} * 3^5 &&= \frac{9}{13}\,,
\end{aligned}
$$

and in Exercise 5.11 you verify that this pattern continues, that is, the sequence of deviations is periodic with period 3. Because the deviations do not decrease below $1/2$, even eventual rounding doesn't occur.

5.3   Estimation of the Roots

5.3.1   Estimation of the dominant root

In Section 5.5 we will show that for nonnegative difference equations, $s_n = \Theta(\lambda_0^n)$ often holds. In fact, when the nonnegative equation is *primitive*, the dominant eigenvalue λ_0 is simple and $\lim_{n \to \infty} (s_n/\lambda_0^n) = \alpha$ holds for some constant α. Because of this, we want to estimate the dominant root of the recurrence in order to find the long-term behavior of solutions.

Lemma 5.3.1. *The root λ_0 depends on $c_1 + c_2 + \cdots + c_k$ in the following way:*

(1) *if $c_1 + \cdots + c_k = 1$, then $\lambda_0 = 1$;*

(2) *if $c_1 + \cdots + c_k > 1$, then $1 < \lambda_0 < c_1 + \cdots + c_k$;*

(3) *if $c_1 + \cdots + c_k < 1$, then $c_1 + \cdots + c_k < \lambda_0 < 1$.*

Proof. We observe that $p(1) = 1 - S$, where $S = c_1 + \cdots + c_k$. Since λ_0 is the only positive root of $p(x)$, then $\lambda_0 = 1$ iff $S = 1$, which proves (1) and allows us to assume $\lambda_0 \neq 1$. Recall that

$$\lambda_0 > 1 \iff 0 > p(1) = 1 - S \iff S > 1,$$

the required relative ordering of λ_0 and 1. Further,

$$\begin{aligned} p(S) &= S^k - c_1 S^{k-1} - \cdots - c_{k-1} S - c_k \\ &= (S - c_1) S^{k-1} - c_2 S^{k-2} - \cdots - c_{k-1} S - c_k \\ &= (c_2 + \cdots + c_k) S^{k-1} - c_2 S^{k-2} - \cdots - c_{k-1} S - c_k \\ &= c_2 S^{k-2}(S - 1) + c_3 S^{k-3}(S^2 - 1) + \cdots + c_{k-1} S(S^{k-2} - 1) \\ &\qquad + c_k(S^{k-1} - 1), \end{aligned}$$

and the facts that each c_i is nonnegative and $c_k > 0$ imply

$$S > 1 \iff p(S) > 0 \iff S > \lambda_0,$$

as required. $\qquad\qquad\qquad\qquad\qquad\qquad\qquad\qquad\qquad\qquad\qquad\qquad\qquad\quad$ $\square$

This lemma gives bounds on the dominant root, positioning it between 1 and $c_1 + \cdots + c_k$. This will be a useful first approximation for Newton's method in Section 5.4.

5.3.2   Estimation of the second root

While estimates of the dominant root of the recurrence are used to find the asymptotic behavior, bounds on the second eigenvalue tell how quickly and in what sense the solution approaches the long-term behavior. When the nonnegative equation is primitive, there exists a polynomial $\beta(x)$ such that

$$|s_n - \alpha\lambda_0^n| < \beta(n)|\lambda_1|^n \,,$$

where λ_1 is the maximum of the nondominant eigenvalues. From this we see that when $|\lambda_1| < 1$, there exist positive constants β, M with $M < 1$ such that

(5.11) $$|s_n - \alpha\lambda_0^n| < \beta M^n \,,$$

and $\langle s_n \rangle$ converges to $\alpha\lambda_0^n$ in the sense of **absolute error**. On the other hand, when $|\lambda_1| \geq 1$ there exist β and M with $1 < M < \lambda_0$ such that

(5.12) $$\left|\frac{s_n}{\lambda_0^n} - \alpha\right| < \beta\left(\frac{M}{\lambda_0}\right)^n \,,$$

and $\langle s_n \rangle$ is said to converge to $\alpha\lambda_0^n$ in the sense of **relative error**.

In Appendix D we describe an algorithm for counting the number of roots of a polynomial within the (complex) unit circle $|z| < 1$. If the algorithm finds that the kth degree characteristic polynomial has at least $k - 1$ roots inside the unit circle, then $|\lambda_1| < 1$, and the above analysis gives absolute error convergence of $\langle s_n \rangle$ to $\alpha\lambda_0^n$. On the other hand, if it's found that there are fewer than $k-1$ roots within the circle, only relative error convergence is obtained from this argument. The method in Appendix D does not actually compute an M with $\lambda_0 > M > |\lambda_1|$. We might instead take the nonnegative polynomial $p(x)$ and its positive root λ_0 and find the least M for which $(x - M)p(x)/(x - \lambda_0)$ is a nonnegative polynomial. Unfortunately, this method may only yield $M = \lambda_0$. Next we prove some upper bounds on the nondominant roots of a nonnegative polynomial.

Theorem 5.3.2. *Consider a polynomial* $p(x) = x^k - c_1x^{k-1} - \cdots - c_k$, *where all* c_i *are strictly positive (and so* $p(x)$ *is primitive.) Then*

$$M = \max\left\{\frac{c_2}{c_1}, \frac{c_3}{c_2}, \ldots, \frac{c_k}{c_{k-1}}\right\}$$

is an upper bound on $|\lambda_1|$.

Proof. Since the conclusion follows if $M > \lambda_0$, we may assume $M \leq \lambda_0$. Multiplying $p(x) = x^k - c_1x^{k-1} - \cdots - c_{k-1}x - c_k$ by $x - M$ gives

$$x^{k+1} - (c_1 + M)x^k + (Mc_1 - c_2)x^{k-1} + \cdots + (Mc_{k-1} - c_k)x + c_kM \,,$$

a polynomial that we call $q(x)$. Since any root of $p(x)$ is also a root of $q(x)$, then $q(\lambda_1) = 0$, which means that

$$\lambda_1^{k+1} + (Mc_1 - c_2)\lambda_1^{k-1} + \cdots + (Mc_{k-1} - c_k)x + c_kM = (c_1 + M)\lambda_1^k \,.$$

The definition of M implies that all coefficients on the left side are positive, and taking absolute values gives

$$|\lambda_1|^{k+1} + (Mc_1 - c_2)|\lambda_1|^{k-1} + \cdots + c_k M \geq (c_1 + M)|\lambda_1|^k,$$

with $q(|\lambda_1|) = (|\lambda_1| - M)p(|\lambda_1|) \geq 0$. Because $p(x)$ is a nonnegative polynomial, its values change from negative to positive at $x = \lambda_0$, and primitivity implies that λ_0 is strictly dominant. This means that $p(|\lambda_1|) < 0$, and $q(|\lambda_1|) \geq 0$ gives $|\lambda_1| \leq M$. □

For example, if $c_1 > c_2 > c_3 > \cdots > c_k > 0$, then $|\lambda_1| < M < 1$, and (5.11) says that any nonnegative solution $\langle s_n \rangle$ has limiting behavior in the *absolute* error sense, that is, $\lim_{n \to \infty} |s_n - \alpha \lambda_0^n| = 0$.

If any of $c_1, \ldots, c_{k-1}$ are zero, then Theorem 5.3.2 cannot be used to get an upper bound on λ_1. Another result is the following.

Theorem 5.3.3. *Let* $p(x) = x^k - c_1 x^{k-1} - \cdots - c_k$ *be a nonnegative polynomial with positive root* λ_0. *If* $H_i = \lambda_0^i - c_1 \lambda_0^{i-1} - \cdots - c_i$ *for all* $i = 1, \ldots, k$ *(as used earlier in (5.7)), then the maximum of the absolute values of the roots of* $p(x)$ *excluding* λ_0 *satisfies*

$$(5.13) \qquad |\lambda_1| \leq \max\left\{ H_1, \frac{H_2}{H_1}, \frac{H_3}{H_2}, \ldots, \frac{H_{k-1}}{H_{k-2}} \right\}.$$

Proof. (In Exercise 5.14 you show that each H_i is positive.) We may assume that the maximum in (5.13) is strictly less than λ_0, since otherwise the conclusion holds. The roots of the polynomial $p(x)/(x - \lambda_0)$ are $\lambda_1, \ldots, \lambda_{k-1}$, and

$$\frac{p(x)}{x - \lambda_0} = x^{k-1} + H_1 x^{k-2} + \cdots + H_{k-1}.$$

Also, for any positive w the polynomial $g(x) = (x - w)\dfrac{p(x)}{x - \lambda_0}$ has only one positive root and

$$g(x) = x^k - (w - H_1)x^{k-1} - (H_1 w - H_2)x^{k-2}$$
$$- \cdots - (H_{k-2}w - H_{k-1})x - w H_{k-1}.$$

Restricting w to

$$w > \max\left\{ H_1, \frac{H_2}{H_1}, \ldots, \frac{H_{k-1}}{H_{k-2}} \right\},$$

only the leading coefficient of $g(x)$ is positive, and the coefficient of x^{k-1} is strictly negative. Therefore, $g(x)$ is a primitive polynomial with strictly dominant root w. Since w can be taken arbitrarily close to the maximal element of (5.13), $|\lambda_1|$ must be less than or equal to this maximum. □

Example 5.3.1. Consider the generalized Fibonacci polynomial $x^k - x^{k-1} - \cdots - 1$, whose dominant root satisfies $\lambda_0 < 2$. (Refer to Exercise 5.4.) To apply the last theorem we need to bound the various H's. It is true that $1 > H_1 > H_2 > \cdots > H_{k-1}$, because $1 > H_1 = \lambda_0 - 1$ (since $2 > \lambda_0$), and $H_i > H_{i+1}$ is equivalent to $\lambda_0^i > \lambda_0^{i+1} - \lambda_0^i$. Finally, $H_1 = \lambda_0 - 1 > \lambda_0 - \frac{1}{H_i} = \frac{H_{i+1}}{H_i}$ because $1 > H_i$. So for the generalized Fibonacci polynomial, the maximum in (5.13) occurs at H_1, and the result gives $|\lambda_1| \leq \lambda_0 - 1$, a slightly better bound than $|\lambda_1| < 1$.

Note that in general, $\frac{H_{i+1}}{H_i} = \lambda_0 - \frac{c_{i+1}}{H_i}$ holds, and this shows:

Corollary 5.3.4. $|\lambda_1| \leq \lambda_0 - \min \left\{ c_1, \dfrac{c_2}{H_1}, \ldots, \dfrac{c_{k-1}}{H_{k-2}} \right\}.$

We can also write the result in yet another form. For this, set

$$g_i = \sum_{j=i}^{k} c_j \lambda_0^{k-j} = c_i \lambda_0^{k-i} + \cdots + c_{k-1} \lambda_0 + c_k,$$

where $0 = p(\lambda_0) = \lambda_0^{k-i} H_i - g_{i+1}$. Therefore,

$$H_i = \frac{g_{i+1}}{\lambda_0^{k-i}} \quad \text{and} \quad \frac{H_{i+1}}{H_i} = \frac{g_{i+2}}{\lambda_0^{k-i-1}} \cdot \frac{\lambda_0^{k-1}}{g_{i+1}} = \frac{g_{i+2}}{g_{i+1}} \lambda_0,$$

and Theorem 5.3.3 can be rewritten as:

Corollary 5.3.5. $|\lambda_1| \leq \lambda_0 \max \left\{ \dfrac{g_2}{\lambda_0^k}, \dfrac{g_3}{g_2}, \dfrac{g_4}{g_3}, \ldots, \dfrac{g_k}{g_{k-1}} \right\}.$

5.4   Calculation of the Roots

As we said earlier, we want an efficient algorithm for computing the dominant root to desired accuracy because the dominant root can be used to find the long-term behavior of solutions. Beginning with any interval $[L, U]$ that contains λ_0, another interval approximation of λ_0 can be obtained by computing $p(M)$ for $M = (U + L)/2$. (Refer to Figure 5.1.) If $p(M) = 0$, then $M = \lambda_0$ and we've located the root. If $p(M) > 0$, we replace U by M and otherwise replace L by M. This method, usually called the **Bisection Method**, is guaranteed (up to round-off error) to produce a sequence of decreasing intervals that always contain λ_0. Since the interval is halved at each step, repeating the bisection method for n steps produces an interval of length $(U - L)/2^n$ that is known to contain λ_0. In order to apply the Bisection Method we need an initial interval that contains the root. For a nonnegative polynomial, the single positive root is between 1 and $c_1 + \cdots + c_k$ (recall Lemma 5.3.1), and so good starting values for U and L are immediately available.

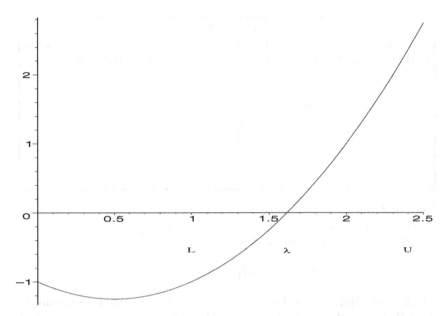

FIGURE 5.1. A nonnegative polynomial with the U and L to be used in calculating the root λ.

Bisection Method

The following code finds an interval of width E that contains the root of the polynomial $p(x)$ when started with an interval $[L, U]$, where $p(L)$ and $p(U)$ have different signs.

```
A := p(U)
B := p(L)
WHILE U − L > E DO
        MID := (U + L)/2
        C := p(MID)
        IF A ∗ C > 0 THEN A := C
                             U := MID
                       ELSE B := C
                             L := MID
```

The decision rule used in our procedure for the Bisection Method requires polynomial evaluation. If direct substitution is used, every evaluation of an arbitrary polynomial of $p(x) = c_0 x^d + c_1 x^{d-1} + \cdots + c_d$ of degree d at some $x = a$ generically requires $\Theta(d^2)$ multiplications. The following code, which is usually called **Horner's Method**, uses only d multiplications.

Horner's Method

A polynomial of the form $p(x) = c_0 x^d + c_1 x^{d-1} + \cdots + c_d$ can be evaluated at $x = a$ by the following code which uses d multiplications and d additions.

$$H := c_0$$
$$\underline{\text{FOR}} \ i := 1 \ \underline{\text{TO}} \ d \ \underline{\text{DO}}$$
$$H := H * a + c_i$$

Note that when the polynomial $p(x)$ is monic (namely, $c_0 = 1$) the terms generated by Horner's Method in the evaluation of $p(x)$ at λ_0 are exactly the H_i defined earlier in (5.7). Also,

$$p(x) = (x - \lambda_0)(x^{d-1} + H_1 x^{d-2} + H_2 x^{d-3} + \cdots + H_{d-1}),$$

which is why Horner's method is sometimes called **synthetic division**. In 1964 Pan [123] showed that Horner's Method uses the fewest number of operations to evaluate a polynomial at an arbitrary value when no precomputing using the coefficients is allowed. This does not preclude a method for evaluating a polynomial at m values using fewer than md operations. (Also refer to Exercise 38 in Section 4.6.4 in Knuth [88].)

Because a polynomial of degree d can be evaluated using Horner's Method in d multiplications and d additions, the Bisection Method will compute the root to an accuracy of $(U - L)/2^n$ in $\Theta(nd)$ steps. Since multiplications often take much more time than additions, the computing time is usually given as only the number of multiplications. Therefore, if E is the designated error bound for the root and E_0 the initial error bound, the root can be found within error E using $d \log(E_0/E)$ multiplications.

While the Bisection Method works reasonably well, there is another method, **Newton's Method** (sometimes called the **Newton–Raphson method**), which usually locates a root more quickly. Again consider the graph of a nonnegative polynomial, but this time include the tangent to the curve $y = p(x)$ at $x = U$ and extrapolate the tangent until it intersects the x-axis at some x-value, which we name U_1. (Refer to Figure 5.2.) This can be done because (refer to Exercise 5.2) for nonnegative polynomials the tangent line at $x = \lambda$ always has a positive slope when $\lambda \geq \lambda_0$. This also guarantees that $U_1 < U$. (From the graph we see that U_1 is greater than the positive root of $p(x)$, which means that it lies between λ_0 and U.) This construction can be repeated using U_1 in place of U, giving the general iteration

$$(5.14) \qquad\qquad U_{n+1} = U_n - \frac{p(U_n)}{p'(U_n)}$$

because the value of the derivative $p'(U_n)$ equals the slope of the tangent line, $\dfrac{p(U_n) - 0}{U_n - U_{n+1}}$. The formula in (5.14) is a nonlinear recurrence and defines the iterative procedure known as Newton's Method. Although there is little chance that we can write a closed-form solution to this recurrence, we would like to know whether it converges to λ_0 as well as to have an idea of the rate of convergence.

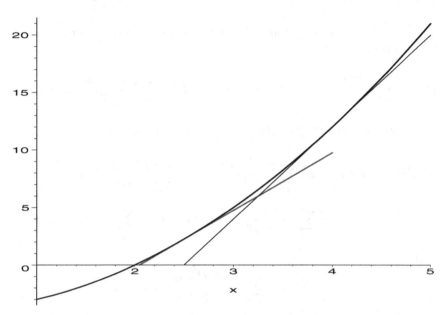

FIGURE 5.2. A nonnegative polynomial with two tangent lines indicated. The tangents cross the axis at the successive Newton approximations to the root of the polynomial.

The following is a standard theorem for Newton's method applied to a general polynomial (for a proof refer to [20, 155]).

Theorem 5.4.1 (Newton's Method). *For any polynomial and any simple real root R there is a small interval around R such that if Newton's method is started at any point within this interval, the approximations found by Newton's method will stay in the interval and converge to R. Further, if the error E_i is the distance from R at the ith step of Newton's method, then $E_{i+1} = O(E_i^2)$—this is called* **quadratic convergence.**

There are several practical problems with this theorem. One is that it only states the existence of the interval and it does not tell how to find it. Also, it only promises rapid *asymptotic* convergence. When the polynomial is nonnegative, the graph in Figure 5.2 suggests that Newton's Method might yield a sequence that is rapidly decreasing to λ_0.

Recall that we plan to initialize Newton's Method with either $U_0 = 1$ or $U_0 = c_1 + \cdots + c_k$, according to whether $p(1)$ is positive or negative. We've shown that the constructed sequence $\langle U_n \rangle$ is decreasing, provided each U_n is greater than λ_0, that is, provided each error $E_n = U_n - \lambda_0$ is positive. The sequence of errors has the form

$$(5.15) \qquad E_{n+1} = U_{n+1} - \lambda_0 = E_n - \frac{p(U_n)}{p'(U_n)} = \frac{E_n p'(U_n) - p(U_n)}{p'(U_n)},$$

and expanding each of $p(x)$ and $p'(x)$ into its Taylor series about λ_0 (refer to Appendix B) gives

$$p(x) = \sum_{i=1}^{d} \frac{p^{(i)}(\lambda_0)}{i!}(x - \lambda_0)^i \quad \text{and also} \quad p'(x) = \sum_{i=0}^{d-1} \frac{p^{(i+1)}(\lambda_0)}{i!}(x - \lambda_0)^i,$$

where $p^{(i)}(x)$ is the ith derivative of $p(x)$. This means that

$$E_n p'(U_n) - p(U_n) = E_n \sum_{i=0}^{d-1} \frac{p^{(i+1)}(\lambda_0)}{i!}(E_n)^i - \sum_{i=1}^{d} \frac{p^{(i)}(\lambda_0)}{i!}(E_n)^i$$

$$= \sum_{i=1}^{d} \left(\frac{1}{(i-1)!} - \frac{1}{i!} \right) p^{(i)}(\lambda_0) E_n^i = \sum_{i=2}^{d} \frac{i-1}{i!} p^{(i)}(\lambda_0) E_n^i$$

and

$$E_{n+1} = \sum_{i=2}^{d} \frac{i-1}{i!} \frac{p^{(i)}(\lambda_0)}{p'(U_n)} E_n^i,$$

where $U_n = \lambda_0 + E_n$. For sufficiently small E_n, $p'(U_n) \approx p'(\lambda_0)$ and $E_{n+1} \approx \frac{p^{(2)}(\lambda_0)}{2p'(\lambda_0)} E_n^2$, and since both $p^{(i)}(\lambda_0)$ and $p'(U_n)$ are positive, E_n and E_{n+1} must have the same sign. Therefore, Newton's method does yield a sequence that is monotonically decreasing to λ_0, provided the initial value is greater than λ_0.

If Newton's method is applied to functions that are not nonnegative polynomials, then various types of odd behavior are possible (refer to [142], [160]). But even for nonnegative polynomials, Newton's method is only guaranteed to converge to the positive root when the initial value is larger than λ_0. For example, consider the nonnegative polynomial $f(x) = x^3 - 5x$, which has the three distinct real roots: $-\sqrt{5}$, 0, and $\lambda_0 = \sqrt{5}$. Newton's formula for this polynomial can be written as

$$N(x) = x - \frac{x(x^2 - 5)}{3x^2 - 5} = \frac{2x^3}{3x^2 - 5},$$

where $\sqrt{5} < N(x) < x$ holds for $x > \sqrt{5}$. When the iteration is started at some $x > \sqrt{5}$, the iteration will monotonically decrease to $\sqrt{5}$. But for instance if the iteration were started at $x = 1$, then

$$N(1) = -1 \quad \text{and} \quad N(-1) = 1,$$

and the Newton iteration oscillates with period 2. Such values of period 2 satisfy $x = N^{(2)}(x)$, the two-fold iteration of N, and so they must be zeros of the polynomial

$$x(x^2 - 1)(x^2 - 5)(4x^4 - 15x^2 + 25),$$

where $x = 0$ and $x = \pm\sqrt{5}$ are fixed points (have period 1). As we saw above, the points $x = \pm 1$ are points of period 2, and this polynomial has four other (nonreal) roots, which are also points of period 2.

It is straightforward but extremely tedious to derive an analogous 27th degree polynomial from $x = N^{(3)}(x)$, the third iterate of N. After eliminating the three points of period 1 there remain 24 points of period 3, some of which might not be real. Similarly, for any n one can derive a polynomial from $x = N^{(n)}(x)$, a polynomial whose roots have period dividing n. Once the points that have a period that is a proper divisor of n are eliminated, all remaining points oscillate with period n under the Newton iteration.

This shows that one should exercise care in picking a good starting point, because even something as seemingly simple as Newton's method applied to a nonnegative polynomial can have complicated behavior. In practice, these complications are unlikely to arise, because round-off error will usually be sufficient to keep the iteration from settling down into periodic behavior. A more practical difficulty is starting at or near a zero of the derivative. This can throw the iteration into some unexpected region and may lead to numerical underflow or overflow. For nonnegative polynomials, starting at $x > \lambda_0$ ensures that the derivatives have non-zero values. In summary, Newton's method is well-behaved when used to find the positive root of a nonnegative polynomial, but you must remember to use a starting value that is larger than the root.

Newton's method can also be used to approximate the other roots of a nonnegative polynomial, but these roots have some extra complications. For instance, we do not know a good initial estimate for the second root. Also, the second root may be nonreal. While there is a variant of Newton's Method that can be used for complex numbers, other methods are more effective [1].

5.4.1 The rate of convergence in Newton's method

To obtain additional insight into the behavior of Newton's method for nonnegative polynomials $p(x)$ we look at the sequence of errors. For this it's helpful to write $E(x) = x - \lambda_0$ and $\hat{E} = E(N(x))$. Suppressing all occurrences of the variable x, the error equation (5.15) becomes

$$\hat{E} = E^2 \alpha,$$

where $\alpha = (Ep' - p)/p'E^2$ is a function of x. Since $E = x - \lambda_0$ is a factor of $p(x)$, there exists a polynomial q such that $p = Eq$, and using the product rule to calculate the derivative $p' = Eq' + q$, we obtain

$$(5.16) \qquad\qquad \alpha = \frac{Ep' - p}{p'E^2} = \frac{q'}{p'} .$$

This allows us to prove the following result.

Theorem 5.4.2. *For a nonnegative polynomial $p(x)$ with positive root λ_0, the error in Newton's method started at any $x > \lambda_0$ obeys*

$$\hat{E} < E^2 \frac{p''(\lambda_0)}{2\,p'(\lambda_0)} .$$

Proof. Since $E(\lambda_0) = 0$ gives $p''(\lambda_0) = 2q'(\lambda_0)$, the result follows from (5.16) if we can show that $\alpha(x)$ is decreasing for $x \geq \lambda_0$. For this we prove that $\alpha'(x)$ is nonpositive for all $x > \lambda_0$, which is equivalent to

$$(5.17) \qquad\qquad p'q'' - p''q' = T_1 q'' - p''T_2 \leq 0 ,$$

where

$$T_1 = p' - \frac{x}{k-1}p'' ; \quad T_2 = q' - \frac{x}{k-1}q''.$$

Recalling that $q(x) = \sum_{i=0}^{k-1} H_i x^{k-1-i}$, where $H_0 = 1, H_1, \ldots, H_{k-1}$ are the positive numbers given by the Horner sequence applied to the evaluation of $p(\lambda_0)$, we see that $q''(x)$ is positive for $x > \lambda_0$. Since $p'(x)$ is a nonnegative polynomial whose dominant root must be less than λ_0, its derivative must also be positive for all $x > \lambda_0$. We complete the proof by showing that $T_1(x) < 0$ and $T_2(x) > 0$ for $x > \lambda_0$. To do this, we note that each of these polynomials has the form $f'(x) - \frac{x}{d-1}f''(x)$ for some polynomial $f(x) = \sum_{j=0}^{d} a_j x^{d-j}$, and

$$f'(x) - \frac{x}{d-1}f''(x)$$

$$= \sum_{j=0}^{d-1}(d-j)a_j x^{d-1-j} - \frac{x}{d-1}\sum_{j=0}^{d-1}(d-j)(d-1-j)a_j x^{d-2-j}$$

$$= \sum_{j=1}^{d-1}(d-j)a_j x^{d-1-j}\left(1 - \frac{d-1-j}{d-1}\right)$$

$$= \sum_{j=1}^{d-1}(d-j)a_j x^{d-1-j}\frac{j}{d-1} .$$

This identity says that $T_1(x) < 0$ and $T_2(x) > 0$ for all positive x, and (5.17) is satisfied for all $x > \lambda_0$. □

This result is helpful because it says that the sequence of errors satisfies $E_i < \alpha(\lambda_0)^{-(2^i+1)}$ after i steps if initially $E_0 < 1/\alpha(\lambda_0)^2$. The difficulty is that we don't know how long it takes for Newton's Method to reduce the error to less than $1/\alpha(\lambda_0)^2$.

Until now we've considered only *absolute* error, that is, the difference (or deviation) of our estimate from the actual value. Often, comparing the size of the actual error with the size of λ_0 is more appropriate, and that's what is meant by the **relative error**, $\delta(x) = E(x)/\lambda_0$. Rephrasing the above result in terms of relative error, we have

$$\hat{\delta}(x) = \frac{\hat{E}(x)}{\lambda_0} = \frac{(\delta\lambda_0)^2\alpha(x)}{\lambda_0} < \delta^2\frac{\lambda_0 p''(\lambda_0)}{2\,p'(\lambda_0)}.$$

The homogeneity of the convergence factor associated with the relative error criterion yields the following theorem.

Theorem 5.4.3. *The relative error in Newton's method for approximating the dominant root of a kth degree nonnegative polynomial obeys*

$$\hat{\delta} < (k-1)\delta^2.$$

In fact, the convergence factor $\lambda_0 p'(\lambda_0)/2p''(\lambda_0)$ is guaranteed to lie in the interval $[(k-1)/2,\, k-1)$.

Proof. Following the proof of Theorem 5.4.2, we note that for any polynomial $f(x) = \sum_{i=0}^{k-1} a_i x^{k-1-i}$ of degree $k-1$ we have

(5.18) $$\qquad\qquad (k-1)f(x) - xf'(x) = \sum_{i=1}^{k-1} a_i i x^{k-1-i}.$$

Using this identity for $f(x) = p'(x) = kx^{k-1} - c_1(k-1)x^{k-2} - \cdots - c_{k-1}$, we see that no coefficient in the polynomial $(k-1)p'(x) - xp''(x)$ is positive, and $x = \lambda_0$ gives the lower bound of $(k-1)/2$ for the convergence factor.

From $p'(\lambda_0) = q(\lambda_0)$ and $p''(\lambda_0) = 2q'(\lambda_0)$ we get another representation of the convergence factor that is not computationally helpful unless the quotient $q(x)$ can be determined. However, this representation can be used to successfully obtain the upper bound we want to prove here, namely that

$$\lambda_0 q'(\lambda_0) < (k-1)q(\lambda_0).$$

Recalling that

$$q(\lambda_0) = \lambda_0^{k-1} + H_1\lambda_0^{k-2} + \cdots + H_{k-1},$$

where each Horner element H_i is positive, (5.18) with $f(x) = q(x)$ gives the required upper bound. $\square$

This result gives doubling convergence for *quadratic* polynomials. It can be extended to polynomials of any degree.

Corollary 5.4.4. *Let δ_0 be the initial relative error when Newton's method is used to approximate the dominant root of a kth degree nonnegative polynomial. If δ_0 satisfies $\delta_0 < 1/k$, then the number of correct digits doubles at each iteration when the computation is done in base $b = k/(k-1)$.*

Proof. From the theorem, $\delta_{i+1} < (k-1)\delta_i$ and

$$\delta_1 < (k-1)/k^2, \quad \delta_2 < \frac{(k-1)^3}{k^4}, \quad \ldots, \delta_r < \frac{(k-1)^{2^r-1}}{k^{2^r}}.$$

$\square$

It remains to locate an initial value for which the relative error is sufficiently small. For this we consider the notion of "scaling". In this context, scaling means replacing x by a new variable y of the form $x = \beta y$. The standard example is approximating a square root. (Refer to Section 9.4.4 for an analysis of the run time when Newton's method is used to calculate a square root.) The square root of $A > 0$ can be computed using Newton's method, since it is the dominant root λ_0 of $p(x) = x^2 - A$. The Newton iteration formula is

$$N(x) = x - \frac{x^2 - A}{2x},$$

or the computationally better

$$N(x) = \frac{x}{2} + \frac{A}{2x}.$$

We scale the expression $x^2 - A$ by replacing x by $x = 2^l y$, where A lies inside the interval $4^{l-1} \le A < 4^l$ (such l might be negative). Then $x^2 - A = 0$ iff $y^2 - \frac{A}{4^l} = 0$. Once we scale, the root $\mu_0 = \sqrt{A}/2^l$ of $h(y) = y^2 - A/4^l$ satisfies $1/2 < \mu_0 < 1$, and when $y = 1$ is chosen as the initial value, we have $E_0 < 1/2$. In this example, the convergence factor for the relative error is

$$\frac{\mu_0 h''(\mu_0)}{h'(\mu_0)} = 1,$$

from which we obtain doubling convergence. Hence, the number of correct bits when Newton's Method is used to find μ_0 at least doubles at each iteration, and $E < 1/2^r$ implies $\hat{E} < 1/2^{2r}$. After the root μ_0 has been determined with desired accuracy, the root λ_0 is easily obtained by multiplying by 2^l, a binary shift. Scaling helped to identify a good initial value for the Newton iteration, and information about the scaled polynomial easily translates back to information about the original polynomial.

In general, if $p(x) = x^k - \sum_{i=1}^{k} c_i x^{k-i}$ is a nonnegative polynomial with dominant root λ_0, scaling it by β yields the new polynomial

$$h(y) = \beta^{-k} \left((\beta y)^k - \sum_{i=1}^{k} c_i (\beta y)^{k-i} \right) = y^k - \sum_{i=1}^{k} \beta^{-i} c_i y^{k-i},$$

with dominant root $\mu_0 = \lambda_0/\beta$. It can be checked that the relative convergence factor remains unchanged, that is,

$$\lambda_0 p''(\lambda_0)/p'(\lambda_0) = \mu_0 h''(\mu_0)/h'(\mu_0).$$

How is this used? First, if it happens that $\sum c_i > 1$, we repeatedly scale by a factor of $\beta = k/(k-1)$ until the coefficient sum either equals 1 (in which case the dominant root equals 1) or is less than 1. The last scaling places the scaled root μ_0 in the interval $((k-1)/k, 1)$. If it happens that μ_0 lies in the subinterval $((k-1)/k, k/(k+1))$ of length $1/k(k+1)$, then $\delta_0 < 1/(k^2-1) < 1/k$ holds for the initial value $U_0 = k/(k+1)$. Otherwise, when $\mu_0 \in (k/(k+1), 1)$, the usual $U_0 = 1$ gives $\delta_0 < 1/(k+1)$. In either case, we've identified an initial value for which the relative error satisfies $\delta_0 < 1/k$ and doubling convergence is ensured. Once the correct degree of accuracy is attained for μ_0, we scale back to λ_0 by a shift in base $\beta = k/(k-1)$. Since our scaled values are less than 1, the relative error is an upper bound on the absolute error, and the number of correct digits really does double at each iteration.

5.5 Asymptotic Size of Solutions

5.5.1 Homogeneous nonnegative recurrences

In this section we consider the asymptotic size of solutions to homogeneous nonnegative recurrences that can be written in the form

(HNN) $s_n = c_1 s_{n-1} + c_2 s_{n-2} + \cdots + c_k s_{n-k}$ where all $c_i \geq 0$ and $c_k \neq 0$.

Since the characteristic polynomial is a nonnegative polynomial, the results of Section 5.1 can be applied, and we let λ_0 be the dominant eigenvalue of the recurrence. Keep in mind that we're also assuming that the initial conditions are nonnegative.

Theorem 5.5.1. *Let λ_0 be the dominant eigenvalue of the homogeneous nonnegative recurrence (HNN). If the initial conditions are nonnegative, then $s_n = O(\lambda_0^n)$. If in addition there are k consecutive positive elements in $\langle s_n \rangle$, then $s_n = \Theta(\lambda_0^n)$.*

Proof. For any natural number N we define the finite set of real numbers

$$S_N = \{s_i/\lambda_0^i : N \leq i < N+k\},$$

and set $\alpha_N = \min(S_N)$, $\beta_N = \max(S_N)$. Then

(5.19) $$\alpha_N \lambda_0^i \leq s_i \leq \beta_N \lambda_0^i \text{ for all } N \leq i < N + k$$

and we prove that

(5.20) $$s_n \leq \beta_0 \lambda_0^n \text{ for all } n \geq 0$$

by induction on n. By construction, (5.20) holds for all $0 \leq n < k$. If (5.20) holds for all $0 \leq n < K$, then

$$s_K = \sum_{i=1}^k c_i s_{K-i} \leq \beta_0 \sum_{i=1}^k c_i \lambda_0^{K-i} = \beta_0 \lambda_0^K,$$

since $\sum_{i=1}^k c_i \lambda_0^{K-i} = \lambda_0^K$. Therefore, $s_K \leq \beta_0 \lambda_0^K$, and we've proved (5.20), and so $s_n = O(\lambda_0^n)$. We can obtain $s_n \geq \alpha_0 \lambda_0^n$ by a similar argument, but α_0 might be zero. When there are k consecutive positive values, say $s_N, \ldots, s_{N+k-1}$, then α_N is positive, and beginning the above argument at $n = N$ instead of $n = 0$ gives

$$s_{N+k} = \sum_{i=0}^{k-1} c_i s_{N+i} \geq \alpha_N \sum_{i=0}^{k-1} c_i \lambda_0^{N+i} = \alpha_N \lambda_0^{N+k}.$$

By induction, $s_n \geq \alpha_N \lambda_0^n$ for all $n \geq N$, and so $s_n = \Theta(\lambda_0^n)$.   □

For example, the Fibonacci recurrence is a homogeneous nonnegative recurrence in which all terms after the first are positive, and this result says that $f_n = \Theta(\lambda_0^n)$, where $\lambda_0 = \frac{1+\sqrt{5}}{2}$ is the unique positive eigenvalue. Another simple example is

$$s_n = s_{n-1} + 2s_{n-2} \text{ with } s_0 = 1,\ s_1 = 2.$$

Since both initial conditions are positive and $ch(x) = x^2 - x - 2 = (x+1)(x-2)$, the solution is $\Theta(2^n)$. In Exercise 5.3 you consider this recurrence under various choices of initial conditions, including when some initial conditions are negative.

Corollary 5.5.2. *Let $\langle s_n \rangle$ be the solution to a homogeneous nonnegative recurrence with nonnegative initial conditions, not all of which are zero. If the characteristic polynomial is primitive, then $\langle s_n \rangle$ is eventually positive and $s_n = \Theta(\lambda_0^n)$.*

Proof. It suffices to prove that the sequence is eventually positive. Let $s_n = c_1 s_{n-1} + \cdots + c_k s_{n-k}$. We may assume that $s_0 > 0$. Setting $G = \{i : c_i > 0\}$, then $s_0 > 0$ implies $s_i > 0$ for all $i \in G$, and so also $s_j > 0$ for all j that are nonnegative sums of elements in G. By primitivity, $\gcd(G) = 1$, and this in turn means that every sufficiently large integer can be expressed as a nonnegative sum of elements of G. (Refer to Exercise 5.5.)   □

What about periodic recurrences? An simple example is

$$s_n = 4s_{n-2} \, ,$$

which has $g = 2$, periodic characteristic polynomial $ch(x) = x^2 - 4$, and dominant root $\lambda_0 = 2$. With initial conditions $s_0 = 1 = s_1$, the solution is

$$s_n = 4^{\lfloor n/2 \rfloor} \, ,$$

and $s_n = \Theta(2^n) = \Theta(\lambda_0^n)$. However, for the initial conditions $s_0 = 1$, $s_1 = 0$, the solution is

$$s_n = \begin{cases} 4^{n/2} & \text{when } n \text{ is even} \, , \\ 0 & \text{when } n \text{ is odd} \, , \end{cases}$$

and $s_n = O(2^n)$. But $s_n \neq \Theta(\lambda_0^n)$, since every odd position in the sequence is zero.

In general, a periodic recurrence whose index of imprimitivity is $g \neq 1$ can be expressed as a system of g primitive equations because

$$s_n = c_g s_{n-g} + c_{2g} s_{n-2g} + \cdots + c_{rg} s_{n-rg}$$

can be written as a primitive system; namely,

$$t_n^{(0)} = c_g t_{n-1}^{(0)} + c_{2g} t_{n-2}^{(0)} + \cdots + c_{rg} t_{n-r}^{(0)}$$
$$t_n^{(1)} = c_g t_{n-1}^{(1)} + c_{2g} t_{n-2}^{(1)} + \cdots + c_{rg} t_{n-r}^{(1)}$$
$$\vdots$$
$$t_n^{(g-1)} = c_g t_{n-1}^{(g-1)} + c_{2g} t_{n-2}^{(g-1)} + \cdots + c_{rg} t_{n-r}^{(g-1)}$$

where the initial conditions for each $\langle t_n^{(j)} \rangle$ are the set of original initial conditions whose subscripts are congruent to j (mod g). If all initial conditions for a particular j are zero, the sequence $\langle t_n^{(j)} \rangle$ is the zero sequence. Otherwise, at least one initial condition for $\langle t_n^{(j)} \rangle$ is positive, and by the corollary, $t_n^{(j)} = \Theta(\lambda_0^{ng})$. Translating this back to the original sequence $\langle s_n \rangle$, we see that for a fixed j the subsequence whose subscripts satisfy the arithmetic progression $n \equiv j$ (mod g) either is the zero sequence or grows like λ_0^n. In some sense we can therefore regard $\langle s_n \rangle$ as a periodic sequence with period g, or under some special circumstances the period of $\langle s_n \rangle$ may be a divisor of g.

5.5.2   Nonhomogeneous nonnegative equations

Here we look at certain *nonhomogeneous* nonnegative equations with one of three types of non-zero input functions.

Theorem 5.5.3. *Consider a nonnegative difference equation*

$$(5.21) \qquad s_n = c_1 s_{n-1} + c_2 s_{n-2} + \cdots + c_k s_{n-k} + g(n) \,,$$

where $g(n)$ is nonnegative for each natural number n. Let λ_0 be the dominant eigenvalue of the recurrence. If $s_n > 0$ for all sufficiently large n, then

- *$s_n = \Theta(\lambda_0^n)$ if $g(n) = O(\lambda_1^n)$ for some positive $\lambda_1 < \lambda_0$;*

- *$s_n = \Theta(g(n))$ if $g(n) = \Theta(\lambda_2^n)$ for some $\lambda_2 > \lambda_0$;*

- *$s_n = \Theta(n^{d+1}\lambda_0^n)$ if $g(n) = f(n)\lambda_0^n$ for a polynomial f with $\deg(f) = d$.*

These three types of nonnegative forcing functions correspond to the situations in which $g(n)$ is much less than λ_0^n, much greater than λ_0^n, and equal to λ_0^n times a polynomial in n. For the last two types of forcing functions the hypothesis requiring s_n to be eventually positive is of course superfluous. In the first type, the required eventual positivity follows if there are k consecutive positive terms, which can result from positive initial conditions or positive $g(n)$ or some combination of these forms of positivity.

Notice that the solutions asymptotically behave about as one expects, except that the behavior in the third case may be a little unexpected. In that case, the response to forcing by a polynomial times λ_0^n results in a polynomial of one higher degree times λ_0^n. The most common occurrence of this is with $g(n) = \lambda_0^n$, and we emphasize this special case because it arises so often in practice.

Corollary 5.5.4. *For any positive constant b, any nonnegative difference equation of the form*

$$s_n = c_1 s_{n-1} + c_2 s_{n-2} + \cdots + c_k s_{n-k} + b\lambda_0^n$$

has solution $s_n = \Theta(n\lambda_0^n)$, where λ_0 is the dominant eigenvalue.

The remainder of this section is occupied with proving the theorem. The proof is a good example of the technical arguments that are often involved in asymptotic analysis.

Proof of the Theorem. Let N be such that $s_n \neq 0$ for all $n \geq N$. Similarly to what was done for the homogeneous case, for any positive λ we define

$$S_\lambda = \{s_i/\lambda^i \ : \ N \leq i < N+k\},$$

and set $\alpha_\lambda = \min(S_\lambda)$, $\beta_\lambda = \max(S_\lambda)$. (Notice that α_λ and β_λ are functions of λ.) In particular,

$$(5.22) \qquad \alpha_\lambda \lambda^i \leq s_i \leq \beta_\lambda \lambda^i \quad \text{for all } N \leq i < N+k.$$

Case A. $g(n) = O(\lambda_1^n)$ *for some positive* $\lambda_1 < \lambda_0$. Setting $\alpha_0 = \alpha_{\lambda_0}$ and $\beta_0 = \beta_{\lambda_0}$,

$$\alpha_0 \lambda_0^n \leq s_n \leq \beta_0 \lambda_0^n \text{ for all } N \leq n < N + k.$$

As in the proof of the homogeneous case, if $s_n \geq \alpha_0 \lambda_0^n$ for all $N \leq n < K$, then

$$s_K = \sum_{i=1}^{k} c_i s_{K-i} + g(n) \geq \alpha_0 \sum_{i=1}^{k} c_i \lambda_0^{K-i} = \alpha_0 \lambda_0^K,$$

since $g(K) \geq 0$ and $ch(\lambda_0) = 0$. Therefore, by induction we have that $s_n \geq \alpha \lambda_0^n$ for all $n \geq N$.

Since $g(n) = O(\lambda_1^n)$, there exist positive C, N_1 such that $0 \leq g(n) < C\lambda_1^n$ for all $n \geq N_1$, and we can increase C if necessary (when $N < N_1$) to obtain

$$0 \leq g(n) < C\lambda_1^n \text{ for all } n \geq N.$$

From the fact that $\lambda_1 < \lambda_0$, then $-C\lambda_1^k/ch(\lambda_1)$ is a *positive* constant, which we call γ_1. (Recall Exercise 5.2.) Setting $\beta_1 = \beta_0 + \gamma_1$, we'll prove that

$$s_n \leq \beta_1 \lambda_0^n - \gamma_1 \lambda_1^n \text{ for all } n \geq N$$

by induction on n and thereby obtain $s_n = \Theta(\lambda_0^n)$. For this, we note that

$$\beta_1 \lambda_0^n - \gamma_1 \lambda_1^n = \beta_0 \lambda_0^n + \gamma_1(\lambda_0^n - \lambda_1^n) > \beta_0 \lambda_0^n \geq s_n \text{ for all } N \leq n < N + k.$$

Suppose we've shown that $s_n \leq \beta_1 \lambda_0^n - \gamma_1 \lambda_1^n$ for all $N \leq n < K$. Then

$$s_K = \sum_{i=1}^{k} c_i s_{K-i} + g(K)$$

$$\leq \sum_{i=1}^{k} c_i(\beta_1 \lambda_0^{K-i} - \gamma_1 \lambda_1^{K-i}) + C\lambda_1^K$$

$$= \beta_1 \lambda_0^{K-k} \sum_{i=1}^{k} c_i \lambda_0^{k-i} - \gamma_1 \lambda_1^{K-k} \sum_{i=1}^{k} c_i \lambda_1^{k-i} + C\lambda_1^K$$

$$= \beta_1 \lambda_0^K + \gamma_1 \lambda_1^{K-k}(ch(\lambda_1) - \lambda_1^k) + C\lambda_1^K$$

$$= \beta_1 \lambda_0^K - \gamma_1 \lambda_1^K + \lambda_1^{K-k}(\gamma_1 ch(\lambda_1) + C\lambda_1^k)$$

$$= \beta_1 \lambda_0^K - \gamma_1 \lambda_1^K,$$

by construction of γ_1. Then $s_n = \Theta(\lambda_0^n)$ does hold.

Case B. $g(n) = \Theta(\lambda_2^n)$ *for* $\lambda_2 > \lambda_0$. Let $\alpha, \beta > 0$ be such that

$$\alpha \lambda_2^n < g(n) < \beta \lambda_2^n, \text{ for all } n \geq N_1.$$

For $n \geq N_1$,

$$s_n = \sum_{i=1}^{k} c_i s_{n-i} + g(n) \geq g(n) \geq \alpha \lambda_2^n,$$

giving a lower bound on s_n. For an upper bound on s_n, from (5.22) with $\lambda = \lambda_2$ we obtain β_2 such that

(5.23) $$s_n \leq \beta_2 \lambda_2^n \text{ for all } N \leq n < N + k.$$

Increase β_2 if necessary to obtain $\beta_2 \geq \lambda_2^k \beta / ch(\lambda_2)$, which is positive because $\lambda_2 > \lambda_0$. Recalling (5.23), we assume $s_n \leq \beta_2 \lambda_2^n$ for all $N \leq n < M$ and prove that this also holds for $n = M$, since

$$s_M = \sum_{i=1}^{k} c_i s_{M-i} + g(M)$$

$$\leq \sum_{i=1}^{k} c_i \beta_2 \lambda_2^{M-i} + \beta \lambda_2^M$$

$$= \beta_2 \lambda_2^{M-k} \sum_{i=1}^{k} c_i \lambda_2^{k-i} + \beta \lambda_2^M$$

$$= \beta_2 \lambda_2^{M-k}(\lambda_2^k - ch(\lambda_2)) + \beta \lambda_2^M$$

$$= \beta_2 \lambda_2^M + \lambda_2^{M-k}(\beta \lambda_2^k - \beta_2 \, ch(\lambda_2)) \leq \beta_2 \lambda_2^M,$$

from the choice of β_2. Therefore, for all $n \geq N$ we have

$$\alpha \lambda_2^n \leq s_n \leq \beta \lambda_2^n,$$

and we've proved that $s_n = \Theta(\lambda_2^n)$.

Case C. $g(n) = \lambda_0^n f(n)$. For this case we divide the recurrence in (5.21) by λ_0^n to get

$$\frac{s_n}{\lambda_0^n} = \sum_{i=1}^{k} \frac{c_i}{\lambda_0^i} \frac{s_{n-i}}{\lambda_0^{n-i}} + f(n).$$

Defining $t_n = s_n/\lambda_0^n$ and $b_i = c_i/\lambda_0^i$, this becomes the new recurrence

$$t_n = \sum_{i=1}^{k} b_i t_{n-i} + f(n),$$

where we note that

$$\sum_{i=1}^{k} b_i = \sum_{i=1}^{k} \frac{c_i}{\lambda_0^i} = \frac{1}{\lambda_0^k} \sum_{i=1}^{k} c_i \lambda_0^{k-i} = 1,$$

since $\lambda_0^k = \sum_{i=1}^{k} c_i \lambda_0^{k-i}$. Because $s_n = t_n \lambda_0^n$, it suffices to prove $t_n = \Theta(n^{d+1})$.

Since $f(n) = \Theta(n^d)$, there exist constants γ_1, γ_2 and positive integer N such that $\gamma_1 n^d \leq f(n) \leq \gamma_2 n^d$ for all $n \geq N$. Define $S = \{t_n/n^{d+1} : N \leq n < N+k\}$ and let $\alpha > 0$ be the minimal element of $S \cup \{\gamma_1/(k+1)^{d+1}\}$ and

β the maximal element of $S \cup \{\gamma_2\}$. We'll prove that $\alpha n^{d+1} \leq t_n \leq \beta n^{d+1}$ by induction on n, which by the construction of α and β holds for all $N \leq n < N + k$. Assuming $\alpha n^{d+1} \leq t_n \leq \beta n^{d+1}$ for all $n < K$, we prove that this inequality also holds for $n = K$. Then

$$\sum_{i=1}^{k} b_i \alpha (K - i)^{d+1} + \gamma_1 K^d \leq t_K = \sum_{i=1}^{k} b_i t_{K-i} + f(K)$$

$$\leq \sum_{i=1}^{k} b_i \beta (K - i)^{d+1} + \gamma_2 K^d,$$

where $K - k \leq K - i \leq K - 1$ and $\sum_{i=1}^{k} b_i = 1$ give

$$\alpha (K - k)^{d+1} + \gamma_1 K^d \leq t_K \leq \beta (K - 1)^{d+1} + \gamma_2 K^d.$$

Therefore, the choice of $\gamma_2 \leq \beta$ implies

$$t_K \leq \beta K^d (K - 1 + 1) = \beta K^{d+1},$$

and $\alpha \leq \gamma_1 / (k + 1)^{d+1}$ gives

$$t_K \geq \alpha ((K - k)^{d+1} + (k + 1)^{d+1} K^d)) \geq \alpha K^{d+1}$$

by Exercise 5.18.   □

There are many variations of the results in Theorem 5.5.3. Because they can be proved in essentially the same way as the results given in the theorem, we simply state them here and leave their proofs as exercises.

- If $g(n) = O(\lambda_1^n)$ for $\lambda_1 < \lambda_0$, then $s_n = O(\lambda_0^n)$.
  - (a) If the difference equation is primitive and $g(n) = O(\lambda_1^n)$ for $\lambda_1 < \lambda_0$, then $s_n = \Theta(\lambda_0^n)$.
  - (b) If $s_n > 0$ for k consecutive values of n and $g(n) = O(\lambda_1^n)$ for $\lambda_1 < \lambda_0$, then $s_n = \Theta(\lambda_0^n)$.
- Let $G(n) = \max\{g(n), \lambda_2 g(n-1), \ldots, \lambda_2^{n-1} g(1)\}$. If $g(n) = \Omega(\lambda_2^n)$ for $\lambda_2 > \lambda_0$, then:
  - (a) $s_n = \Omega(\lambda_2^n)$.
  - (b) $s_n = \Omega(g(n))$.
  - (c) $s_n = O(G(n))$.
  - (d) If $G(n) = O(g(n))$, then $s_n = \Theta(g(n))$.
  - (e) If $g(n) = \Omega(G(n))$, then $s_n = \Theta(G(n))$.
- If $g(n) = h(n)\lambda_0^n$ for some nonnegative $h(n)$, then $s_n = O(ng(n))$ and $s_n = \Omega(g(n))$.
- If $g(n) = h(n)\lambda_0^n$ for some nonnegative $h(n)$ that is nondecreasing, then $s_n = \Theta(\sum_{j=1}^{n} \lambda_0^{n-j} g(j))$.

5.6   Exercises

Ex 5.1. Let $p(x)$ be a polynomial, and let λ be a root of $p(x)$ whose multiplicity is $m \geq 2$. Show that λ is a root of $p'(x)$ with multiplicity $m - 1$.

Ex 5.2. Show that if $p(x)$ is a nonnegative polynomial with positive root λ_0, then the values of $p(x)$ change from negative to positive at $x = \lambda_0$. From this conclude that for $x > 0$,

$$ch(x) > 0 \iff x > \lambda_0$$

by showing that $p'(x)$ is positive for all $x > \lambda_0$.

Ex 5.3. Consider the recurrence $s_n = s_{n-1} + 2s_{n-2}$.
  (a) Show that $s_n = \Theta(2^n)$ for $s_0 = 0$, $s_1 = 15$.
  (b) When $s_0 = -4$ and $s_1 = 7$, show that $s_n > 0$ for all $n \geq 3$, and $s_n = \Theta(2^n)$ still holds.
  (c) When $s_0 = -1$, $s_1 = 1$, show that $s_n \neq \Theta(2^n)$.

Ex 5.4. In this problem we consider the kth order Fibonacci recurrence

$$f_n = f_{n-1} + f_{n-2} + \cdots + f_{n-k}$$

with primitive characteristic polynomial and dominant root λ_0.
  (a) Use Newton's Method initialized at $x_0 = 2$ to show that $\lambda_0 \leq 2 - 1/(2^k - 1)$.
  (b) Show that $\lambda_0 \geq 2 - 2/(k+1)$
  (c) (HARDER) Show that $\lambda_0 > 2 - 2/2^k$.

Ex 5.5. Let a and b be fixed positive integers.
  (a) Show that an integer n is expressible as a nonnegative combination of a and b (that is, there exist *nonnegative* integers i, j such that $n = ia + jb$) iff n is in one of the following sequences:

$$0, b, 2b, \ldots ,$$
$$a, b + a, 2b + a, \ldots ,$$
$$2a, b + 2a, 2b + 2a, \ldots ,$$
$$\ldots$$
$$(b-1)a, b + (b-1)a, 2b + (b-1)a, \ldots .$$

  (b) When $\gcd(a, b) = 1$, show that each of the above sequences is in a different congruence class modulo b. Use this to show that every integer n greater than $(a-1)(b-1)$ is in one of these sequences and so can be written as a nonnegative combination of a and b.
  (c) If G is any finite set of positive integers with $\gcd(G) = 1$, show that every sufficiently large integer can be written as a nonnegative sum of elements in G.

Ex 5.6. Let G be a finite set of positive integers and let $g = \gcd(G)$ be the **greatest common divisor** (gcd) of the elements of G. Then ζ is an ith root of unity for every $i \in G$ iff ζ is a gth root of unity.

Ex 5.7. Show that
   (a) $p(x) = (x - 3)(x + 1)^3$ is a primitive nonnegative polynomial;
   (b) $p(x) = x^4 - x^3 - 5x^2 - x - 6$ is a primitive nonnegative polynomial with two roots on the unit circle.

Ex 5.8. Consider

$$p(x) = x^5 - \frac{1}{2}x^3 - \frac{9}{2}x^2 - \frac{55}{16}x - \frac{25}{8} .$$

Show that $\lambda_0 = 2$ is the unique positive root of $p(x)$ and that there exists a quadratic polynomial $q(x)$ such that $p(x) = (x - 2)(q(x))^2$, where every root λ of $q(x)$ satisfies $|\lambda| = \sqrt{5}/2$. With this you see that even a primitive nonnegative polynomial can have subdominant eigenvalues that are not simple.

Ex 5.9. If $\sum_{i=1}^{k} |b_i| \leq 1$, show that every solution to

$$y_n = b_1 y_{n-1} + \cdots + b_k y_{n-k}$$

satisfies $|y_n| \leq \max\{|y_0|, |y_1|, \ldots, |y_{k-1}|\}$.

Ex 5.10. In this problem we consider

$$s_n = 2s_{n-1} + 2s_{n-2} + 3s_{n-3}.$$

Use the method of Exercise 4.22 to show that the coefficient of the dominant eigenvalue in the closed form of the solution is $\alpha = (s_0 + s_1 + s_2)/13$.

Ex 5.11. Show that the sequence of deviations is periodic for the solution to $s_n = 2s_{n-1} + 2s_{n-2} + 3s_{n-3}$ with $s_0 = 0, s_1 = 3, s_2 = 9$.

Ex 5.12. Show that the nonnegative solutions to

$$x_n = 5x_{n-1} + 4x_{n-2}$$

converge to $\alpha \lambda_0^n$ in the absolute value sense. If the initial conditions are nonnegative integers, will $x_n = \text{Round}(\alpha \lambda_0^n)$?

Ex 5.13. Let $p(x) = x^k + c_1 x^{k-1} + \cdots + c_{k-1}x + c_k$ be a polynomial with integer coefficients.
   (a) Show that any real root of $p(x)$ is either an integer or is an irrational number.
   (b) Give an efficient algorithm to determine whether or not the dominant root of a nonnegative polynomial with integer coefficients is an integer.

Ex 5.14. Let $p(x)$ be a monic nonnegative polynomial with dominant root λ_0. Show that each H_i generated by Horner's Method in the evaluation of $p(x)$ at λ_0 is positive.

Ex 5.15. Show that the polynomial $f(x) = x^3 + 3x^2 - 13x + 17$ has one real root and it is negative. Try various positive initial conditions to see if Newton's method converges to the root. Conclude from this that the assumption of nonnegativity is needed to ensure that any initialization of Newton's method above the root will converge to the root.

Ex 5.16. Show that there exists an initial value such that Newton's Method applied to $f(x) = x^3 + 3x^2 - 13x + 17$ oscillates with period 3.

Ex 5.17. Show that applying Newton's Method to the Fibonacci polynomial $x^2 - x - 1$ with initial condition 2 produces the sequence $\langle f_{2^j+1}/f_{2^j} \rangle$ converging to the dominant root $(1 + \sqrt{5})/2$.

Ex 5.18. Write n^{d+1} as $[(n - k) + k]^{d+1}$ and use the binomial expansions of this and of $(k + 1)^{d+1}$ to show that

$$(n - k)^{d+1} \geq n^{d+1} - (k + 1)^{d+1} n^d .$$

Ex 5.19. Find the asymptotic size of solutions to

$$s_n = 2s_{n-1} + 2s_{n-2} + 3s_{n-3} + g(n)$$

with nonnegative initial conditions for each of the following choices of $g(n)$:

$$g(n) = 2^n ; \quad g(n) = 4^n ; \quad g(n) = n3^n .$$

Ex 5.20. Find the asymptotic size of the solution to

$$f_n = f_{n-1} + f_{n-2} + g(n) \quad f_0 = 0, f_1 = 1 ,$$

where $g(n)$ satisfies the recurrence

$$g(n) = g(n - 1) + g(n - 2) \quad g(0) = 0, g(1) = 1 .$$

Ex 5.21. Let $\langle s_n \rangle$ be the solution to a nonnegative recurrence

$$s_n = c_1 s_{n-1} + c_2 s_{n-2} + \cdots + c_k s_{n-k} + g(n) .$$

Show that:
 (a) If $g(n) = O(\lambda_1^n)$ for $\lambda_1 < \lambda_0$, then $s_n = O(\lambda_0^n)$.
 (b) If the recurrence is primitive and $g(n) = O(\lambda_1^n)$ for $\lambda_1 < \lambda_0$, then $s_n = \Theta(\lambda_0^n)$.
 (c) If $s_n > 0$ for k consecutive values of n and $g(n) = O(\lambda_1^n)$ for $\lambda_1 < \lambda_0$, then $s_n = \Theta(\lambda_0^n)$.

Ex 5.22. Let $\langle s_n \rangle$ be a solution of the nonnegative recurrence

$$s_n = c_1 s_{n-1} + c_2 s_{n-2} + \cdots + c_k s_{n-k} + g(n).$$

Let $G(n) = \max\{g(n), \lambda_2 g(n-1), \ldots, \lambda_2^{n-1} g(1)\}$. If $g(n) = \Omega(\lambda_2^n)$ for $\lambda_2 > \lambda_0$, show that:
  (a) $s_n = \Omega(\lambda_2^n)$.
  (b) $s_n = \Omega(g(n))$.
  (c) $s_n = O(G(n))$.
  (d) If $G(n) = O(g(n))$, then $s_n = \Theta(g(n))$.
  (e) If $g(n) = \Omega(G(n))$, then $s_n = \Theta(G(n))$.

Ex 5.23. Let $\langle s_n \rangle$ be a solution of the nonnegative recurrence

$$s_n = c_1 s_{n-1} + c_2 s_{n-2} + \cdots + c_k s_{n-k} + g(n).$$

Show that:
  (a) If $g(n) = h(n)\lambda_0^n$ for some nonnegative $h(n)$, then $s_n = O(ng(n))$ and $s_n = \Omega(g(n))$.
  (b) If $g(n) = h(n)\lambda_0^n$ for some nonnegative $h(n)$ that is nondecreasing, then $s_n = \Theta(\sum_{j=1}^{n} \lambda_0^{n-j} g(j))$.

6
Leslie's Population Matrix Model

6.1 Leslie's Model

In the Fibonacci story (see Chapter 1) the rabbits are immortal, but with a few exceptions, like the Energizer Bunny, we know that real rabbits have small finite lifetimes. We can make this model more reasonable by reinterpreting the meaning of the age classes while still maintaining the mathematical form of the model. Recalling the original model, if A_t is the number of adult pairs at time t, and Y_t is the number of young (not yet breeding) pairs at time t, then

$$\begin{pmatrix} A_{t+1} \\ Y_{t+1} \end{pmatrix} = \begin{bmatrix} 1 & 1 \\ 1 & 0 \end{bmatrix} \begin{pmatrix} A_t \\ Y_t \end{pmatrix}.$$

The assumed immortality of the adults is represented by the 1 in the upper left entry of the matrix. The young disappear by becoming adults, but they are replaced by each adult pair producing a new young pair. What happens if we simply interchange adults and young, leaving the matrix alone? We get

$$\begin{pmatrix} Y_{t+1} \\ A_{t+1} \end{pmatrix} = \begin{bmatrix} 1 & 1 \\ 1 & 0 \end{bmatrix} \begin{pmatrix} Y_t \\ A_t \end{pmatrix},$$

but our interpretation is now quite different. New young are produced by both young and adults. The adults disappear, but they are replaced by youths who grow into adulthood. So by reinterpreting this model we can replace the unrealistic assumption of immortality with the biologically realistic assumptions of aging and death. As a leading biologist said at a recent meeting, "The facts are always changing, but a good model goes on forever."

The reinterpreted model still has some unrealistic features. Should every youth survive to adulthood? Should the number of offspring produced by an adult pair and a young pair be the same? The model can be generalized to avoid these unrealistic features by replacing the 1's in the matrix with other constants. Since the lower left 1 represents a youth surviving to become an adult, we could replace this 1 by s, which is called the **survival rate**, and reasonably assume that $1 \geq s > 0$. The 1's in the top row represent the number of offspring produced by a youth and an adult respectively. We can replace these 1's with the **fertility rates** f_1 and f_2 and simply assume that $f_1 \geq 0$ and $f_2 \geq 0$. The generalized model is

$$\begin{pmatrix} Y_{t+1} \\ A_{t+1} \end{pmatrix} = \begin{bmatrix} f_1 & f_2 \\ s & 0 \end{bmatrix} \begin{pmatrix} Y_t \\ A_t \end{pmatrix}.$$

Several special cases suggest themselves. What happens if $f_1 = f_2 = 0$? Then, if we start with the population $\begin{pmatrix} Y_0 \\ A_0 \end{pmatrix}$, we next get $\begin{pmatrix} 0 \\ sY_t \end{pmatrix}$ and then $\begin{pmatrix} 0 \\ 0 \end{pmatrix}$. So as expected, if the fertility rates are 0, the population dies out, and in this model the population dies out in two time steps. What happens if $f_2 = 0$, but $f_1 \neq 0$? Then starting with $\begin{pmatrix} Y_t \\ A_t \end{pmatrix}$, we get $\begin{pmatrix} f_1 Y_t \\ sY_t \end{pmatrix}$ and then $\begin{pmatrix} Y_t f_1^2 \\ Y_t s f_1 \end{pmatrix}$. These last two vectors can be rewritten as $Y_t \begin{pmatrix} f_1 \\ s \end{pmatrix}$ and $Y_t f_1 \begin{pmatrix} f_1 \\ s \end{pmatrix}$. Notice that these are both multiples of the vector $\begin{pmatrix} f_1 \\ s \end{pmatrix}$, and that neither depends on A_t. These observations agree with the expectations that the individuals past reproductive age eventually have no effect on the composition of the population, and the eventual shape of the population depends on the structural parameters f_1, f_2, and s, and not on the initial population. What happens if $f_1 = 0$ and $f_2 \neq 0$? Then, starting with $\begin{pmatrix} Y_t \\ A_t \end{pmatrix}$, we get $\begin{pmatrix} f_2 A_t \\ sY_t \end{pmatrix}$ and then $\begin{pmatrix} sf_2 Y_t \\ sf_2 A_t \end{pmatrix}$. In contrast to the other cases, the shape of the population oscillates between multiples of $\begin{pmatrix} Y_t \\ A_t \end{pmatrix}$ and $\begin{pmatrix} f_2 A_t \\ sY_t \end{pmatrix}$. In the second and third cases the population may increase or decrease depending on whether the multiplier, f_1 or sf_2, is or is not larger than 1. These cases show that this simple model does have some features that we want in a population model. The population size may increase or decrease. The population may die out. The shape of the population may approach a stable form, or the shape of the population may oscillate. By doing some simple calculations on the parameters, we would like to be able to determine which of these various situations occurs.

We can generalize the Fibonacci model with two age classes to a model with k age classes. In population biology the model with k age classes is usually called **Leslie's model**. In 1945, Leslie [95] published one of the most influential papers in population biology. In it he introduced a generation of biologists to vectors and matrices. The model Leslie described is quite similar to the renewal model [62] and the method of life tables, which were devised in the 1700's by such mathematicians as the Bernoullis, Euler, and Halley. (For more on the history, consult Boyer's *History of Mathematics* [12].) These methods form the basis of an entire industry— insurance. Further, these methods were not unknown to earlier biologists. For example, Lotka's 1925 book, *Elements of Mathematical Biology* [101], devotes an entire chapter to them. On the other hand, Leslie presented a concise formulation of the model, and the mid-twentieth century generation of biologists was a receptive audience.

The model can be concisely stated as

$$X_{t+1} = LX_t,$$

where X_t and X_{t+1} are population vectors and L is a Leslie matrix (which is explicitly defined below.) The model assumes discrete time, and the time unit must be chosen appropriately for the organism being modeled. For bacteria the time unit might be 20 minutes. For many insects an appropriate time unit might be one week. For many vertebrates a one-year time unit might be used. For human populations a five-year time unit is often used. The population vectors have a number of components. Each component is the number of individuals of a particular age. For example, if

$$X_t = (x_1, x_2, x_3, x_4)^T,$$

then at time t, x_1 is the number of individuals in the first age class, x_2 is the number of individuals in the second age class, and so forth. Said another way, x_i is the number of individuals of age $i - 1$, because newborns are usually assigned age 0 and not age 1. If one is not really happy with the discrete-time assumption and assumes that discrete time is only an approximation to an underlying continuous time, then one could say that x_i represents the individuals with ages a for which $i - 1 \leq a < i$. One could instead include in x_i the ages for which $i - 1 < a \leq i$, with newborns included in x_1. Or one could say that ages for which $i - 1 \leq a \leq i$ are represented by x_i and claim that there is no ambiguity because there are no individuals of exactly age i for any i. The main point is that the width of an age class is one time unit. If the data are arranged for five-year age classes, this model will not calculate the population even one year in the future.

Generalizing from the Fibonacci example, a **Leslie matrix** contains both survival rates and fertility rates; specifically,

$$
L = \begin{bmatrix}
f_1 & f_2 & \cdots & \cdots & f_k \\
s_1 & 0 & \cdots & \cdots & 0 \\
 & s_2 & & & \vdots \\
 & & \ddots & & \vdots \\
 & & & s_{k-1} & 0
\end{bmatrix}.
$$

The first row of L consists of fertility rates, where f_i is the number of offspring (newborn) produced by an individual of age class i in one time unit, and the subdiagonal of L contains the survival rates, where s_i is the probability that an individual in age class i will survive to age class $i+1$. All other entries in L are zero.

The usual assumptions on these parameters are that for each i, $0 < s_i \leq 1$ and $f_i \geq 0$. The first assumption makes sense if one interprets s_i as a probability and assumes that there is some possibility for an individual to survive a particular age class into the next age class. Further, if any s_i were zero, then in $k - i$ steps the population would become a population in which the last $k - i$ age classes are empty, and all future developments occur within and depend only on the first i age classes. For similar reasons, one usually assumes that $f_k \geq 0$. That is, if one or several of the oldest age classes have zero fertility, then the composition of these older age classes has no effect on the rest of the population, and in a small number of steps the composition of these age classes is determined by the younger age classes with no effect from the original composition of these oldest age classes.

An extra assumption made in Leslie's original model and often used in demographic applications is that at least two adjacent fertility rates are positive. This assumption is often enforced by averaging fertilities. That is, the number of offspring from females in each age class is measured, but a fraction of these are attributed to females in the next age class because the females are assumed to be aging as the measurements are taken. A mathematically more appropriate assumption, which includes the Leslie assumption as a special case, is that there is a power of the Leslie matrix that is strictly positive. Luckily, this can be checked easily using the greatest common divisor of the indices of positive fertility rates. Later, we will prove the following result.

Theorem 6.1.1. *Let L be a Leslie matrix. Then there exists $m \geq 0$ with $L^m \gg 0$ iff $\gcd\{i \mid f_i > 0\} = 1$, where $A \gg 0$ means that every entry in the matrix A is strictly positive.*

This theorem says that there is a power of L that is a strictly positive matrix. The following theorem gives bounds on the least exponent m for which the power L^m is strictly positive.

Theorem 6.1.2. *Let L be an $n \times n$ Leslie matrix with $\gcd\{i \mid f_i > 0\} = 1$. Let m_0 be the least nonnegative integer such that $L^{m_0} \gg 0$, and let l be the least positive integer with $f_l > 0$. Then*

$$m_0 \leq l(n-2) + n \leq (n-1)^2 + 1.$$

We will prove both of these theorems in Chapter 7 in the more general context of nonnegative matrices (which might not be in Leslie form.) The bounds given in Theorem 6.1.2 are tight. For instance, for any $n \geq 2$ consider a Leslie matrix in which $f_1 = f_2 = \cdots = f_{n-2} = 0$, $f_{n-1} > 0$, and $f_n > 0$. Here, $\gcd(n, n-1) = 1$, and it is relatively easy to show that $m_0 = (n-1)^2 + 1$. (See Exercise 6.8.)

These two theorems can be generalized to apply to all nonnegative matrices, but the gcd condition must be generalized. For this we use a graphical interpretation. A nonnegative matrix A is called **primitive** if there exists positive m with $A^m \gg 0$. A graph G can be associated with A in which vertices v_i and v_j have an edge $v_i \leftarrow v_j$ iff $a_{ij} > 0$. (We chose this direction, but the other ordering, $v_i \rightarrow v_j$, could be used.) A graph is **strongly connected** if for each pair of vertices v_i and v_j there is a directed path from v_i to v_j (and so also a directed path from v_j to v_i.) When the directed edges are represented by arrows, a directed path always follows the arrows from tail to head. The generalized theorems are as follows:

Theorem 6.1.3. *A nonnegative $n \times n$ matrix A is primitive iff the associated graph G is strongly connected and the greatest common divisor of the cycle lengths in G is 1.*

Theorem 6.1.4. *If A is a primitive $n \times n$ matrix, then m_0, the least nonnegative integer such that $A^{m_0} \gg 0$, obeys*

$$m_0 \leq l(n-2) + n \leq (n-1)^2 + 1,$$

where l is the length of the shortest cycle in the associated graph G.

We defer the proofs of these theorems to Chapter 7, where we will also describe the graph associated with a nonnegative matrix in more detail.

6.1.1   How to tell whether a Leslie matrix is primitive

How does one determine whether a nonnegative matrix has a power that is strictly positive? For a Leslie matrix, the strong connectedness in Theorem 6.1.3 holds, since all survival rates s_i are positive. Also, from the form of Leslie matrices we see that every cycle in the graph contains the vertex 1, which means that the necessary and sufficient condition for primitivity becomes $\gcd\{i \mid f_i > 0\} = 1$ for Leslie matrices.

Notice that if L^m is positive, then L^{m_1} is positive for all $m_1 > m$. So one way to test whether a Leslie matrix is primitive is to raise the matrix to the

$(n-1)^2 + 1$ power and then check whether this matrix is strictly positive. This technique will require a number of matrix multiplications and hence may not be very efficient. A more efficient algorithm is based on the gcd condition, where $\gcd\{i | f_i > 0\}$ is computed using the **Euclidean greatest common divisor algorithm**. It can be shown (see Exercise 6.2) that the Euclidean Algorithm computes the gcd of two numbers that are less than n using $O(\log n)$ arithmetic operations. Since the Euclidean Algorithm is used at most $O(n)$ times, the whole algorithm takes $O(n \log n)$. The algorithm for primitivity is given in the next two boxes.

**Recursive Form of the Euclidean Algorithm
for Greatest Common Divisor**

$$\gcd(a, b) = \begin{cases} b & \text{if } a = 0, \\ \gcd(b \bmod a, \ a) & \text{if } a \neq 0. \end{cases}$$

Algorithm to determine whether a Leslie Matrix is primitive

Let $\{i_1, i_2, \ldots, i_r\} = \{i | f_i > 0\}$
  $G := i_1$
  FOR $J := 2$ TO r DO
      $G := \gcd(G, i_J)$
  ENDFOR
  IF $G = 1$ THEN Primitive
              ELSE Not Primitive

6.2   Leslie's Convergence Theorem

The content of Leslie's Convergence Theorem is that a multi-dimensional model asymptotically becomes a one-dimensional model. Specifically, the theorem states that under reasonable circumstances the population distribution will converge to a single distribution regardless of the initial distribution. In general, this convergence is convergence in the sense of relative error. With an additional assumption, the convergence can become convergence in the sense of absolute error.

The asymptotic growth rate of the population is determined by a single number, λ_0, which depends on the fertility rates and survival rates but is independent of the initial population distribution. There is no closed-form formula for λ_0, but it can be calculated quickly to high numerical accuracy using Newton's method (see Section 5.3).

Will the population be asymptotically increasing or decreasing? Obviously, this depends on λ_0, but a highly accurate estimate of λ_0 is not needed to answer this question. All one needs to know is whether $\lambda_0 > 1$ or $\lambda_0 < 1$, and this can be determined in $O(n)$ arithmetic operations.

What will the asymptotic distribution, usually called the **stable age distribution**, look like? This distribution can be given as an explicit formula in λ_0, the fertility rates, and the survival rates. Further, assuming that $\lambda_0 > 1$ holds, this stable age distribution has an "inverted pyramid" form. That is, the largest age class is the newborns, and the size of each age class decreases as age increases. Here is the theorem.

Theorem 6.2.1 (Leslie's Convergence Theorem). *Let L be an $n \times n$ Leslie matrix in which $\gcd\{i \mid f_i > 0\} = 1$, and let X be a nonnegative vector. Then there is a unique positive eigenvalue λ_0 of L such that*

$$\lim_{t \to \infty} \frac{L^t X}{\lambda_0^t} = \gamma \begin{pmatrix} \frac{c_1}{\lambda_0} \\ \vdots \\ \frac{c_i}{\lambda_0^i} \\ \vdots \\ \frac{c_n}{\lambda_0^n} \end{pmatrix},$$

where

$$c_i = \begin{cases} 1 & \text{if } i = 1, \\ s_1 s_2 \cdots s_{i-1} & \text{if } i = 2, \ldots, n, \end{cases}$$

and

$$\gamma = \sum_{j=1}^{n} \frac{\lambda_0^{j-1} g_j(\lambda_0) x_j}{ch'_L(\lambda_0) c_j},$$

where

$$ch_L(\lambda) = \lambda^n - \sum_{i=1}^{n} c_i f_i \lambda^{n-i}$$

and

$$g_j(\lambda) = \sum_{i=j}^{n} c_i f_i \lambda^{n-i}.$$

This form of the theorem is slightly stronger than Leslie's original theorem, because he assumed that there was some i such that both fertility rates f_i and f_{i+1} are positive. Clearly his assumption implies the gcd condition, but there are Leslie matrices that satisfy the gcd condition and do not have two adjacent positive fertility rates. We will not prove the theorem now. Rather, it will follow as a corollary to some more general theorems that we will prove later in this chapter. (See Section 6.6.) The above form for the limiting vector was chosen to display the inverted pyramid form.

Corollary 6.2.2. *If $\lambda_0 > 1$ and for each i, either $s_i \leq 1$, or $\lambda_0 \geq 1$ and $s_i < 1$, then the limiting vector has the inverted pyramid form in which the entries of the vector decrease as one goes from top to bottom.*

The other important point is that $\gamma > 0$ if $X > 0$. This follows because $ch'_L(\lambda_0) > 0$, and for each j both c_j and $g_j(\lambda_0)$ are positive. (For this, refer to Exercise 5.2. Also notice that if $X = 0$, then $\gamma = 0$ and $L^t X$ is always 0.) The convergence is in the relative error sense, since the theorems on nonnegative polynomials (Section 5.1) guarantee only that λ_0 is greater than the absolute value of the other eigenvalues. For convergence in the absolute error sense, one wants $\lambda_0 \geq 1$ and that all other eigenvalues satisfy $|\lambda| < 1$. For example, the Fibonacci matrix F satisfies these hypotheses, and so not only does

$$\lim_{t \to \infty} \frac{F^t X}{\lambda_0^t} = \gamma \begin{pmatrix} 1/\lambda_0 \\ 1/\lambda_0^2 \end{pmatrix} = \hat{\gamma} \begin{pmatrix} \lambda_0 \\ 1 \end{pmatrix}$$

hold, but also there exists a constant $\hat{\gamma}$ depending only on X such that

$$\lim_{t \to \infty} \left| F^t X - \hat{\gamma}\lambda_0^t \begin{pmatrix} \lambda_0 \\ 1 \end{pmatrix} \right| = 0 \,,$$

where here the absolute value of a vector means the maximum absolute value of its coordinates.

6.3   Imprimitive Leslie Matrices

Not all Leslie matrices converge. Some can oscillate. The idea of using oscillating matrices for populations even predates Leslie's paper. For example, Bernadelli [8] described a hypothetical population of beetles that he modeled with an oscillating matrix. The periodic cicadas (the famous 17-year locusts) can also be described by the oscillating matrices that we will cover in this section.

6.3.1   A simple example

Consider the two-dimensional Leslie matrix

$$M = \begin{bmatrix} 0 & 8 \\ 1/2 & 0 \end{bmatrix}.$$

There is only one positive fertility rate, f_2, and so $g = 2$, and we expect an oscillation of period 2. The characteristic polynomial is $\lambda^2 - 4$, and the eigenvalues are ± 2, which gives

$$M^{2m} = 4^m I \quad \text{and} \quad M^{2m+1} = 4^m M \,.$$

So in some sense the powers of M are periodic. To be more precise, we can normalize by dividing M by its positive eigenvalue, $\lambda_0 = 2$, giving

$$\frac{M}{2} = \begin{bmatrix} 0 & 4 \\ 1/4 & 0 \end{bmatrix},$$

and clearly,

$$\left(\frac{M}{2}\right)^{2m} = I \quad \text{and} \quad \left(\frac{M}{2}\right)^{2m+1} = \frac{M}{2},$$

and thus $\frac{M}{2}$ is periodic with period 2. In general, we say that a matrix A is **periodic with period** p if $A^{p+i} = A^i$ for all $i \geq 0$, and p is the smallest positive integer that makes the equation true. While in our example it is true that $\left(\frac{M}{2}\right)^{512+i} = \left(\frac{M}{2}\right)^i$ for all i, we do *not* say that $\left(\frac{M}{2}\right)$ has period 512, because the periodicity equations are also satisfied by $p = 2$. When we say that the period p equals 2, we are also implying that $A^{1+i} \neq A^i$; that is, the matrix does not have period 1.

6.3.2   A special case: Only one positive fertility rate

Theorem 6.3.1. Let M be an n-dimensional Leslie matrix with f_n its only non-zero fertility rate. If λ_0 is the positive eigenvalue of M, then M/λ_0 is periodic with period n. Also, $\lambda_0 = (s_1 \cdots s_{n-1} f_n)^{1/n}$.

Proof. By direct calculation, the characteristic polynomial of M is the non-negative polynomial $\lambda^n - s_1 \cdots s_{n-1} f_n$. So $\lambda_0 = (s_1 \cdots s_{n-1} f_n)^{1/n}$ is the (unique) positive eigenvalue. By the Cayley–Hamilton Theorem (refer to Appendix C), $M^n = \lambda_0^n I$ and n satisfies the periodicity equations. To see that n is the smallest such positive integer, consider the orbit of e_n under M: $e_n, Me_n, M^2 e_n, \ldots, M^{n-1} e_n$, where $e_1, e_2, \ldots, e_n$ are the standard coordinate vectors. For $j > 0$,

$$M^j e_n = s_1 \cdots s_{j-1} f_n e_j,$$

implying that the period divides n. Now if M/λ_0 were periodic with period p, then for every vector X, $M^p X$ would be a scalar multiple of X. But no two e_j's are scalar multiples of one another. Therefore, for any $p < n$, $M^p e_n$ is not a scalar multiple of e_n, so M cannot have period less than n. (Note that our proof basically shows that for such M the minimal and characteristic polynomials are equal.)   □

6.3.3   Asymptotically periodic Leslie matrices

Fortunately or unfortunately, most matrices—even most Leslie matrices—are not truly periodic. Even imprimitive Leslie matrices are usually not periodic. (Recall that a matrix is primitive when there is a power of the matrix that has only positive entries.) All powers of an imprimitive Leslie matrix contain some zeros, and these zeros move through the powers of the matrix in a periodic fashion. The non-zero entries also appear periodically, but the values of the non-zero entries are changing, preventing true periodicity. If an $n \times n$ imprimitive Leslie matrix has at least two positive fertility rates, then g is less than n, and the matrix has g eigenvalues of largest magnitude and $n - g$ eigenvalues of smaller magnitude. These smaller magnitude eigenvalues prevent the matrix from being truly periodic, but we can expect the contribution from these smaller eigenvalues to disappear (at least in a ratio sense) as we take higher powers of the matrix.

For example, consider the Leslie matrix

$$A = \begin{bmatrix} 0 & 1 & 0 & 2 \\ 1/2 & 0 & 0 & 0 \\ 0 & 1 & 0 & 0 \\ 0 & 0 & 1/2 & 0 \end{bmatrix},$$

with characteristic polynomial $\lambda^4 - \frac{1}{2}\lambda^2 - \frac{1}{2}$, eigenvalues $\pm 1, \pm i/\sqrt{2}$, and $g = 2$. In this example we expect that taking high powers of A leads to a matrix that is close to a periodic matrix of period 2, and we can use the characteristic polynomial to display this behavior. Since $A^4 = \frac{1}{2}(A^2 + I)$,

$$A^{K+2} - A^K = A^{K-2}(A^4 - A^2) = A^{K-2}(\frac{1}{2}A^2 + \frac{1}{2}I - A^2)$$

$$= -\frac{1}{2}A^{K-2}(A^2 - I) = -\frac{1}{2}(A^K - A^{K-2}).$$

Iterating this formula gives

$$A^{K+2} - A^K = \begin{cases} (-\frac{1}{2})^{K/2}(A^2 - I) & \text{if } K \text{ is even}, \\ (-\frac{1}{2})^{\lfloor K/2 \rfloor} A(A^2 - I) & \text{if } K \text{ is odd}. \end{cases}$$

In either case, $A^2 - I$ or $A(A^2 - I)$ are fixed matrices independent of K. Since $(-\frac{1}{2})^{K/2}$ is decreasing exponentially to 0,

$$\lim_{K \to \infty} (A^{K+2} - A^K) = 0,$$

where this convergence is componentwise, which means that each entry in the matrix $A^{K+2} - A^K$ goes to 0. (Other notions of convergence are possible for matrices.)

Using this example as our paradigm, we say that A is an **asymptotically periodic matrix** with period g if

$$\lim_{K \to \infty} (A^{K+g} - A^K) = 0$$

and g is the smallest positive integer that satisfies this equation.

To show that our example A has asymptotic period 2, we must show that A cannot have a smaller period. If A had asymptotic period 1, then for every vector X, we would have

$$\lim_{K \to \infty} (A^{K+1} - A^K)X = 0.$$

(The 0 on the right side is a vector, not a matrix.) Consider the vector

$$X = (4, -2, 2, -1)^T.$$

Calculation shows that AX equals $-X$. Hence $A^K X = (-1)^K X$, and

$$(A^{K+1} - A^K)X = (-1)^{K+1}X - (-1)^K X = 2(-1)^{K+1}X,$$

which does not go to zero. This means that A does have asymptotic period 2.

6.4 Companion Matrices

To set the stage for our proof of Leslie's Convergence Theorem we take a detour through companion matrices. Our intention is to prove convergence results about *matrices* and from these to derive convergence results for *vectors*. We generalize Leslie's theorem by considering both primitive and imprimitive matrices.

Let A be a matrix in **companion** form; that is,

$$A = \begin{bmatrix} c_1 & c_2 & \cdots & c_n \\ 1 & 0 & \cdots & 0 \\ & & \ddots & \\ 0 & & 1 & 0 \end{bmatrix},$$

where $c_n \neq 0$. Companion matrices have features in common with Leslie matrices, and the relationship between these two types of matrices, will be explained a little later. (See Section 6.6.) Companion matrices are called *companion* because they are closely associated with polynomials of the form $\lambda^n - c_1\lambda^{n-1} - c_2\lambda^{n-2} - \cdots - c_n$. This companionship is not just one-way, since the characteristic polynomial of A is, up to sign, the polynomial to which A is companion. This last remark can be verified easily by writing down the determinant and expanding it via its last column:

$$ch_A(\lambda) = \det(A - \lambda I) = \begin{vmatrix} c_1 - \lambda & c_2 & \cdots & \cdots & c_n \\ 1 & -\lambda & & & \\ & 1 & \ddots & & \\ & & \ddots & \ddots & \\ & & & 1 & -\lambda \end{vmatrix}$$

$$= -\lambda \begin{vmatrix} c_1 - \lambda & c_2 & \cdots & \cdots & c_{n-1} \\ 1 & -\lambda & & & \\ & 1 & \ddots & & \\ & & \ddots & \ddots & \\ & & & 1 & -\lambda \end{vmatrix} - (-1)^n c_n \begin{vmatrix} 1 & -\lambda & & \\ & 1 & -\lambda & \\ & & \ddots & \ddots \\ & & & \ddots & -\lambda \\ & & & & 1 \end{vmatrix}$$

$$= \ \cdots$$

$$= -\lambda(-1)^{n-1}(\lambda^{n-1} - c_1\lambda^{n-2} - \cdots - c_{n-1}) - (-1)^n c_n$$

$$= (-1)^n[\lambda^n - c_1\lambda^{n-1} - \cdots - c_{n-1}\lambda - c_n],$$

where as usual, the ellipses, $\ldots$, indicate an inductive argument. In what follows we will blithely ignore the $(-1)^n$, and simply say that $ch_A(\lambda) = \lambda^n - c_1\lambda^{n-1} - \cdots - c_n$. In fact, we will usually drop the subscript A and just say $ch(\lambda) = \lambda^n - c_1\lambda^{n-1} - \cdots - c_n$.

The Cayley–Hamilton Theorem (see Appendix C) says that every matrix satisfies $ch(A) = A^n - c_1 A^{n-1} - c_2 A^{n-2} - \cdots - c_n I = 0$, giving a relationship between a polynomial and its companion matrix. We can use this relation to compute powers of the matrix A. From the relation

$$(6.1) \qquad A^n = c_1 A^{n-1} + c_2 A^{n-2} + \cdots + c_n I,$$

so that

$$A^{n+1} = c_1 A^n + c_2 A^{n-1} + \cdots + c_n A,$$

and using (6.1), we obtain

$$A^{n+1} = (c_1^2 + c_2)A^{n-1} + (c_1 c_2 + c_3)A^{n-2} + \cdots$$
$$+ (c_1 c_{n-1} + c_n)A + c_1 c_n I.$$

Extending this example, all powers of A can be expressed as linear combinations of the 0th through $n - 1^{st}$ power of A. For any k there are coefficients $\alpha_1(k), \ldots, \alpha_n(k)$ such that

$$A^k = \alpha_1(k)A^{n-1} + \alpha_2(k)A^{n-2} + \cdots + \alpha_n(k)I.$$

This formula says that the problem of computing the powers of a matrix can be reduced to computing n scalar functions. Further, the formulas for computing these functions have the following simple form:

$$\alpha_1(k+1) = c_1 \cdot \alpha_1(k) + \alpha_2(k),$$
$$\alpha_2(k+1) = c_2 \cdot \alpha_1(k) + \alpha_3(k),$$
$$\vdots$$
$$\alpha_{n-1}(k+1) = c_{n-1} \cdot \alpha_1(k) + \alpha_n(k),$$
$$\alpha_n(k+1) = c_n \cdot \alpha_1(k).$$

This is a system of n coupled difference equations, which can be separated by converting to an nth order difference equation, for example,

$$\alpha_1(k+1) = c_1\alpha_1(k) + c_2\alpha_1(k-1) + \cdots + c_n\alpha_1(k-n+1).$$

If one knew the solution to this difference equation, then the other scalar functions could be computed by

$$\alpha_2(k+1) = c_2\alpha_1(k) + c_3\alpha_1(k-1) + \cdots + c_n\alpha_1(k-n+2),$$
$$\vdots$$
$$\alpha_n(k+1) = c_n\alpha_1(k).$$

If one desired, these could also be replaced by n uncoupled difference equations

$$\alpha_1(k+1) = c_1\alpha_1(k) + c_2\alpha_1(k-1) + \cdots + c_n\alpha_1(k-n+1),$$
$$\alpha_2(k+1) = c_1\alpha_2(k) + c_2\alpha_2(k-1) + \cdots + c_n\alpha_2(k-n+1),$$
$$\vdots$$
$$\alpha_n(k+1) = c_1\alpha_n(k) + c_2\alpha_n(k-1) + \cdots + c_n\alpha_n(k-n+1).$$

Notice that these are really the same difference equation, but each copy of the equation has, in general, a different solution because the initial conditions are different for each copy.

Although the above formulas give computationally efficient methods to compute powers of a matrix, we want a single method that gives a good estimate of the asymptotic behavior of the sequence of powers. What does A^k look like when k is large? In general, the answer is easy. Either all the entries in A^k are large—that is, grow exponentially in magnitude as a function of k—or all entries of A^k are small—that is, decrease exponentially in magnitude as a function of k. Of course, there are A's that are exceptions to this general case. In fact, we would like to change A so that the overall growth or decline trend is normalized out, and then see how A^k behaves. To do so we assume that there is a single number λ_0 that is the largest number (in magnitude) such that $A - \lambda_0 I$ is singular and then study the behavior of $(A/\lambda_0)^k$. We expect the growth trend to be normalized out, so that $(A/\lambda_0)^k$ has a limit, and we want to compute this limit. It is important to determine or at least estimate λ_0, and we return to this estimation later.

First, let us look for an **eigenmatrix** P_λ whose form is preserved but whose magnitude is multiplied by λ when P_λ is multiplied by A. That is, we want to solve the equation

$$(6.2) \qquad AP_\lambda = \lambda P_\lambda .$$

Since we are interested in computing powers of A, we assume that P_λ can be represented as a polynomial in A. Assuming

$$P_\lambda = p_0 A^{n-1} + p_1 A^{n-2} + \cdots + p_{n-1} I ,$$

then

$$
\begin{aligned}
(6.3) \qquad \lambda P_\lambda = AP_\lambda &= p_0 A^n + p_1 A^{n-1} + \cdots + p_{n-1} A \\
&= (p_0 c_1 + p_1) A^{n-1} + (p_0 c_2 + p_2) A^{n-2} + \cdots \\
&\quad + (p_0 c_{n-1} + p_{n-1}) A + p_0 c_n I .
\end{aligned}
$$

Equating the coefficients of corresponding powers of A on each side of (6.3) gives

$$p_0 c_n = p_{n-1} \lambda ,$$
$$p_0 c_{n-1} + p_{n-1} = p_{n-2} \lambda ,$$
$$\vdots$$
$$p_0 c_2 + p_2 = p_1 \lambda ,$$
$$(6.4) \qquad p_0 c_1 + p_1 = p_0 \lambda .$$

Assuming $\lambda \neq 0$, these equations can be solved as

$$
\begin{aligned}
p_{n-1} &= \frac{p_0 c_n}{\lambda} , \\
(6.5) \quad p_{n-2} &= \frac{p_0 c_{n-1}}{\lambda} + \frac{p_{n-1}}{\lambda} &= p_0 \left(\frac{c_{n-1}}{\lambda} + \frac{c_n}{\lambda^2} \right) \\
&\;\;\vdots &\vdots \;\;, \\
p_1 &= \frac{p_0 c_2}{\lambda} + \frac{p_2}{\lambda} &= p_0 \left(\frac{c_2}{\lambda} + \frac{c_3}{\lambda^2} + \cdots + \frac{c_n}{\lambda^{n-1}} \right) .
\end{aligned}
$$

The last equation and (6.4) give

$$p_0 c_1 + p_1 = p_0 \left(c_1 + \frac{c_2}{\lambda} + \cdots + \frac{c_n}{\lambda^{n-1}} \right) = p_0 \lambda .$$

This is a valid equation for all values of p_0 if $\lambda^n = c_1 \lambda^{n-1} + \cdots + c_n$, which is true for every λ that is a root of the characteristic polynomial. Of course, if λ is not a root of the characteristic polynomial, then this last equation implies that $p_0 = 0$, and hence by (6.5) that $p_i = 0$ for all i.

The eigenmatrices P_λ can also be expressed in terms of the values of the polynomials $g_i(\lambda)$, defined as

$$g_1(\lambda) = \lambda - c_1 \, ,$$
$$g_2(\lambda) = \lambda^2 - c_1 \lambda - c_2 = \lambda g_1(\lambda) - c_2 \, ,$$
$$\vdots$$
$$g_i(\lambda) = \lambda^i - c_1 \lambda^{i-1} - \cdots - c_i = \lambda g_{i-1}(\lambda) - c_i \, ,$$
$$g_{n-1}(\lambda) = \lambda^{n-1} - c_1 \lambda^{n-2} - \cdots - c_{n-1} \, ,$$
$$g_n(\lambda) = \lambda^n - c_1 \lambda^{n-1} - \cdots - c_{n-1} \lambda - c_n = ch(\lambda) \, ,$$

which form the sequence of polynomials from Horner's method (see Section 5.3). So up to a scalar multiplier P_λ is

$$(6.6) \qquad P_\lambda = A^{n-1} + g_1(\lambda) A^{n-2} + g_2(\lambda) A^{n-2} + \cdots + g_{n-1}(\lambda) I \, ,$$

which follows by using $ch(\lambda) = 0$ and rearranging the formulas from (6.5).

In the simple case in which $ch_A(\lambda)$ has n distinct roots $\lambda_1, \lambda_2, \ldots, \lambda_n$, we will show that it is easy to represent powers of A in terms of the P_λ. Assume that there are n scalars $\alpha_1, \alpha_2, \ldots, \alpha_n$ such that

$$(6.7) \qquad I = \sum_{i=1}^{n} \alpha_i P_{\lambda_i} \, .$$

Then

$$A^k = A^k \sum_{i=1}^{n} \alpha_i P_{\lambda_i} = \sum_{i=1}^{n} \alpha_i A^k P_{\lambda_i} = \sum_{i=1}^{n} \alpha_i \lambda_i^k P_{\lambda_i} \, .$$

Further, if λ_1 is a strictly dominant eigenvalue then

$$\lim_{k \to \infty} \frac{A^k}{\lambda_1^k} = \lim_{k \to \infty} \sum_{i=1}^{n} \alpha_i \left(\frac{\lambda_i}{\lambda_1} \right)^k P_{\lambda_i} = \alpha_1 P_{\lambda_1} \, .$$

We are left with calculating α_1. Luckily, this is not difficult. Since $A P_{\lambda_i} = \lambda_i P_{\lambda_i}$, then $(A - \lambda_i I) P_{\lambda_i} = 0$, and for any scalars a, b,

$$(A - \lambda_i I)(A - \lambda_j I)(a P_{\lambda_i} + b P_{\lambda_j})$$
$$= a(A - \lambda_j I)(A - \lambda_i I) P_{\lambda_i} + b(A - \lambda_i I)(A - \lambda_j I) P_{\lambda_j}$$
$$= a(A - \lambda_j I) 0 + b(A - \lambda_i I) 0 = 0 \, .$$

So the matrix polynomial $(A - \lambda_2 I)(A - \lambda_3 I) \cdots (A - \lambda_n I)$ annihilates every linear combination of $P_{\lambda_2}, \ldots, P_{\lambda_n}$; that is,

$$(A - \lambda_2 I) \cdots (A - \lambda_n I) \sum_{i=2}^{n} b_i P_{\lambda_i} = 0 \, .$$

Now the matrix polynomial $(A - \lambda_2 I) \cdots (A - \lambda_n I)$ can be written as $\frac{ch(A)}{A - \lambda_1 I}$ because the polynomial in A in the denominator exactly divides the polynomial in A in the numerator. Next, consider the action of $(A - \lambda_j I)$ on P_{λ_1}. We see that

$$(A - \lambda_j I)P_{\lambda_1} = AP_{\lambda_1} - \lambda_j P_{\lambda_1} = \lambda_1 P_{\lambda_1} - \lambda_j P_{\lambda_1} = (\lambda_1 - \lambda_j)P_{\lambda_1}.$$

So

$$(A - \lambda_i I)(A - \lambda_j I)P_{\lambda_1} = (\lambda_1 - \lambda_i)(\lambda_1 - \lambda_j)P_{\lambda_1}$$

and

$$\frac{ch(A)}{A - \lambda_1 I}P_{\lambda_1} = (\lambda_1 - \lambda_2)(\lambda_1 - \lambda_3)\ldots(\lambda_1 - \lambda_n)P_{\lambda_1}$$

$$= \lim_{\lambda \to \lambda_1} \frac{ch(\lambda)}{\lambda - \lambda_1}P_{\lambda_1}$$

$$= \lim_{\lambda \to \lambda_1} \frac{ch(\lambda) - ch(\lambda_1)}{\lambda - \lambda_1}P_{\lambda_1}$$

$$= ch'(\lambda_1)P_{\lambda_1}.$$

Actually, in this calculation we made use of no special property of the eigenvalue λ_1, so

$$\frac{ch(A)}{A - \lambda_i I}P_{\lambda_i} = ch'(\lambda_i)P_{\lambda_i}$$

holds for each eigenvalue λ_i.

We now want to use this result on the equation

$$I = \sum_{i=1}^{n} \alpha_i P_{\lambda_i}$$

to compute the α_i's. So

(6.8) $$\frac{ch(A)}{A - \lambda_j I}I = \sum_{i=1}^{n} \alpha_i \frac{ch(A)}{A - \lambda_j I}P_{\lambda_i} = \alpha_j \, ch'(\lambda_j)P_{\lambda_j}$$

where

$$P_{\lambda_j} = A^{n-1} + g_1(\lambda_j)A^{n-2} + \cdots + g_{n-1}(\lambda_j)I.$$

Since the coefficient of A^{n-1} in $\frac{ch(A)}{A - \lambda_j I}$ is 1, equating the coefficients gives $\alpha_j = 1/ch'(\lambda_j)$. In fact, we also have $P_{\lambda_j} = \frac{ch(A)}{A - \lambda_j I}$, which one can easily verify is the same matrix polynomial as the one we computed in (6.6). These computations demonstrate the following theorem.

Theorem 6.4.1. *If A is a companion matrix with n distinct eigenvalues and λ_1 is a strictly dominant eigenvalue, then*

$$\lim_{k \to \infty} \frac{A^k}{\lambda_1^k} = \frac{1}{ch'(\lambda_1)}P_{\lambda_1}, \quad where \quad P_{\lambda_1} = \frac{ch(A)}{A - \lambda_1 I}.$$

A few comments on this theorem are necessary. First, if the eigenvalues are distinct, isn't there only one *largest* eigenvalue? No, there may still be several eigenvalues with the same magnitude. For instance,

$$A = \begin{bmatrix} 0 & 1 \\ 1 & 0 \end{bmatrix}$$

has $ch_A(\lambda) = \lambda^2 - 1$, and the two eigenvalues, $\lambda_1 = 1$ and $\lambda_2 = -1$, have the same magnitude. In this case, it is easy to see that

$$A^K = \begin{cases} I & \text{if } K \text{ is even,} \\ A & \text{if } K \text{ is odd,} \end{cases}$$

and A^K/λ_1^K has no limit. Taking absolute values of matrix elements does not help, because

$$|A^K/\lambda_1^K| = |A^K/\lambda_2^K| = \begin{cases} I & \text{if } K \text{ is even,} \\ A & \text{if } K \text{ is odd.} \end{cases}$$

It is possible to get a convergence theorem when absolute values are replaced by **matrix norms**. Recall that absolute value takes a complex number, which may be a negative real number or have a non-zero imaginary part, and returns a positive real number, the magnitude of the complex number. In a similar fashion, it is possible to define a norm that takes a matrix and returns a positive real number, the magnitude of the matrix, but we do not pursue matrix norms at this point. Second, why must λ_1 be distinct from the other eigenvalues? Because the limit contains $1/ch'(\lambda_1)$, and if λ_1 were a multiple eigenvalue, $ch'(\lambda_1)$ would be zero, and the claimed limit would not be defined. Also, there would be an inconsistency in (6.8), and the scalars in (6.7) could not be found.

There is a logical hole in what we have just done. We assumed that solving for P_λ could be carried out by equating the coefficients of A^j for each j on either side of the equation $AP_\lambda = \lambda P_\lambda$. (See equation (6.2).) This is tantamount to assuming that the powers of A are **linearly independent**; that is, if

$$b_1 A^{n-1} + b_2 A^{n-2} + \cdots + b_n I = 0,$$

then $b_1 = b_2 = \cdots = b_n = 0$. This is equivalent to the situation in which the minimal polynomial equals the characteristic polynomial, and for general matrices this assumption is not valid. For example, if $A = I$, then $1 \cdot A + (-1) \cdot I = 0$, and the powers of A are not linearly independent. Fortunately, we have an extra assumption about the matrices we are using. We have assumed that they are in **companion form**; that is,

$$A = \begin{bmatrix} c_1 & c_2 & \cdots & c_n \\ 1 & 0 & \cdots & 0 \\ & & \ddots & \\ 0 & & 1 & 0 \end{bmatrix}$$

with $c_n \neq 0$.

We now show that if A is in companion form, then $A^{n-1}, A^{n-2}, \ldots, I$ are linearly independent, which means that the only choice of $b_1, b_2, \ldots, b_n$ with

(6.9) $$b_1 A^{n-1} + b_2 A^{n-2} + \cdots + b_n I = 0$$

has all b_i equal to 0. This equation (6.9) is a matrix equation, and for it to hold as a matrix equation it must hold for each column. In particular, $A^i e_n$ is the last column of A^i, and (6.9) implies that

(6.10) $$b_1 A^{n-1} e_n + b_2 A^{n-2} e_n + \cdots + b_n I e_n = 0.$$

We claim that $A^i e_n$ has the form

$$\begin{pmatrix} - \\ \vdots \\ - \\ c_n \\ 0 \\ \vdots \\ 0 \\ 0 \end{pmatrix} \begin{matrix} \left.\vphantom{\begin{pmatrix}-\\ \vdots \\ -\end{pmatrix}}\right\} & i \\ \\ \left.\vphantom{\begin{pmatrix}0\\ \vdots \\ 0\end{pmatrix}}\right\} & n - i \, ; \end{matrix}$$

that is, the ith entry is c_n and the last $n - i$ entries are 0. This claim is valid for $i = 1$, since the last column of A is simply $\begin{pmatrix} c_n \\ 0 \\ \vdots \\ 0 \end{pmatrix}$. Assuming that the last column has the claimed form for A^i, then the last column of A^{i+1} is given by

$$A^{i+1} e_n = A \, A^i e_n = \begin{bmatrix} c_1 & c_2 & \cdots & c_n \\ 1 & 0 & \cdots & 0 \\ & & \ddots & \\ 0 & & 1 & 0 \end{bmatrix} \begin{pmatrix} - \\ \vdots \\ - \\ c_n \\ 0 \\ \vdots \\ 0 \end{pmatrix} = \begin{pmatrix} - \\ \vdots \\ - \\ c_n \\ 0 \\ \vdots \\ 0 \end{pmatrix} ,$$

and the claim holds for $i = 1$ through $i = n - 1$. For $i = 0$, we want the last column of I, which is e_n. For (6.10) to be valid, b_n must be 0, since all of the other vectors being added have a 0 as the last component. Hence, (6.10) becomes $b_1 A^{n-1} e_n + \cdots + b_{n-1} A\, e_n = 0$, and similarly we can see that $b_{n-1} = 0$. Continuing this argument, $b_1 = b_2 = \cdots = b_n = 0$. So the first n powers of a companion matrix *are* linearly independent, and equating coefficients of these powers was a valid operation and there is no logical hole in the proof of Theorem 6.4.1.

6.4.1   Matrices with repeated eigenvalues

Unfortunately, even companion matrices can fail to have distinct eigenvalues, and since we get only one P_λ for each λ, it may not be possible to represent I as a linear combination of the P_λ's. But the last theorem suggests a natural generalization. If λ_1 is an eigenvalue of multiplicity m, then $ch(A)$ is divisible by $(A - \lambda_1 I)^m$. So instead of having a single P_{λ_1} given by $\frac{ch(A)}{A - \lambda_1 I}$, we can have instead m matrices $P_{\lambda_1}^{(1)}, P_{\lambda_1}^{(2)}, \ldots, P_{\lambda_1}^{(m)}$ with

$$P_{\lambda_1}^{(j)} = \frac{ch(A)}{(A - \lambda_1 I)^j}.$$

This form has the pleasant consequence that

$$(A - \lambda_1 I) P_{\lambda_1}^{(j)} = P_{\lambda_1}^{(j-1)} \text{ for } j > 1,$$

and, of course, $(A - \lambda_1 I) P_{\lambda_1}^{(1)} = 0$.

As before, to study the behavior of powers of A we look at an expansion of I. To keep the notation simple, we write the P_λ's without superscripts, and have

$$I = \sum_{i=1}^{n} \alpha_i P_{\lambda_i},$$

which looks like the expansion we used before. The extra complication is that the formula for the α_i's is slightly more complex. If λ_j is a root of multiplicity 1, consider

$$\frac{ch(A)}{A - \lambda_j I} I = \frac{ch(A)}{A - \lambda_j I} \sum \alpha_i P_{\lambda_i} = \alpha_j \frac{ch(A)}{A - \lambda_j I} P_{\lambda_j} = \alpha_j ch'(\lambda_j) P_{\lambda_j},$$

and so $\alpha_j = \frac{1}{ch'(\lambda_j)}$, since $P_{\lambda_j} = \frac{ch(A)}{A - \lambda_j I}$. If λ_i is a root of multiplicity m, consider

$$\frac{ch(A)}{(A - \lambda_i I)^m} I = \frac{ch(A)}{(A - \lambda_i I)^m} \sum \alpha_j P_{\lambda_j}$$

$$= \frac{ch(A)}{(A - \lambda_i I)^m} [\alpha_1 P^{(1)} + \alpha_2 P^{(2)} + \cdots + \alpha_m P^{(m)}]$$

where $P^{(1)}, \ldots, P^{(m)}$ are the eigenmatrices associated with λ_i. Of course, the eigenmatrices associated with other eigenvectors are annihilated by this operator. For

$$r(A) = \frac{ch(A)}{(A - \lambda_i I)^m}$$

we have

$$P^{(m)} = r(A)[\alpha_1 P^{(1)} + \alpha_2 P^{(2)} + \cdots + \alpha_m P^{(m)}].$$

Now

$$AP^{(1)} = \lambda_i P^{(1)},$$
$$AP^{(2)} = \lambda_i P^{(2)} + P^{(1)}, \quad A^2 P^{(2)} = \lambda_i^2 P^{(2)} + 2\lambda_i P^{(1)}.$$

So

$$A^l P^{(2)} = \lambda_i^l P^{(2)} + l \cdot \lambda_i^{l-1} P^{(1)},$$

and

$$A^l P^{(j)} = \lambda_i^l P^{(j)} + \binom{l}{1} \lambda_i^{l-1} P^{(j-1)} + \binom{l}{2} \lambda_i^{l-2} P^{(j-2)} + \cdots.$$

Hence,

$$
\begin{aligned}
P^{(m)} = {} & \alpha_1 r(\lambda_i) P^{(1)} \\
& + \alpha_2 [r(\lambda_i) P^{(2)} + r'(\lambda_i) P^{(1)}] \\
& \quad \vdots \\
& + \alpha_m [r(\lambda_i) P^{(m)} + r'(\lambda_i) P^{(m-1)} + r''(\lambda_i) P^{(m-2)} + \cdots].
\end{aligned}
$$

Equating the coefficients of the $P^{(i)}$'s gives

$$
\begin{aligned}
1 &= \alpha_m r(\lambda_i), \\
0 &= \alpha_m r'(\lambda_i) + \alpha_{m-1} r(\lambda_i), \\
0 &= \alpha_m r''(\lambda_i) + \alpha_{m-1} r'(\lambda_i) + \alpha_{m-2} r(\lambda_i), \\
&\quad \vdots \\
0 &= \alpha_m r^{m-1}(\lambda_i) + \alpha_{m-1} r^{m-2}(\lambda_i) + \cdots + \alpha_1 r(\lambda_i).
\end{aligned}
$$

Since $r(\lambda_i)$ is non-zero, this is an invertible triangular system of m linear equations in m unknowns; it has a unique solution which can be found by back substitution. We can compute powers of A by

$$A^k = \sum_{i=1}^{n} \alpha_i A^k P_{\lambda_i},$$

but we may have that $AP_{\lambda_i}^{(2)} = \lambda_i P_{\lambda_i}^{(2)} + P_{\lambda_i}^{(1)}$ if there are eigenvalues with multiplicity greater than 1. But these "extra terms" can only grow as a polynomial in k times λ_i^k. Thus, if we assume that λ_1 is the eigenvalue of greatest magnitude and that λ_1 is *not* a multiple root, then $\lim_{k\to\infty} \frac{A^k}{\lambda_1^k}$ does not have any extra terms, because $(\lambda_i/\lambda_1)^k$ times any polynomial in k still goes to 0 as k increases. This gives an improved form of Theorem 6.4.1.

Theorem 6.4.2. *If A is a companion matrix and the dominant eigenvalue (the eigenvalue with largest magnitude) λ_1 is simple, then*

$$\lim_{k\to\infty} \frac{A^k}{\lambda_1^k} = \frac{1}{ch'(\lambda_1)} P_{\lambda_1}, \ \text{ where } P_{\lambda_1} = \frac{ch(A)}{A - \lambda_1 I},$$

and this limiting matrix has rank 1. That is, there is some vector X such that the limiting matrix can be written as

$$(X \quad X \quad \cdots \quad X) \begin{bmatrix} b_1 & 0 & \cdots & 0 \\ & b_2 & & \\ & & \ddots & \\ 0 & \cdots & 0 & b_n \end{bmatrix},$$

or said another way, every column of the limiting matrix is a scalar multiple of any other column of the limiting matrix.

Proof. If Z is the limiting matrix, then it does satisfy the equation $AZ = \lambda_1 Z$. Each column Z_i of Z also satisfies $AZ_i = \lambda_1 Z_i$, and so

$$c_1 Z_{i\cdot 1} + c_2 Z_{i\cdot 2} + \cdots + c_n Z_{i\cdot n} = \lambda_1 Z_{i\cdot 1},$$
$$Z_{i\cdot 1} = \lambda_1 Z_{i\cdot 2},$$
$$\vdots$$
$$Z_{i\cdot n-1} = \lambda_1 Z_{i\cdot n}$$

where $Z_{i\cdot j}$ means the jth component of the vector Z_i. If $Z_{i\cdot n}$ were known, then $Z_{i\cdot n-1}, \ldots, Z_{i\cdot 1}$ could be computed. The top equation is redundant and merely says that λ_1 satisfies the characteristic polynomial. Thus, up to one unknown multiplier $Z_{i\cdot n}$, this system of equations has a unique solution, and the theorem follows. $\square$

6.5 Nonnegative Companion Matrices

Although companion matrices capture the *structure* of Leslie matrices, they do not take into account the Leslie assumptions of nonnegativity. In this section we consider the extra properties that follow when all entries in a companion matrix are nonnegative.

The characteristic polynomial of either a Leslie matrix or a nonnegative companion matrix can be written in the form

$$ch(\lambda) = \lambda^n - \sum_{i=1}^{n} c_i \lambda^{n-i},$$

where each $c_i \geq 0$ and $c_n > 0$. As shown in Section 5.1, such nonnegative polynomials are of two types: the primitive type, which has $\gcd\{i|c_i > 0\} = 1$, and the periodic type, which has $g = \gcd\{i|c_i > 0\} > 1$.

A primitive polynomial has exactly one positive real root λ_0, and this root is dominant in the sense that $\lambda_0 > |\lambda_i|$ for any other root λ_i. A periodic polynomial with period g has non-zero coefficients only at positions that are multiples of g, which means that λ appears in the polynomial only as powers of λ^g. Therefore, the characteristic polynomial of a periodic nonnegative companion matrix can be viewed as a polynomial $p(x)$, where $x = \lambda^g$ and $p(x)$ is now a *primitive* polynomial with a unique positive real root x_0. The positive root of $ch(\lambda)$ is the positive solution to $\lambda^g = x_0$, and since λ_0 is positive, then $\lambda_0 = x_0^{1/g}$. This λ_0 is not strictly dominant, because $\lambda^g = x_0$ has g roots all of the same magnitude, and these equal-magnitude roots give rise to oscillations and asymptotic periodic behavior of period g. We consider these possibilities in the next subsection. Oscillations are also possible in companion matrices that have some negative entries. (Are these non-nonnegative?) We do not consider such matrices in detail, but in Exercise 6.19 there is an example of a companion matrix that behaves periodically but its characteristic polynomial is not periodic.

Since a nonnegative companion matrix with a primitive characteristic polynomial has a dominant eigenvalue, we can prove a stronger version of Theorem 6.4.2 for such matrices.

Theorem 6.5.1. *If A is a nonnegative companion matrix with a primitive characteristic polynomial, then there is a power of A that is strictly positive, A has a strictly dominant positive eigenvalue λ_0,*

$$\lim_{K \to \infty} \frac{A^K}{\lambda_0^K} = \frac{1}{ch'(\lambda_0)} P_{\lambda_0}, \quad where \quad P_{\lambda_0} = \frac{ch(A)}{A - \lambda_0 I},$$

and this limiting matrix is a strictly positive matrix with rank 1.

Proof. Nonnegativity implies that λ_0 is the dominant eigenvalue, and from Exercise 5.2 we know that $ch'(\lambda_0) > 0$. Theorem 6.4.2 implies convergence. We are left with showing that the limiting matrix is strictly positive. By the division algorithm, $ch(\lambda)/(\lambda - \lambda_0) = \lambda^{n-1} + g_1\lambda^{n-2} + \cdots + g_{n-1}$, where

the coefficients are

$$g_1 = \lambda_0 - c_1,$$
$$g_2 = \lambda_0^2 - c_1\lambda_0 - c_2,$$
$$\vdots$$
$$g_{n-1} = \lambda_0^{n-1} - c_1\lambda_0^{n-2} - \cdots - c_{n-1} = c_n/\lambda_0.$$

Clearly, $g_{n-1} > 0$, and this implies that g_{n-2} through g_1 are also positive. (Also, refer to Exercise 5.14.) So $P_{\lambda_0} = A^{n-1} + g_1 A^{n-2} + \cdots + g_{n-1}I$, which is a positive sum of nonnegative matrices. But can it happen that none of these powers of A is strictly positive? If we look at the graph G_A corresponding to A (see Section 7.3 for more details on the correspondence between a nonnegative matrix and its graph), then A_{ij}^r, the $(i,j)^{\text{th}}$ entry of the rth power of A, is positive when there is a path from v_j to v_i of length r in G_A. The graph for a nonnegative companion matrix contains at least the edges in the following diagram

so there is a path between any pair of vertices, and the length of this path is between 0 and $n - 1$. Hence, A_{ij}^r is positive for some $0 \leq r \leq n - 1$, which means that the positive sum P_{λ_0} of these powers of A is strictly positive. □

6.5.1 Periodic nonnegative companion matrices

Now that we understand the convergence properties of *primitive* companion matrices (that is, companion matrices whose characteristic polynomials are primitive), we would like to consider nonnegative companion matrices that are periodic in the sense that their characteristic polynomials can be written in the form

$$ch(\lambda) = (\lambda^g)^{\frac{n}{g}} - c_g(\lambda^g)^{\frac{n-g}{g}} - c_{2g}(\lambda^g)^{\frac{n-2g}{g}} - \cdots - c_{g\,n/g},$$

and these can be viewed as polynomials in λ^g rather than as polynomials in λ. We want to reduce the periodic case to the primitive case. Here "reduce" means that we would like to represent or think about an imprimitive matrix as several primitive matrices, apply the theorems about primitive matrices, and then put the results together to obtain an analysis of imprimitive matrices. This idea of **reduction** is a central unifying concept in all of mathematics, and is also the basis for divide-and-conquer algorithms [102] in computer science. (See also Section 9.4.)

Assume that we have a matrix L with $g = 2$, where this g is, of course, the greatest common divisor of the cycle lengths in the graph of the matrix. This quantity is also called the **index of imprimitivity**, but it is simply easier to call it g. We can associate a graph $G(L)$ with the matrix

$$L = \begin{bmatrix} 0 & c_2 & 0 & \dots & 0 & c_n \\ 1 & 0 & 0 & \dots & 0 & 0 \\ & & & \dots & & \\ 0 & 0 & 0 & \dots & 1 & 0 \end{bmatrix}.$$

We've chosen to use the convention that if $L_{i,j} > 0$, there is an edge from v_i to v_j. This is the opposite direction to that used in the previous diagram. There is an edge from v_n to v_1 and a cycle of length n passing through this edge. The assumption that $g = 2$ implies that n is even. So the graph looks like this:

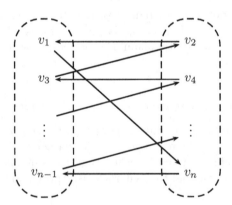

There may be some other edges from v_1, but they all go to vertices in the second column, vertices with an even index because $g = 2$. Now let's form a new graph whose edges are the paths of length 2 in $G(L)$; a part of this graph looks like this:

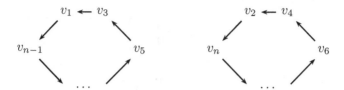

The old edges from v_1 to v_i now appear as edges from v_1 to v_{i-1} and as edges from v_2 to v_i, but there are *no* edges from an even vertex to an odd vertex, and there are *no* edges from an odd vertex to an even vertex. Notice that the original connected graph has been split into two graphs with no edges between them. Further notice that two new graphs are identical (isomorphic). By the usual reasoning about adjacency matrices, L^2 is the matrix that corresponds to this graph with two parts. If we permute the columns of L^2 so that the odd positions appear first and we permute the rows in the same way, we obtain a matrix of the form

$$\begin{bmatrix} A & 0 \\ 0 & A \end{bmatrix}, \quad \text{where} \quad A = \begin{bmatrix} f_2 & f_4 & \cdots & f_n \\ 1 & & & 0 \\ & \ddots & & \vdots \\ & & 1 & 0 \end{bmatrix}.$$

So, pleasantly enough, A is a companion matrix, and in fact a *primitive* companion matrix because using paths of length two divides the cycle lengths by 2, and converts $g = 2$ for L into $g = 1$ for A.

At this point we should recall a few facts about permutation matrices. A **permutation matrix** Π has exactly one 1 in each row and each column and 0's elsewhere. If A is a matrix and Π is a permutation matrix, then $A\Pi$ is a matrix with the columns of A permuted. The position of the 1 in the ith column of Π determines which column of A appears as the ith column of $A\Pi$. Similarly, ΠA is a matrix with the rows permuted. The position of the 1 in the ith row of Π determines which row of A appears as the ith row of ΠA. Now consider the effect of applying Π^T, the **transpose** of Π, the matrix whose rows are the columns of Π. Then $\Pi^T A\Pi$ is the matrix that results from permuting the rows of $A\Pi$ in the same way as the columns of A were permuted in $A\Pi$. This is exactly the kind of operation we want to carry out on imprimitive matrices. Finally, consider $\Pi^T\Pi$. The single 1 in the ith row of Π^T matches the single 1 in the ith column of Π, and since there are no other 1's in the same row as this 1, the product matrix has in its ith row a 1 only at position i. Hence, $\Pi^T\Pi$ equals the identity matrix, and we conclude that the transpose of a permutation matrix is the inverse of the permutation matrix.

For our example matrix L, the appropriate Π permutes the odd columns so that they lie in front of the even columns, and

$$\Pi^T L^2 \Pi = \begin{bmatrix} A & 0 \\ 0 & A \end{bmatrix}$$

and

$$L^2 = \Pi \begin{bmatrix} A & 0 \\ 0 & A \end{bmatrix} \Pi^T .$$

Now consider $(\Pi^T L^2 \Pi)(\Pi^T L^2 \Pi) = \Pi^T L^4 \Pi$, since $\Pi^T \Pi = I$. This gives

$$\Pi^T L^{2K} \Pi = \begin{bmatrix} A^K & 0 \\ 0 & A^K \end{bmatrix} .$$

Of course, A satisfies the hypotheses of Theorem 6.5.1, so

$$\lim_{K \to \infty} \frac{A^K}{\mu_0^K} = \frac{1}{ch_A'(\mu_0)} \frac{ch_A(A)}{A - \mu_0 I} ,$$

where $ch_A(\lambda)$ is the characteristic polynomial of A and μ_0 is the unique positive root of this nonnegative polynomial. Most conveniently, $ch_A(\lambda^2) = ch_L(\lambda)$, and replacing λ^2 by λ gives the characteristic polynomial for A. This also means that the dominant eigenvalue μ_0 for A is the square of the dominant eigenvalue λ_0 for L. Hence,

$$\lim_{K \to \infty} \frac{L^{2K}}{\lambda_0^{2K}} = \Pi \begin{bmatrix} \lim_{K \to \infty} \dfrac{A^K}{\mu_0^K} & 0 \\ 0 & \lim_{K \to \infty} \dfrac{A^K}{\mu_0^K} \end{bmatrix} \Pi^T$$

$$= \frac{1}{ch_A'(\mu_0)} \Pi \left\{ \frac{ch_A(A)}{A - \mu_0 I} \begin{bmatrix} I & 0 \\ 0 & I \end{bmatrix} \right\} \Pi^T ,$$

where I is the $\frac{n}{2} \times \frac{n}{2}$ identity matrix. While this formula gives an answer, the matrix without the Π's might be more intuitive:

$$\lim_{K \to \infty} \frac{L^{2K}}{\lambda_0^{2K}} = \begin{bmatrix} \alpha_1 & 0 & \alpha_2 & 0 & \cdots & \alpha_{n/2} & 0 \\ 0 & \alpha_1 & 0 & \alpha_2 & \cdots & 0 & \alpha_{n/2} \\ \alpha_1 B_2 & 0 & \alpha_2 B_2 & 0 & \cdots & \alpha_{n/2} B_2 & 0 \\ 0 & \alpha_1 B_2 & 0 & \alpha_2 B_2 & \cdots & 0 & \alpha_{n/2} B_2 \\ \vdots & \vdots & \vdots & \vdots & & \vdots & \vdots \\ \alpha_1 B_{n/2} & 0 & \alpha_2 B_{n/2} & 0 & \cdots & \alpha_{n/2} B_{n/2} & 0 \\ 0 & \alpha_1 B_{n/2} & 0 & \alpha_2 B_{n/2} & \cdots & 0 & \alpha_{n/2} B_{n/2} \end{bmatrix} .$$

Here we are trying to explicitly display two facts: that the limiting matrix for A has rank one and that the permutations Π and Π^T spread out the two copies of the limiting matrix.

Let us call this limiting matrix L_∞ and look at the action of L on L_∞. Clearly, either by the derivation or direct calculation

$$\frac{L^2}{\lambda_0^2} L_\infty = L_\infty.$$

But what about $\frac{L}{\lambda_0} L_\infty$? This operation results in a matrix that is in a sense complementary to L_∞. That is, the produced matrix, let us call it L_w, has 0's where L_∞ has positive entries, and L_w has positive entries where L_∞ has 0's. Unfortunately, the values are not quite so nice. In L_∞, each column is either a multiple of the first column or a multiple of the first column shifted down one component. In L_w, the odd-position columns all start with 0 and are multiples of the first column of L_w. The even-position columns are all multiples of the second column of L_w and this second column begins with a positive entry, but it is not necessary that the second column be a multiple of a shifted copy of the first column.

The limiting behavior for general g follows immediately from our example with $g = 2$.

Theorem 6.5.2. *For a nonnegative companion matrix L with period $g \geq 2$,*

$$\lim_{K \to \infty} \frac{L^{gK}}{\lambda_0^{gK}} = L_0 = \frac{1}{ch_A'(\lambda_0^g)} \Pi \left\{ \frac{ch_A(A)}{A - \lambda_0^g I} \begin{bmatrix} I & & 0 \\ & \ddots & \\ 0 & & I \end{bmatrix} \right\} \Pi^T,$$

where

$$A = \begin{bmatrix} f_g & f_{2g} & \cdots & f_n \\ 1 & & & 0 \\ & \ddots & & \vdots \\ & & 1 & 0 \end{bmatrix}, \quad ch_A(\lambda) = \lambda^{\frac{n}{g}} - f_g \lambda^{\frac{n}{g}-1} - \cdots - f_n,$$

and Π is the permutation matrix that moves the columns with indices $\equiv 1 \bmod g$ before those with indices $\equiv 2 \bmod g, \ldots,$ before those with indices $\equiv 0 \bmod g$. Also:

(a) L_0 has rank g.

(b) If $i \equiv j \pmod{g}$, the ith column of L_0 has positive entries in positions $j, j+g, j+2g, \ldots$ and zeros elsewhere.

(c) Every column of L_0 is a shifted multiple of the first column.

Let $L_r = \left(\frac{L}{\lambda_0}\right)^r L_0$. Then:

(a) $L_0, L_1, \ldots, L_{g-1}$ are linearly independent.

(b) If $i + r \equiv j \pmod{g}$, the ith column of L_r has positive entries in positions $j, j+g, j+2g, \ldots$ and zeros elsewhere.

(c) $\left(\frac{L}{\lambda_0}\right)^g L_r = L_r$.

6.6   Back to Leslie Matrices

Up to this point we have used companion matrices rather than Leslie matrices because the survival rates in the Leslie matrices complicate things slightly. If B is a Leslie matrix,

$$B = \begin{bmatrix} f_1 & f_2 & \cdots & & f_n \\ s_1 & & & & 0 \\ & \ddots & & & \vdots \\ & & s_{n-1} & & 0 \end{bmatrix},$$

then B can be converted to the companion matrix L via the diagonal matrix S by $L = S^{-1}BS$, where

$$S = \begin{bmatrix} 1 & & & & \\ & s_1 & & & \\ & & s_1 s_2 & & \\ & & & \ddots & \\ & & & & s_1 \ldots s_{n-1} \end{bmatrix},$$

$$S^{-1} = \begin{bmatrix} 1 & & & & \\ & 1/s_1 & & & \\ & & 1/s_1 s_2 & & \\ & & & \ddots & \\ & & & & 1/s_1 \ldots s_{n-1} \end{bmatrix},$$

and

$$L = \begin{bmatrix} f_1 & s_1 f_2 & s_1 s_2 f_3 & \cdots & s_1 \ldots s_{n-1} f_n \\ 1 & & & & 0 \\ & 1 & & & \vdots \\ & & \ddots & & \vdots \\ & & & 1 & 0 \end{bmatrix}.$$

This claim can be easily verified by direct calculation. Of course, we also have $B = SLS^{-1}$ and $B^K = SL^K S^{-1}$. Hence, the asymptotic behavior of B can be directly calculated from the asymptotic behavior of the companion matrix L. Each entry of B^K can be obtained from the corresponding entry of L^K by multiplying and dividing by products of the survival rates. Specifically, the $(i,j)^{\text{th}}$ entry is obtained by multiplying by $s_1 \cdots s_{i-1}$ and dividing by $s_1 \cdots s_{j-1}$. So the $(i,j)^{\text{th}}$ multiplier is

$$\frac{s_1 \cdots s_{i-1}}{s_1 \cdots s_{j-1}} = \begin{cases} s_j \cdots s_{i-1} & \text{if } i > j, \\ 1 & \text{if } i = j, \\ 1/s_i \cdots s_{j-1} & \text{if } i < j. \end{cases}$$

If B is a *primitive* Leslie matrix, the corresponding companion matrix L is also primitive and there is an analog to Theorem 6.5.1. But S and S^{-1} can be factored across powers and limits and then applied to the limiting matrix. Since the limiting matrix is a polynomial in the companion matrix, applying S and S^{-1} converts this to a polynomial in the Leslie matrix. Hence the convergence theorem for *primitive* Leslie matrices is identical to the convergence theorem for primitive companion matrices:

Theorem 6.6.1. *If A is a primitive Leslie matrix, then A has a strictly dominant positive eigenvalue λ_0,*

$$\lim_{K \to \infty} \frac{A^K}{\lambda_0^K} = \frac{1}{ch'(\lambda_0)} P_{\lambda_0}, \quad where \quad P_{\lambda_0} = \frac{ch(A)}{A - \lambda_0 I},$$

and this limiting matrix is a strictly positive matrix with rank 1.

6.6.1 Periodic Leslie matrices

For periodic Leslie matrices we can obtain a convergence theorem by taking Theorem 6.5.2 and multiplying the companion matrices by S and S^{-1}. In a moment we will restate that theorem, this time for Leslie matrices rather than companion matrices. An obvious question to ask is, why did we bother to analyze companion matrices rather than looking directly at Leslie matrices? The answer is that if we took a Leslie matrix with $g > 1$, took the gth power, and then permuted rows and columns, the resulting matrix would become

$$\begin{bmatrix} A_1 & & & \\ & A_2 & & \\ & & \ddots & \\ & & & A_g \end{bmatrix},$$

where the matrices $A_1, \ldots, A_g$ are Leslie matrices but they are not necessarily identical. The complication comes from the fact that the survival rates can be different. (The matrices A_i are similar, but that is somewhat messy to check.) By using companion matrices, the corresponding A_i are identical because their subdiagonals are strings of ones.

Theorem 6.6.2. *For a Leslie matrix L with index of imprimitivity g,*

$$\lim_{K \to \infty} \frac{L^{gK}}{\lambda_0^{gK}} = L_0 = \frac{1}{ch'_A(\lambda_0^g)} S\Pi \left\{ \frac{ch_A(A)}{A - \lambda_0^g I} \begin{bmatrix} I & & 0 \\ & \ddots & \\ 0 & & I \end{bmatrix} \right\} \Pi^T S^{-1},$$

where

$$A = \begin{bmatrix} s_1 \cdots s_{g-1} f_g & s_1 \cdots s_{2g-1} f_{2g} & \cdots & s_1 \cdots s_{n-1} f_n \\ 1 & & & 0 \\ & \ddots & & \vdots \\ & 1 & & 0 \end{bmatrix},$$

$$ch_A(\lambda) = \lambda^{\frac{n}{g}} - s_1 \cdots s_{g-1} f_g \lambda^{\frac{n}{g}-1} - \cdots - s_1 \cdots s_{n-1} f_n,$$
$$ch_A(\lambda_0^g) = 0,$$
$$S = \mathrm{diag}(1, s_1, s_1 s_2, \ldots, s_1 \cdots s_{n-1}),$$

and Π is the appropriate permutation matrix. Also:

(a) L_0 has rank g.
(b) If $i \equiv j \pmod{g}$, the ith column of L_0 has positive entries in positions $j, j+g, j+2g, \ldots$.
(c) Every column of L_0 is a shifted multiple of the first column.

Let $L_r = \left(\frac{L}{\lambda_0}\right)^r L_0$, $0 \le r < g$. Then:
(a) $L_0, L_1, \ldots, L_{g-1}$ are linearly independent.
(b) If $i + r \equiv j \pmod{g}$, the ith column of L_r has positive entries in positions $j, j+g, j+2g, \cdots$.
(c) $\left(\frac{L}{\lambda_0}\right)^g L_r = L_r$.

Figures 6.1 and 6.2 are from the Census Bureau and give the predicted number of males and females in the United States for two different years. Both of them are projections; that is, they are estimates based on a model and data gathered at an earlier date. Figure 6.1 shows the estimate for 2000, and Figure 6.2 shows the estimate for 2050. The age classes begin with the youngest on the bottom and proceed to the oldest on the top. In this form, we might expect the graphs to have a pyramid form rather than the inverted pyramid form expected when the oldest age class is on the bottom. Figure 6.1 shows that the U.S. population is not in equilibrium, because the "baby boom" is still passing through the population. Figure 6.2 shows a prediction that the population in 2050 will have reached a distribution that has a rough pyramid shape. We should mention that the U.S. Census Bureau uses models that are more complicated than the simple Leslie model; for example, their models include immigration. In spite of this, an approach to a pyramid shape is still predicted. A conclusion like this, which does not depend on the details of the mathematical model, is called **robust**. Scientists generally have more confidence in robust predictions because they know that their model omits many details or variables.

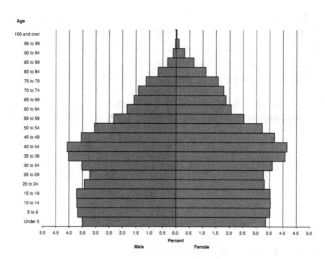

FIGURE 6.1. Projected U.S. Population in 2000.

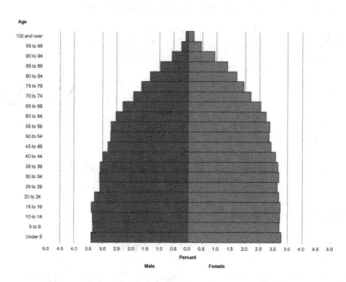

FIGURE 6.2. Projected U.S. Population in 2050.

6.6.2 Averaging

As we have just seen, the asymptotic behavior of a periodic Leslie matrix is more complicated than the straightforward convergence result for primitive matrices. Here we want to look at the "average" behavior of a periodic matrix over its period, and we show that a convergence theorem similar to that for primitive matrices can be obtained.

Theorem 6.6.3 (Averaging Theorem). *If L is an $n \times n$ Leslie matrix (or a nonnegative companion matrix) with period $g > 1$, then*

$$\lim_{t \to \infty} \frac{1}{g} \sum_{i=1}^{g} \frac{L^{t+i}}{\lambda_0^{t+i}} = \frac{1}{ch'(\lambda_0)} P_{\lambda_0} = \frac{1}{ch'(\lambda_0)} \frac{ch(L)}{L - \lambda_0 I},$$

where

$$P_{\lambda_0} = L^{n-1} + \cdots + \lambda_0^{g-1} L^{n-g} + (\lambda_0^g - c_g) L^{n-g-1} + \cdots + (\lambda_0^{n-1} - c_g \lambda_0^{n-g-1} - \ldots) I$$

is a strictly positive matrix of rank 1.

Proof. As we saw in Theorem 5.1.4, L has g roots of maximum modulus; namely, $\lambda_0, \omega\lambda_0, \ldots, \omega^{g-1}\lambda_0$, where ω is a principal g^{th} root of unity. For each of these simple roots there is an eigenmatrix P_j such that $LP_j = \lambda_0 \omega^j P_j$. As in our previous arguments, we can expand I in terms of these eigenmatrices, which are again linearly independent. Applying $(L/\lambda_0)^t$ and taking the limit, all terms except those corresponding to the maximal-modulus eigenvalues disappear. So we need only look at

$$\frac{1}{g} \sum_{i=0}^{g-1} \left(\frac{L}{\lambda_0}\right)^i \sum_{j=0}^{g-1} \alpha_j P_j = \frac{1}{g} \sum_{j=0}^{g-1} \alpha_j \sum_{i=0}^{g-1} \left(\frac{L}{\lambda_0}\right)^i P_j$$

$$= \frac{1}{g} \sum_{j=0}^{g-1} \alpha_j \sum_{i=0}^{g-1} \omega^{ij} P_j = \frac{1}{g} \sum_{j=0}^{g-1} \alpha_j P_j \sum_{i=0}^{g-1} \omega^{ij}.$$

Now, since ω is a principal g^{th} root of unity,

$$\sum_{i=0}^{g-1} \omega^{ij} = \begin{cases} 0 & \text{if } j \not\equiv 0 \pmod{g}, \\ g & \text{if } j \equiv 0 \pmod{g}, \end{cases}$$

and the above sum becomes

$$\frac{1}{g} \alpha_0 P_0 g = \alpha_0 P_0.$$

By the argument leading up to Theorem 6.4.1, $\alpha_0 = 1/ch'(\lambda_0)$. The argument for Theorem 6.4.2 shows that the limiting matrix has rank 1, and the argument for Theorem 6.5.1 shows that the limiting matrix is strictly positive. $\square$

6.7 The Limiting Effect of L on Nonnegative Vectors

Now that we know the limiting form of Leslie matrices, we are in a position to ask about the longterm predictions made by the Leslie model. Specifically, suppose one measures a population and represents this population by the nonnegative vector X_0. If one also measures the fertility rates and survival rates and uses these measurements to construct the Leslie matrix L, then under the assumption that these rates remain constant, one can predict that after k time intervals the population should be $L^k X_0$. Of course, it is relatively easy to calculate $L^k X_0$ even for very large k by using a computer. On the other hand, one hopes that the theory developed in the earlier sections of this chapter will lead to predictions that do not require much computation.

Leslie's Convergence Theorem for nonnegative vectors (Theorem 6.2.1) follows immediately from Theorem 6.6.1 on the convergence of primitive matrices. Provided X_0 is nonnegative with at least one positive component, X_t converges to a positive multiple of the stable age distribution. The convergence is in the absolute error sense if all eigenvalues, except perhaps λ_0, are less than 1 in absolute value. In general, convergence is only in the relative error sense. As long as at least one component of X_0 is positive, then $L^k X_0$ converges to a positive multiple of the unique positive eigenvector that corresponds to λ_0. Further, if $\lambda_0 > 1$, the positive eigenvector has decreasing components and thus displays the inverted pyramid form. Each age class in this eigenvector is less than $1/\lambda_0$ times the previous age class. If $\lambda_0 = 1$, then the components of the eigenvector are nonincreasing, but one needs the assumption that the survival rates are less than 1 to get decreasing age classes. If $\lambda_0 < 1$, then the form of the eigenvector depends on the ratios between the survival rates and λ_0. In the "usual" case, one assumes $\lambda_0 > 1$ (allowing other eigenvalues to have absolute values greater than 1) and concludes that $L^k X_0$ converges in a relative error sense to an inverted pyramid form.

Theorem 6.5.1 also shows that there is convergence even if X_0 is not nonnegative, but the convergence may be to a negative or zero multiple of the stable age distribution. The coefficient γ calculated in Leslie's Convergence Theorem is the appropriate multiplier.

We now want to consider what happens when we have an imprimitive Leslie matrix. Are oscillations possible? Do oscillations *necessarily* occur? To see what happens, we apply L_0 from Theorem 6.6.2 to X_0. First, S^{-1} is applied to X_0, which changes the values by dividing them by positive quantities. So a positive component of X_0 is not changed into a 0 component, and a 0 component is not changed into a positive component. Next, Π^T acts by gathering components into g subvectors. Each subvector consists of those components whose indices were congruent modulo g in the original X_0. Because of the decomposed structure of the matrix, each of the subvectors is now treated separately. In fact, each of the subvectors is now acted on by the matrix

$$\frac{ch_A(A)}{ch'_A(\lambda_0^g)(A - \lambda_0^g I)},$$

which is the limiting form for the primitive Leslie matrix A. So each sub-vector that contains a non-zero component is taken to a multiple of the unique positive eigenvector of A. Next, the permutation matrix Π undoes the permutation caused by Π^T. Finally, the diagonal matrix S multiplies each component by the appropriate product of survival rates.

There are several immediate conclusions obtained from this computation. Provided there is no $i \in \{1, \ldots, g\}$ for which X_i, X_{i+g}, X_{i+2g}, $X_{i+\lfloor \frac{n}{g} \rfloor g}$ are all zero, $L_0 X_0$ is strictly positive. This means that as few as g components of X_0 need to be positive to force $L_0 X_0$ to be positive. On the other hand, if there is such an i, then $L_0 X_0$ will have zeros in the locations whose indices are congruent to $i \bmod g$.

While it is easy to compute the asymptotic period of a Leslie matrix, it is a little more difficult to compute the asymptotic period of the vector X_t because this vector's period depends not only on whether a component is zero, but also on the actual numerical values of its components. For example, if $L = \begin{bmatrix} 0 & 0 & 1 \\ 1 & 0 & 0 \\ 0 & 1 & 0 \end{bmatrix}$, then $L^t X$ has period 3 for most vectors X, but $L^t X$ has period 1 if all the components of X are the same. For instance,

$$L^3 \begin{pmatrix} 3 \\ 2 \\ 1 \end{pmatrix} = L^2 \begin{pmatrix} 1 \\ 3 \\ 2 \end{pmatrix} = L \begin{pmatrix} 2 \\ 1 \\ 3 \end{pmatrix} = \begin{pmatrix} 3 \\ 2 \\ 1 \end{pmatrix}, \qquad \text{but} \quad L \begin{pmatrix} 1 \\ 1 \\ 1 \end{pmatrix} = \begin{pmatrix} 1 \\ 1 \\ 1 \end{pmatrix}.$$

For this example, $X_0 = (x_1, x_2, x_3)^T$ leads to a cycle of period 3 iff at least one of $(1, \omega, \omega^2)X_0 = x_1 + \omega x_2 + \omega^2 x_3$ and $(1, \omega^2, \omega)X_0 = x_1 + \omega^2 x_2 + \omega x_3$ is non-zero, where ω is a primitive third root of unity.

More generally, using the partial sums from Horner's Method,

$$h_0(\lambda) = 1,$$
$$h_1(\lambda) = \lambda - f_1,$$
$$h_2(\lambda) = \lambda^2 - f_1\lambda - f_2 s_1,$$
$$\vdots$$
$$h_{n-1}(\lambda) = \lambda^{n-1} - f_1\lambda^{n-2} - f_2 s_1 \lambda^{n-3} - \cdots - f_{n-1} s_1 \cdots s_1,$$

define the row vectors R_j for $j = 0, \ldots, g - 1$ by

$$R_0 = \left(1, \frac{h_1(\lambda_0)}{s_1}, \frac{h_2(\lambda_0)}{s_1 s_2}, \ldots, \frac{h_{n-1}(\lambda_0)}{s_1 \cdots s_{n-1}}\right)$$

$$R_j = \left(1, \frac{h_1(\lambda_0 \omega^j)}{s_1}, \frac{h_2(\lambda_0 \omega^j)}{s_1 s_2}, \ldots, \frac{h_{n-1}(\lambda_0 \omega^j)}{s_1 \cdots s_{n-1}}\right).$$

It is easy to check that $R_j L = \lambda_0 \omega^j R_j$ and that $R_j L^g = \lambda_0^g R_j$. Now, $R_j L^t X_0 / \lambda_0^t = (R_j \cdot X_0) \omega^{jt}$, a periodic function of t. Let us define per(j) as the least $p > 0$ such that $\omega^{jp} = 1$. In particular, if $j = 0$, per$(j) = 1$ and the above function has period 1; that is, it is a constant function. Notice also that per$(j) \le g$ and per(j) always divides g. Asymptotically, $L^t X_0 / \lambda_0^t$ is a sum of periodic functions, and its period is the least common multiple of the periods of the functions in the sum. These considerations lead to the following theorem.

Theorem 6.7.1. *If L is a Leslie matrix with period g, then asymptotically X_t / λ_0^t is a periodic function of t with period $\mathrm{lcm}\{\,\mathrm{per}(j) \mid (R_j \cdot X_0) \ne 0\,\}$. This period is a divisor of g. (Here, $\mathrm{per}(j)$ is the least $p > 0$ such that $\omega^{jp} = 1$, where ω is a principal g^{th} root of unity.)*

6.7.1   The period of the total population

Now that we know that X_t / λ_0^t becomes periodic, we would like to know whether this periodicity is visible if one observes only the total population rather than each component of the population vector. Here the **total population** is the sum of the components of X_t, and can be written as the inner product $TOTAL = (E, X_t)$, where E is the vector of all 1's. (The standard **inner product** of two n-component vectors A and B is the sum of the products of the corresponding components; that is, $\sum_{i=1}^n A_i B_i$. Here we use (A, B) to represent the inner product of A and B.)

Since we are interested in the asymptotic period, let $\overline{X}_t = \sum_{j=0}^{g-1} \alpha_j \omega^{jt} C_j$, where λ_0 is the positive real eigenvalue of L, ω is a principal g^{th} root of unity, and for $\lambda_j = \lambda_0 \omega^j$,

$$C_j = (\lambda_j^{n-1}, \lambda_j^{n-2} s_1, \lambda_j^{n-3} s_1 s_2, \ldots, s_1 \cdots s_{n-1})^T.$$

Note that $\overline{X}_t$ is a periodic function of t, and X_t / λ_0^t approaches $\overline{X}_t$. The period of $\overline{X}_t$ divides g and depends on which of the α's are non-zero, namely, $\mathrm{per}(\overline{X}_t) = \mathrm{lcm}\{\mathrm{per}(j) \mid \alpha_j \ne 0\}$. Similarly,

$$\mathrm{per}(TOTAL) = \mathrm{per}((E, \overline{X}_t)) = \mathrm{lcm}\{\mathrm{per}(j) \mid \alpha_j (E, C_j) \ne 0\}$$

and

$$\mathrm{per}(\overline{X}_t) = \mathrm{per}(TOTAL) \text{ for every } X_0 \iff (E, C_j) \ne 0 \text{ for all } j = 0, \ldots, g-1.$$

To see when this will occur, let us define the polynomial $e(\lambda)$ by

$$e(\lambda) = \sum_{i=1}^{n} s_1 \cdots s_{i-1} \lambda^{n-i} = \lambda^{n-1} + s_1 \lambda^{n-2} + \cdots + s_1 \cdots s_{n-1},$$

so that $e(\lambda_j) = (E, C_j)$. To show that the inner product (E, C_j) is non-zero, we need only show that $e(\lambda)$ does not have λ_j as a root. Since each λ_j has modulus λ_0, if we can show that $e(\lambda)$ has no root of modulus λ_0, then we can conclude that no λ_j is a root of $e(\lambda)$. For this, we use the fairly standard trick of multiplying $e(\lambda)$ by $\lambda - \lambda_0$ and find that

$$e(\lambda)(\lambda - \lambda_0) = \lambda^n - (\lambda_0 - s_1)\lambda^{n-1} - (\lambda_0 - s_2)s_1\lambda^{n-1}$$
$$- \cdots - (\lambda_0 - s_{n-1})s_1 \cdots s_{n-2}\lambda - s_1 \cdots s_{n-1}\lambda_0,$$

which is a primitive nonnegative polynomial when $\lambda_0 > s_i$ for all $i = 1, \ldots, n-1$. Under this assumption, by Corollary 5.1.5 the polynomial $p(\lambda) = e(\lambda)(\lambda - \lambda_0)$ has a unique positive root, which is strictly dominant. Since $\lambda_0 > 0$ is a root of $p(\lambda)$, λ_0 must be this root. Hence, all other roots of $p(\lambda)$, which are exactly the roots of $e(\lambda)$, are strictly less than λ_0 in complex modulus. From this we obtain the following theorem.

Theorem 6.7.2. *For a Leslie model the period of total population equals the period of the population vector if*

$$\lambda_0 > \max\{s_1, s_2, \ldots, s_{n-1}\}.$$

For Leslie models we assume that all survival rates s_i are less than or equal to 1, and in most applications the further restriction $\lambda_0 > 1$ applies. So, this theorem suggests that we should see the same period of oscillation in both the population vector and the total population. Because this theorem gives only a *sufficient* condition for the periods to be equal, they may be the same even if the condition of the theorem is not satisfied. The following results, which are proved in Cull and Vogt [44], give some other sufficient conditions.

Corollary 6.7.3. *Other sufficient conditions for the period of total population to equal the period of the population vector are:*
   (a) $\lambda_0 \geq \max\{s_1, s_2, \ldots, s_{n-1}\}$ and $\gcd(\{j | \lambda_0 > s_j\} \cup \{g\}) = 1$;
   (b) $\lambda_0 < \min\{s_1, s_2, \ldots, s_{n-1}\}$;
   (c) $\lambda_0 \leq \min\{s_1, s_2, \ldots, s_{n-1}\}$ and $\gcd(\{j | \lambda_0 < s_j\} \cup \{g\}) = 1$.

In actual practice, even the total population is difficult to observe, but a weighted total population might be observed. For example, the probability of observing an organism might be correlated with its size and age. So older age classes might be more heavily represented in a sample than younger age classes. Such observations can be modeled by replacing E with an arbitrary nonnegative weight vector W. Arguments similar to the above can show that W must satisfy a very restrictive set of equations to give a period different from that of the population vector. Hence it is reasonable to expect that the asymptotic periods of the population vector and the weighted total population are identical.

6.8   Afterword

Let us review briefly what we have done in this chapter. We started with the Fibonacci model and generalized it to the Leslie model by allowing more than two age classes. Two complications arose: the problem of periodicity, and the difficulty of handling survival rates. We first showed that a simple gcd condition eliminates the periodic case. For survival rates, we switched to companion matrices, analyzed their behavior, and then used similarity to transform back to Leslie matrices.

Our major result is the asymptotic convergence of powers of Leslie matrices. We showed that these limiting forms are simple enough to be written as closed-form expressions. For primitive matrices this form is particularly simple, and the limiting matrix is one-dimensional.

Of course, biologists are more interested in the population vectors that they observe than the matrices that they infer. The convergence of population vectors follows from the convergence of powers of the matrices. For primitive matrices, the population vector converges to the stable age distribution, and under reasonable circumstances this distribution has the inverted pyramid form. In general, population vectors will have an oscillating limit whose period depends on the period of the matrix and on the initial population vector. Analyses of the the Leslie model in terms of vectors rather than matrices can be found in Cull and Vogt [42, 43].

It is worthwhile noting that results in this chapter are closely related to results in other chapters. In Chapter 5, we discussed nonnegative difference equations. The companion matrices of this chapter are the matrices that correspond to nonnegative difference equations. Leslie matrices are slightly more general, but since they are related to companion matrices by a similarity transformation, Leslie matrices are closely related to nonnegative difference equations. In Chapter 7, we will discuss matrix difference equations, and Leslie matrices will turn out to be a form of nonnegative matrices. The graph techniques that we used for Leslie matrices will also be used to analyze nonnegative matrices.

The Leslie model has some limitations. It is a linear model. If, as we suspect, biology is nonlinear, we should be cautious about predictions from linear models. For example, if a population is growing, the Leslie model predicts that the population will grow exponentially. Such growth will eventually deplete a population's resources, and so we do not expect exponential growth to continue indefinitely. On the other hand, it may be reasonable to think of the Leslie model as a first–order approximation to a nonlinear growth model. In this case, the linear model *may* give good predictions in the short term, even if its eventual predictions are nonsense.

As we mentioned, the original Leslie model forces the matrix to be primitive, and thus it cannot model periodic populations. In Section 6.7.1, based on work of Cull and Vogt [44], we considered whether periodic behavior would be visible in population totals. We concluded that under reasonable circumstances, periodic behavior would still be seen in population totals. Since periodicities are not seen in most populations, there is some question about how relevant periodic Leslie models are to biology. Fortunately, the example of the periodic cicadas (17-year locusts) shows that there are some periodic populations. In general, for periodic models we showed that a periodic limit is expected (refer to Theorem 6.6.2). But we also showed in the Averaging Theorem (Theorem 6.6.3) that simple Leslie convergence can be obtained by taking a suitable average over a period.

To apply the Leslie model, one must choose a time unit. For some populations, the yearly cycle of the environment gives a natural time unit, but for other populations the appropriate unit is not obvious. In Cull [31], we looked at this problem and showed that inappropriate time units can lead to very bad predictions of population growth or decline.

Finally, we should mention that this chapter only scratches the surface of the myriad uses of matrices in population biology. For further information, Caswell's book [25] is a good place to start.

6.9   Exercises

Ex 6.1. Give an inductive proof that the Euclidean Algorithm correctly computes the greatest common divisor. You should assume that a and b are nonnegative integers. You can use one of these two numbers as your induction variable.

Ex 6.2. Show that for $a \leq n$ and $b \leq n$, the Euclidean Algorithm computes the gcd in time $O(\log n)$.
Hint: Assume that a is the smaller of the two numbers. Consider the situation

$$\gcd(a, b) = \gcd(b \bmod a, a) = \gcd(a \bmod [b \bmod a], b \bmod a)$$

and show that $a \bmod [b \bmod a] \leq \frac{a}{2}$.

Ex 6.3. Show that the bound in the previous exercise is worst case optimal by giving an infinite sequence of pairs of integers such that the Euclidean Algorithm uses $\Omega(\log b)$ steps on the pair (a, b).
Hint: Think Fibonacci.

Ex 6.4. Investigate the use of absolute values with powers of a matrix by considering

$$A = \begin{bmatrix} -1/2 & 1/2 \\ 1 & 0 \end{bmatrix}$$

and showing which of the following have limits: A^k, $|A^k|$, A^k/λ_1^k, $|A^k/\lambda_1^k|$, $|A^k|/\lambda_1^k$, $A^k/|\lambda_1^k|$, $|A^k|/|\lambda_1^k|$. How close is $|A^8/\lambda_1^8|$ to its limit?

Ex 6.5. For the Leslie matrix

$$L = \begin{bmatrix} 0 & 2 & 0 & 0 & 5 \\ 1 & 0 & 0 & 0 & 0 \\ 0 & 1 & 0 & 0 & 0 \\ 0 & 0 & 1 & 0 & 0 \\ 0 & 0 & 0 & 1 & 0 \end{bmatrix},$$

determine the least m_0 such that $L^{m_0} \gg 0$. Compare your m_0 to the upper bound computed from the formula in Theorem 6.1.2.

Ex 6.6. For the Leslie matrix

$$L = \begin{bmatrix} 0 & 2 & 5 & 3 & 4 \\ 1 & 0 & 0 & 0 & 0 \\ 0 & 1/2 & 0 & 0 & 0 \\ 0 & 0 & 1/2 & 0 & 0 \\ 0 & 0 & 0 & 1 & 0 \end{bmatrix},$$

determine the least m_0 such that $L^{m_0} \gg 0$. Compare your m_0 to the upper bound from the formula in Theorem 6.1.2. How does this differ from Exercise 6.5?

Ex 6.7. For the Leslie matrix

$$L = \begin{bmatrix} 0 & 0 & 0 & 3 & 5 \\ 1 & 0 & 0 & 0 & 0 \\ 0 & 1/2 & 0 & 0 & 0 \\ 0 & 0 & 1 & 0 & 0 \\ 0 & 0 & 0 & 1/2 & 0 \end{bmatrix},$$

determine the least m_0 such that $L^{m_0} \gg 0$. Compare your m_0 to the upper bound given by Theorem 6.1.2.

Ex 6.8. Show that if L is a Leslie matrix with only $f_n > 0$ and $f_{n-1} > 0$, then the least m_0 such that $L^{m_0} \gg 0$ is $m_0 = (n-1)^2 + 1$.

Ex 6.9. Create a Leslie matrix L in which $f_i \cdot f_{i+1} = 0$ for all i, but $\gcd\{i \mid f_i > 0\} = 1$. Compute the least power m_0 such that $L^{m_0} \gg 0$.

Ex 6.10. Show that the assumption that λ_1 is unique is necessary for convergence by considering

$$A = \begin{bmatrix} 2 & -1 \\ 1 & 0 \end{bmatrix}$$

and showing that $\lambda_1 = 1$ is a *double* eigenvalue of A. Further, show that

$$A^K = \begin{bmatrix} K+1 & -K \\ K & -(K-1) \end{bmatrix} = K \begin{bmatrix} 2 & -1 \\ 1 & 0 \end{bmatrix} - (K-1) \begin{bmatrix} 1 & 0 \\ 0 & 1 \end{bmatrix},$$

and so A^K / λ_1^K does not have a finite limit.

Ex 6.11. Let

$$F = \begin{bmatrix} 1 & 1 \\ 1 & 0 \end{bmatrix}$$

be the Fibonacci matrix.
   (a) Compute $ch_F(\lambda)$.
   (b) Find the eigenvalue of largest magnitude, λ_1.
   (c) Find $P_{\lambda_1} = ch_F(\lambda)/(F - \lambda_1 I)$.
   (d) Use the formula from Theorem 6.4.1 to compute

$$\lim_{k \to \infty} \frac{F^k}{\lambda_1^k}.$$

   (e) Use $\lambda_1 + \lambda_2 = 1$ and $\lambda_1 \lambda_2 = -1$ to eliminate λ_2 from your formula.
   (f) The limiting matrix should be

$$\frac{1}{\sqrt{5}} \begin{bmatrix} \lambda_1 & 1 \\ 1 & \frac{1}{\lambda_1} \end{bmatrix}.$$

   Is this consistent with what you know about the asymptotic form for the Fibonacci numbers?
   (g) Calculate the rank of the limiting matrix.
   (h) Find a form for the matrix that clearly displays the rank. (Notice that you can multiply each row of a matrix by a different scalar by premultiplying the matrix by a diagonal matrix with the scalar multipliers down the diagonal. Similarly, the columns of a matrix can be scaled by postmultiplying by a diagonal matrix.)

Ex 6.12. Consider running the Fibonacci sequence in reverse. Use the inverse of the Fibonacci matrix.
   (a) Show that the inverse Fibonacci matrix can be transformed by permuting rows and columns to the companion form

$$\begin{bmatrix} -1 & 1 \\ 1 & 0 \end{bmatrix}.$$

   (b) Show that the eigenvalue with largest magnitude is negative.
   (c) Let A be the matrix from (a), and let λ_1 be the eigenvalue of largest magnitude. Does

$$\lim_{k \to \infty} \frac{A^k}{\lambda_1^k}$$

   exist?

(d) Compare

$$\lim_{k \to \infty} \frac{A^k}{\lambda_1^k} \quad \text{to} \quad \lim_{k \to \infty} \frac{F^k}{\lambda_1^k}.$$

(e) What does (d) tell you about the growth in magnitude of the Fibonacci numbers with negative indices?

Ex 6.13. The **trace** of a square matrix A is defined to be the sum of the entries on its main diagonal. If $G(A)$ is the graph associated with a nonnegative matrix A, show that the trace of A^k is non-zero iff there is a cycle of length k in the graph $G(A)$.

Ex 6.14. Assume that you have an $O(n^3)$ algorithm that computes the characteristic polynomial of an $n \times n$ matrix A. Assume that A is a nonnegative matrix and the characteristic polynomial is $\lambda^n + C_1 \lambda^{n-1} + \cdots + C_{n-1}\lambda + C_n$. Use the following lemma to construct an $O(n^3)$ algorithm that determines whether A is primitive.

Leverier's Lemma. *Let $C_1, \ldots, C_n$ be the coefficients of the characteristic polynomial of the matrix A. For each i, let S_i be the trace of A^i. Then,*

$$
\begin{bmatrix}
1 & 0 & 0 & \cdots & 0 \\
S_1 & 2 & 0 & \cdots & 0 \\
S_2 & S_1 & 3 & \ddots & \vdots \\
\vdots & \ddots & \ddots & \ddots & 0 \\
S_{n-1} & \cdots & S_2 & S_1 & n
\end{bmatrix}
\begin{pmatrix}
C_1 \\
C_2 \\
\vdots \\
\vdots \\
C_n
\end{pmatrix}
= -
\begin{pmatrix}
S_1 \\
S_2 \\
\vdots \\
\vdots \\
S_n
\end{pmatrix}.
$$

Ex 6.15. Prove the Cayley–Hamilton Theorem for companion matrices. That is, show that a companion matrix satisfies its characteristic polynomial. Further, show that this is the lowest-degree polynomial satisfied by the matrix.

Ex 6.16. Consider the Leslie matrix

$$\begin{bmatrix} 1 & 2 \\ 1 & 0 \end{bmatrix}.$$

Find the eigenvectors and the stable age distribution. Does every initial population converge to the stable age distribution? Consider both absolute error and relative error, and show that certain vectors converge in one sense but not in the other.

Ex 6.17. Show by example that there are Leslie matrices L such that the entries of the limiting matrix $\lim_{k \to \infty} L^k/\lambda_0^k$ are *not* rational functions of the the entries in L.

Ex 6.18. Many of the results demonstrated for Leslie matrices also hold for positive matrices, matrices in which every entry is strictly positive.

Consider the following two positive matrices

$$A = \begin{bmatrix} 6 & 5 \\ 10 & 11 \end{bmatrix} \quad \text{and} \quad B = \begin{bmatrix} 3 & 2 \\ 1 & 2 \end{bmatrix}.$$

Show by calculation that for each matrix, nonnegative initial vectors converge to the positive eigenvector of the matrix. Is this convergence in the absolute value or the relative value sense? Does either or both of these matrices display the inverted pyramid form expected for Leslie matrices?

Ex 6.19. Show that the matrix

$$A = \begin{bmatrix} 1 & -1 & 1 \\ 1 & 0 & 0 \\ 0 & 1 & 0 \end{bmatrix}$$

has period 4, but its characteristic polynomial is *not* periodic. Further, show that for most vectors X, $A^K X$ has period 4 but there are vectors X for which $A^K X$ has period 1. Show that there are no vectors X such that $A^K X$ has period 2.

Ex 6.20. Prove the results about the period of the total population vector given in Corollary 6.7.3.

Ex 6.21. Consider the Leslie matrix

$$L = \begin{bmatrix} 0 & f_2 \\ s_1 & 0 \end{bmatrix} \quad \text{with} \quad s_1 f_2 = 1.$$

Find conditions on X_0 such that $L^t X_0$ oscillates with period 2. Show that there is a weight vector W such that $W L^t X_0$ always has period 1 even if $L^t X_0$ oscillates with period 2.

7

Matrix Difference Equations

As we saw in Chapter 6 with the Leslie model, elements of a sequence can be vectors instead of the usual real or complex numbers. In this chapter we consider linear difference equations with matrix multipliers whose solutions are sequences of vectors. Such equations are often called matrix difference equations because the equations can be written using matrix and vector notation. We look at homogeneous equations, including the special case in which the matrix is nonnegative, which has applications to Markov chains. We discuss the behavior of primitive matrices and use graph theory to give an efficient algorithm to determine whether a matrix is primitive. After that, we look at nonhomogeneous matrix difference equations and see how to reduce them to one–dimensional difference equations.

7.1   Homogeneous Matrix Equations

Let us return to the simplest difference equation, $x_{t+1} = ax_t$. When we considered such equations before, $\langle x_t \rangle$ was usually a sequence of complex numbers and a was a complex constant. It is usual to use capital letters for vectors and matrices instead of the lowercase letters we use for scalars. So, our first–order homogeneous matrix difference equation is

$$(7.1) \qquad\qquad X_{t+1} = MX_t .$$

For this equation to make sense, the sizes of the vectors and matrices must agree. Accordingly, the X's are $k \times 1$ column vectors and M is a $k \times k$ matrix, and (7.1) will be called a k–**dimensional matrix difference equation**.

In printing it is often more convenient to write row vectors, so we also use X^T (the transpose of X) as the row vector corresponding to the column vector X. To write the components of a vector we use

$$X = (x_1, x_2, \ldots, x_k)^T.$$

You may notice that here subscripts serve double duty, both to indicate a component of a vector and to indicate the position of a vector in a sequence. When there is a possibility of confusion we will use function notation for the sequence, for example,

$$X(t) = (x_1(t), x_2(t), \ldots, x_k(t))^T$$

and

$$X(t+1) = (x_1(t+1), x_2(t+1), \ldots, x_k(t+1))^T.$$

Written in component form, equation (7.1) becomes

$$x_1(t+1) = m_{11}x_1(t) + m_{12}x_2(t) + \cdots + m_{1k}x_k(t),$$
$$x_2(t+1) = m_{21}x_1(t) + m_{22}x_2(t) + \cdots + m_{2k}x_k(t),$$

$$\cdots$$

$$x_k(t+1) = m_{k1}x_1(t) + m_{k2}x_2(t) + \cdots + m_{kk}x_k(t).$$

As you can see, in this form there is one equation for each component, and the next value for each component can depend on the values of *all* components. (This is in contrast to Chapter 2 and Chapter 6, where a companion matrix was always used for the multiplier.) While the component form makes the equations appear complicated, the matrix form makes the equation appear simple, and in matrix form the solution is also simple,

$$(7.2) \qquad\qquad X_t = M^t X_0.$$

Unfortunately, this solution doesn't tell us very much, but it does reduce the problem of solving a difference equation to the problem of matrix multiplication.

There are at least two natural ways to compute the solution to (7.2). One way is to start with X_0, compute X_1 by the multiplication MX_0, compute X_2 by MX_1, and so forth. This procedure requires about $t\,k^2$ arithmetic operations to compute X_t, because k^2 operations are used to compute each matrix–vector product. The other technique is to compute M^t and then compute the one matrix–vector product $M^t X_0$. Using the classical matrix multiplication method (by this we mean using the definition of matrix multiplication directly), computing the product of two matrices uses about k^3 operations. (Refer to Chapter 9 for a discussion of faster algorithms for matrix multiplication.) Thus it might seem to take $t\,k^3$ operations to compute M^t, but the technique of **repeated squaring** (also called fast exponentiation) can save many of these operations. For example, M^{16} can be computed by computing M^2, then squaring to get M^4, then squaring M^4 to get M^8, and squaring M^8 to get M^{16}. In general, M^t can be computed using at most $2\log t$ (here log means the base-2 logarithm) matrix products by using repeated squaring, and as soon as $2\log t$ is less than t, repeated squaring uses fewer arithmetic operations. Further, $2k^3 \log t < k^2 t$ for large enough t, and for such t, matrix exponentiation by repeated squaring uses fewer operations than the matrix–vector product method.

Another method for computing M^t comes from the Cayley–Hamilton Theorem (refer to Appendix C), which says that the characteristic polynomial $ch_M(x)$ is a polynomial whose degree is k such that $ch_M(M) = 0$. This implies that there is a polynomial $p(x)$ of degree at most k such that $p(M) = 0$. Such a polynomial with lowest degree is called the **minimal polynomial** for M, and we symbolize this polynomial as $\min_M(x)$. For some matrices (like companion matrices) the characteristic polynomial is the minimal polynomial, but for other matrices (like the identity matrices) the minimal polynomial may have much lower degree than the characteristic polynomial.

We can use M's minimal polynomial to compute the powers M^n. Setting $\min_M(x) = x^d - c_1 x^{d-1} - \cdots - c_{d-1}x - c_d$, we have

$$M^d = c_1 M^{d-1} + \cdots + c_{d-1}M + c_d I \,,$$

and obviously all powers have the form

$$M^n = c_1 M^{n-1} + \cdots + c_{d-1}M^{n-(d-1)} + c_d M^{n-d}.$$

The key point here is that this formula can be interpreted as the dth order one–dimensional difference equation

$$x_n = c_1 x_{n-1} + \cdots + c_{d-1}x_{n-(d-1)} + c_d x_{n-d}$$

with initial conditions $x_0 = I$, $x_1 = M$, $x_2 = M^2$, ..., $x_{d-1} = M^{d-1}$. As in Chapter 2, the solution to this difference equation can be written in the form

$$M^n = \sum_{i=1}^{d} A_i \, \phi_i(n) \,,$$

where the $\phi_i(n)$ are d linearly independent solutions to the recurrence. What are the A_i's? In this case, the A_i's must be matrices, and further, they must be linear combinations of $I, M, \ldots, M^{d-1}$.

We consider two examples that may help clarify the foregoing argument. As in Chapter 6, we can represent the Fibonacci recurrence in matrix form as

$$\begin{pmatrix} f_n \\ f_{n-1} \end{pmatrix} = \begin{bmatrix} 1 & 1 \\ 1 & 0 \end{bmatrix} \begin{pmatrix} f_{n-1} \\ f_{n-2} \end{pmatrix}$$

and treat this as a matrix difference equation. We want to compute powers of

$$M = \begin{bmatrix} 1 & 1 \\ 1 & 0 \end{bmatrix}.$$

Here it is easy to check that $M^2 = M + I$, and so $M^n = M^{n-1} + M^{n-2}$. Since $\lambda^2 - \lambda - 1$ has two distinct roots, λ_1 and λ_2, we can write the solution of this difference equation for M^n as

$$M^n = A_1 \lambda_1^n + A_2 \lambda_2^n$$

with initial conditions

$$I = M^0 = A_1 + A_2,$$
$$M = M^1 = A_1 \lambda_1 + A_2 \lambda_2,$$

which can be solved to give

$$A_1 = \frac{1}{\lambda_1 - \lambda_2}(M - \lambda_2 I),$$
$$A_2 = \frac{-1}{\lambda_1 - \lambda_2}(M - \lambda_1 I),$$

or

$$M^n = \frac{1}{\lambda_1 - \lambda_2}[\lambda_1^n(M - \lambda_2 I) - \lambda_2^n(M - \lambda_1 I)]$$
$$= \frac{\lambda_1^n - \lambda_2^n}{\lambda_1 - \lambda_2}M + \frac{\lambda_1^{n-1} - \lambda_2^{n-1}}{\lambda_1 - \lambda_2}I$$
$$= f_n M + f_{n-1} I.$$

As another example of this method, we use the matrix

$$M = \begin{bmatrix} -4 & 9 \\ -4 & 8 \end{bmatrix},$$

where here $\min_M(x) = ch_M(x) = x^2 - 4x + 4$, which means that we want (any) two linearly independent solutions of

$$x_n = 4x_{n-1} - 4x_{n-2}.$$

It's easy to verify that

$$x_1(n) = n2^n \quad \text{and} \quad x_2(n) = (n-1)2^{n-1}$$

are two linearly independent solutions to our difference equation. Next we want to find the coefficient matrices A_1 and A_2 such that

$$M^n = n2^n A_1 + (n-1)2^{n-1} A_2.$$

For $n = 1$, we have $M = 2A_1$, which gives $A_1 = M/2$. For $n = 2$, we have

$$M^2 = 4M - 4I = 2 \cdot 2^2 A_1 + 1 \cdot 2^1 A_2 = 4M + 2A_2,$$

which gives $A_2 = -2I$. So,

$$M^n = n2^{n-1}M - (n-1)2^n I,$$

(7.3)
$$M^n = \begin{bmatrix} (-3n+1)2^n & 9n2^{n-1} \\ -n2^{n+1} & (3n+1)2^n \end{bmatrix}.$$

Another way to write M^n comes from the **Jordan Canonical Form**, whose salient properties are stated in the following theorem (also refer to Appendix C).

Theorem 7.1.1. *For any complex $k \times k$ matrix M, there is a nonsingular matrix P such that*

$$P^{-1}MP = \begin{bmatrix} J_1 & & 0 \\ & \ddots & \\ 0 & & J_j \end{bmatrix},$$

where each block J_i is bidiagonal of the form

$$\begin{bmatrix} \lambda_i & 1 & 0 & & \cdots & 0 \\ 0 & \lambda_i & 1 & 0 & \cdots & 0 \\ & & \ddots & \ddots & & \\ & & & & & 1 \\ 0 & 0 & 0 & 0 & \cdots & \lambda_i \end{bmatrix},$$

where λ_i is an eigenvalue of M. (In the simplest case each J_i is 1×1 and $P^{-1}MP$ is a diagonal matrix.) Moreover, the nth power of an $r \times r$ Jordan block J is the upper triangular matrix

(7.4)
$$J^n = \begin{bmatrix} \lambda^n & \binom{n}{1}\lambda^{n-1} & \binom{n}{2}\lambda^{n-2} & \cdots & \binom{n}{r}\lambda^{n-r} \\ 0 & \lambda^n & \binom{n}{1}\lambda^{n-1} & \cdots & \binom{n}{r-1}\lambda^{n-(r-1)} \\ & & \ddots & & \vdots \\ 0 & 0 & 0 & \cdots & \lambda^n \end{bmatrix},$$

where the $(i,j)^{\text{th}}$ entry is $(J^n)_{ij} = \lambda^{n-(j-i)}\binom{n}{j-i}$ for $j \geq i$ and $(J^n)_{ij} = 0$ for $j < i$. (Here $\binom{n}{r}$ is the binomial coefficient $\frac{n!}{r!(n-r)!}$.)

We will not prove this theorem here. Instead, we focus on how this result can be used to compute M^n. If D is a matrix with Jordan blocks $J_1, J_2, \ldots, J_r$ on its diagonal, then

$$D^n = \begin{bmatrix} J_1^n & & & \\ & J_2^n & & \\ & & \ddots & \\ & & & J_r^n \end{bmatrix},$$

and we can write down the exact form of D^n. Also, $P^{-1}MP = D$, and rearranging this gives $M = PDP^{-1}$. So to compute M^n we compute $(PDP^{-1})^n$. This is easy because

$$(PDP^{-1})^2 = PDP^{-1}PDP^{-1} = PD^2P^{-1}$$

and

$$(PDP^{-1})^n = PD^nP^{-1}.$$

From above we know how to compute D^n.

Notice that P and P^{-1} do *not* depend on n, and so all dependence on n is in the powers of the Jordan blocks. Thus the entries of M^n are linear combinations of the entries of the powers of the Jordan blocks. In general, the Jordan form is difficult to calculate. For most matrices it cannot be computed exactly. and the approximate calculation is numerically unstable. In spite of these caveats, the Jordan form method for M^n is useful because it can be computed for many common examples, and because it allows us to estimate the growth of M^n.

The Fibonacci example is simple because the matrix $M = \begin{bmatrix} 1 & 1 \\ 1 & 0 \end{bmatrix}$ has two distinct eigenvalues, which means that its Jordan form is a diagonal matrix. The columns of the P matrix are the corresponding eigenvalues, found by solving the equations

$$M \begin{pmatrix} P_{11} \\ P_{21} \end{pmatrix} = \lambda_1 \begin{pmatrix} P_{11} \\ P_{21} \end{pmatrix} \qquad \text{and} \qquad M \begin{pmatrix} P_{12} \\ P_{22} \end{pmatrix} = \lambda_2 \begin{pmatrix} P_{12} \\ P_{22} \end{pmatrix}$$

which give

$$\begin{pmatrix} P_{11} \\ P_{21} \end{pmatrix} = \begin{pmatrix} \lambda_1 \\ 1 \end{pmatrix} \qquad \text{and} \qquad \begin{pmatrix} P_{12} \\ P_{22} \end{pmatrix} = \begin{pmatrix} \lambda_2 \\ 1 \end{pmatrix}.$$

So,

$$M^n = PD^nP^{-1}$$

$$= \begin{bmatrix} \lambda_1 & \lambda_2 \\ 1 & 1 \end{bmatrix} \begin{bmatrix} \lambda_1^n & 0 \\ 0 & \lambda_2^n \end{bmatrix} \begin{bmatrix} \frac{1}{\lambda_1 - \lambda_2} & \frac{-\lambda_2}{\lambda_1 - \lambda_2} \\ \frac{-1}{\lambda_1 - \lambda_2} & \frac{\lambda_1}{\lambda_1 - \lambda_2} \end{bmatrix}$$

$$= \frac{1}{\lambda_1 - \lambda_2} \begin{bmatrix} \lambda_1^{n+1} - \lambda_2^{n+1} & \lambda_1^n - \lambda_2^n \\ \lambda_1^n - \lambda_2^n & \lambda_1^{n-1} - \lambda_2^{n-1} \end{bmatrix} = \begin{bmatrix} f_{n+1} & f_n \\ f_n & f_{n-1} \end{bmatrix}.$$

(Note that we have used $\lambda_1 \lambda_2 = -1$ to simplify the expression.) It is easy to check that this gives the same values as the previous method, although the expressions may look a little different.

For our second matrix,

$$M = \begin{bmatrix} -4 & 9 \\ -4 & 8 \end{bmatrix},$$

the characteristic polynomial is $ch(x) = x^2 - 4x + 4$, which has the double root $\lambda = 2$. We will show that M is similar to the Jordan block matrix

$$J = \begin{bmatrix} 2 & 1 \\ 0 & 2 \end{bmatrix}.$$

To show this, we solve the eigenvector equation $Mz = 2z$ to get $z_0 = (3, 2)^T$. Since every eigenvector is a multiple of z_0, we need to find a **generalized eigenvector** y, satisfying $My = 2y + z_0$. This equation has a solution $y = (1, 1)^T$, and so for

$$P = \begin{bmatrix} 3 & 1 \\ 2 & 1 \end{bmatrix}$$

we have $MP = PJ$, and $M^n = PJ^nP^{-1}$, where $P^{-1} = \begin{bmatrix} 1 & -1 \\ -2 & 3 \end{bmatrix}$. As in (7.4), we know that

$$J^n = \begin{bmatrix} 2^n & n2^{n-1} \\ 0 & 2^n \end{bmatrix}.$$

Multiplying PJ, J^n, and P^{-1} gives

$$M^n = \begin{bmatrix} (-3n+1)2^n & 9n2^{n-1} \\ -n2^{n+1} & (3n+1)2^n \end{bmatrix},$$

which is consistent with our calculation in (7.3).

A comment is in order. These two methods rely on different polynomials. The first method uses the **minimal polynomial**, the lowest–degree polynomial that maps the matrix M to the 0 matrix, while the Jordan method uses the **characteristic polynomial**. The characteristic polynomial for a $k \times k$ matrix always has degree k, while the minimal polynomial has degree at most k. (In fact, the minimal polynomial is a divisor—a factor—of the characteristic polynomial.) It happens that these two polynomials are different when there are at least two Jordan blocks with the same value of λ. (Refer to [78, Section 7.3].) At this point, a simple example may clarify this. Consider the $k \times k$ identity matrix I, which is already in Jordan form with k Jordan blocks and each block has $\lambda = 1$, the only eigenvalue. Here the characteristic polynomial is $(x - 1)^k$, but the minimal polynomial is $\min_I(x) = x - 1$, a *proper* divisor of the characteristic polynomial when $k > 1$.

We are now ready to estimate the growth of a solution to a matrix difference equation. For this we use the Jordan form. When J is a Jordan block of size r and eigenvalue λ, the largest entry of J^n is $\binom{n}{r}\lambda^{n-r}$, which has growth order $\Theta(n^r|\lambda|^{n-r})$. The Jordan form of a matrix consists of diagonal Jordan blocks, and each block is raised to the nth power independently of the other blocks. When D is the Jordan form, D^n has an entry that grows as $\Theta(n^r|\lambda_1|^{n-r})$, where the absolute value of λ_1 is maximum among the absolute values of the eigenvalues and r is the size of the largest Jordan block that has an eigenvalue of absolute value $|\lambda_1|$. From $M^n = PD^nP^{-1}$, the power M^n consists of entries that are linear combinations of the entries of D^n, and no entry of M^n can grow faster than the fastest–growing entry in D^n. Unfortunately, this argument does not ensure that M^n has an entry that grows like the largest entry in D^n because there may be cancellations that occur in the linear combinations. On the other hand, the largest entry in D^n is also a linear combination of entries of M^n, so by reversing the argument we see that at least one of the entries in this sum must be as large as the largest entry in D^n. This observation gives the following result.

Lemma 7.1.2. *The largest entry in M^n obeys $\max |(M^n)_{ij}| = \Theta(n^{r_1}|\lambda_1|^n)$, where λ_1 is the maximum eigenvalue in absolute value and r_1 is the size of the largest Jordan block that has an eigenvalue of absolute value $|\lambda_1|$. This means the $(i,j)^{\mathrm{th}}$ entry satisfies $|(M^n)_{ij}| = O(n^{r_1}|\lambda_1|^n)$.*

Let us return to solving equation (7.1). We know that the solution is $X_t = M^t X_0$, and what does this say about the growth of X_t? The actual growth depends on X_0. In essence, X_0 can pick out some or all of the Jordan blocks, and so detailed knowledge is necessary to get the actual growth rate. But X_t cannot grow faster than M^t, and we have the following theorem.

Theorem 7.1.3. *The growth of a solution to a matrix difference equation can be bounded from above so that $|x_i(t)|$, $\max |x_i(t)|$, and $|\sum \alpha_i x_i(t)|$ are all $O(n^{r_1}|\lambda_1|^n)$, where λ_1 is the maximum eigenvalue in absolute value and r_1 is the size of the largest Jordan block that has an eigenvalue of absolute value $|\lambda_1|$.*

7.2   Nonnegative Matrix Equations

For the special type of a nonnegative matrix M, the matrix difference equation $X_{t+1} = MX_t$ is as simple as the Leslie equations of Chapter 6, and so they are not much more complicated than scalar equations. These special nonnegative matrices have the property that there is a power M^t in which every entry is strictly positive, which we will write as $M^t \gg 0$, and M is called **primitive**. The associated matrix equation is relatively simple to solve because of the Perron–Frobenius Theorem.

Theorem 7.2.1 (Perron–Frobenius). *If M is a primitive matrix, then:*

   (a) M has a positive real eigenvalue λ_0 of maximum modulus;
   (b) λ_0 is a simple root of the characteristic polynomial;
   (c) for every other eigenvalue λ_i, $\lambda_0 > |\lambda_i|$ (it is strictly dominant);
   (d) $\min_i \sum_j m_{i,j} < \lambda_0 < \max_i \sum_j m_{i,j}$,
        $\min_j \sum_i m_{i,j} < \lambda_0 < \max_j \sum_i m_{i,j}$;

(e) *the row and column eigenvectors associated with λ_0 are strictly positive;*

(f) *the sequence M^t is asymptotically one–dimensional, its columns converge to the column eigenvector associated with λ_0, and its rows converge to the row eigenvector associated with λ_0;*

(g) $\lambda_0 = \max |Mx|/|x|$, *where $|x|$ is the* **Euclidean norm** $|x| = \sqrt{\sum x_i^2}$.

We will not prove this theorem; proofs appear in many places, including [7, Chapter 2] and [146, Chapter 1].

7.2.1 Applications to Markov chains

An important application of nonnegative matrices is Markov chains, which are used as models in the biological, physical, and social sciences. The idea of a **Markov chain** is simple: It has a finite set of states with a set of probabilities that describe how a state transitions to other states. Often, the states are assumed to have certain initial probabilities, and one asks how this probability distribution evolves in time and what its asymptotic distribution will be.

We illustrate this with a Markov chain that has three states called 1, 2, and 3. The probability of being in state i is p_i, and the probability of transitioning among the states is given by

$$P(t+1) = \begin{bmatrix} 1/2 & 1/3 & 0 \\ 1/2 & 1/3 & 1/2 \\ 0 & 1/3 & 1/2 \end{bmatrix} P(t) = M\,P(t)\,,$$

where $P(t)$ is the current vector of probabilities and $P(t+1)$ is the next vector of probabilities. Notice that each column of the matrix M sums to 1, and that ensures that the probabilities in each $P(t)$ sum to 1. These transitions can also be represented by the following labeled graph:

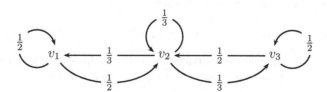

We want to find the long-term behavior of this chain, that is, to calculate $\lim_{t\to\infty} M^t P(0)$. Since we expect this behavior to be independent of the initial distribution $P(0)$, we compute $\lim_{t\to\infty} M^t$.

It is a simple matter to compute the characteristic polynomial for this M, $ch_M(\lambda) = \lambda^3 - \frac{8}{6}\lambda^2 + \frac{1}{4}\lambda + \frac{1}{12}$. By the Cayley–Hamilton Theorem, M satisfies its characteristic polynomial, so for general t we have

$$M^t = \frac{1}{12}(16M^{t-1} - 3M^{t-2} - M^{t-3}),$$

a linear homogeneous difference equation with eigenvalues 1, $1/2$, $-1/6$. Therefore, any solution M^t can be expressed as a (matrix) linear combination of powers of these roots,

$$M^t = C_1 \ (1)^t + C_2 \ (1/2)^t + C_3 \ (-1/6)^t,$$

where the matrices C_1, C_2, C_3 depend on the initial conditions. Since the initial conditions are I, M, and M^2, we expect C_1, C_2, C_3 to be linear combinations of I, M, and M^2. Writing these out in expanded form, we have

$$\begin{aligned}
M^t = \ &(c_{12}M^2 + c_{11}M + c_{10}I)(1)^t \\
&+ (c_{22}M^2 + c_{21}M + c_{20}I)(1/2)^t \\
&+ (c_{32}M^2 + c_{31}M + c_{30}I)(-1/6)^t,
\end{aligned}$$

where the c_{ij}'s are scalars. At first glance it seems we have a problem, because there are 9 unknown coefficients but only 3 initial-condition equations. However, the initial conditions are matrices, and when I, M, and M^2 are linearly independent, we really have 9 equations in scalars. For example, the matrix equation from the first initial condition gives

$$\begin{aligned}
I = M^0 &= 0 \cdot M^2 + 0 \cdot M + 1 \cdot I \\
&= (c_{12}M^2 + c_{11}M + c_{10}I)(1)^0 \\
&\quad + (c_{22}M^2 + c_{21}M + c_{20}I)(1/2)^0 + (c_{32}M^2 + c_{31}M + c_{30}I)(-1/6)^0,
\end{aligned}$$

where the linear independence of I, M, M^2 in this example allows us to equate coefficients of like powers to get

$$0 = c_{12} + c_{22} + c_{32}, \qquad 0 = c_{11} + c_{21} + c_{31}, \qquad 1 = c_{10} + c_{20} + c_{30}.$$

Therefore, each equation in matrices becomes, in this example, three equations in scalars. It is easy to solve for the c_{ij}'s and obtain

$$\begin{aligned}
M^t = \ &\frac{1}{7}(12M^2 - 4M - I) \\
&- \left(\frac{1}{2}\right)^{t+1}(6M^2 - 5M - I) \\
&+ \frac{1}{14}\left(\frac{-1}{6}\right)^t(18M^2 - 27M + 9I).
\end{aligned}$$

Taking limits drives both $(\frac{1}{2})^{t+1}$ and $(\frac{-1}{6})^t$ to 0, and so

$$\lim_{t \to \infty} M^t = \frac{1}{7}(12M^2 - 4M - I),$$

which is

$$\lim_{t \to \infty} M^t = \frac{1}{7}\begin{bmatrix} 2 & 2 & 2 \\ 3 & 3 & 3 \\ 2 & 2 & 2 \end{bmatrix}.$$

Several important results should be noted:

1. The limiting M^t is a matrix of rank 1.

2. Regardless of the initial probability distribution, the limiting distribution is $(2/7, 3/7, 2/7)^T = (p_1, p_2, p_3)^T$.

3. $\lambda_0 = 1$ is a strictly dominant positive eigenvalue of M.

4. The column eigenvector corresponding to 1 is $(2/7, 3/7, 2/7)^T$.

5. The row eigenvector corresponding to 1 is $(1, 1, 1)$.

Since $M^2 \gg 0$, many of these results could have been obtained directly from the Perron–Frobenius Theorem.

7.3 Graphs and Matrices

Many problems about nonnegative matrices can be solved by translating the problem to a problem about graphs. Graphs are objects that consist of vertices and edges between some of these vertices. Small graphs can be represented by diagrams that can be visually inspected for various properties. Often these visual operations can be codified as algorithms, which then can be used to determine properties of graphs that are too large to be conveniently drawn. In many cases, these graph-based algorithms can determine properties of the original matrix more quickly than algorithms based on matrix operations.

More formally, a **(directed) graph** $G = (V, E)$ consists of a finite set V of vertices and an edge set E, where $E \subseteq V \times V$, and the **(directed) edge** (v_i, v_j) goes from vertex v_i to vertex v_j. The **graph associated with the nonnegative matrix** M of size $n \times n$ is $G(M)$ with vertex set $V = \{v_1, v_2, \ldots, v_n\}$ where the directed edge (v_j, v_i) is in E exactly when $M_{i,j} \neq 0$ [1]. A nonnegative matrix M in which all non-zero entries are replaced by 1's is called the **adjacency matrix** for $G(M)$.

---

[1] While (v_j, v_i) corresponding to $M_{i,j}$ may seem backwards, it is the natural definition when matrix M times vector X is computed as MX.

EXAMPLES:

$$M_1 = \begin{bmatrix} 0 & 1 \\ 0 & 0 \end{bmatrix}, \qquad G(M_1) = \quad v_1 \longleftarrow v_2$$

$$M_2 = \begin{bmatrix} 1 & 0 \\ 1 & 2 \end{bmatrix}, \qquad G(M_2) = \quad v_1 \longrightarrow v_2$$

This correspondence between a nonnegative matrix and a graph gives a correspondence between matrix multiplication and paths in the graph, where a **path** from v_i to v_j is a sequence of vertices $v_i, v_{i1}, v_{i2}, \ldots, v_{ir}, v_j$ that starts at v_i, ends at v_j, and for each vertex in the sequence there is an edge to the next vertex in the sequence. The **path length** is the number of edges in the path, which is one less than the number of vertices in the path. For the above graphs, in $G(M_1)$ there is a path of length 1 from v_2 to v_1. In $G(M_2)$ there is a path v_1, v_1, v_2, v_2 of length 3 from v_1 to v_2. Notice also that $G(M_1)$ has no paths of length 2 or greater. Our paths allow a vertex or an edge to be repeated even several times. Paths that do not have repeated vertices are called **simple paths**.

Let us consider matrix multiplication. The row times column rule tells us that if $C = AB$, then

$$c_{ij} = \sum_k a_{ik} b_{kj}.$$

Specializing this formula to the special case in which both A and B are the same nonnegative matrix M gives

$$c_{ij} = \sum_k m_{ik} m_{kj}.$$

When is c_{ij} non-zero? By our assumption that M is nonnegative, each summand is the product of two nonnegative numbers and hence is nonnegative. So the only way for c_{ij} to be 0 is for every summand to be 0. In our graph interpretation this means that there is an edge $v_i \leftarrow v_j$ in $G(C)$ exactly when there is at least one k such that there is an edge $v_i \leftarrow v_k$ and an edge $v_k \leftarrow v_j$ in $G(M)$. Said another way, there is an edge $v_i \leftarrow v_j$ in $G(C)$ exactly when there is a path of length 2 from v_j to v_i in $G(M)$.

This observation tells us that the powers of the matrix M contain information about the existence or nonexistence of paths in $G(M)$. An even stronger relationship can be shown if we assume that M is a **0-1 matrix**, a nonnegative matrix in which all positive entries are 1. For a 0-1 matrix M, $\sum_k m_{ik} m_{kj}$ counts the number of paths of length 2 from v_j to v_i in $G(M)$. This result can be generalized as stated in the following lemma.

Lemma 7.3.1. *Let M be a 0–1 matrix. Then the $(i,j)^{\text{th}}$ entry of the Lth power of M counts the number of paths of length L from v_j to v_i in the associated graph $G(M)$.*

Notice that *paths* rather than *simple paths* are counted. For example, for the graph

$$v_1 \;\rightleftarrows\; v_2 \;\rightleftarrows\; v_3$$

we have

$$M = \begin{bmatrix} 0 & 1 & 0 \\ 1 & 0 & 1 \\ 0 & 1 & 0 \end{bmatrix}, \qquad M^3 = \begin{bmatrix} 0 & 2 & 0 \\ 2 & 0 & 2 \\ 0 & 2 & 0 \end{bmatrix},$$

and M^3 counts the paths of length 3. In particular, $(M^3)_{21} = 2$ says that there are two paths of length 3 from v_1 to v_2. These (nonsimple) paths are

$$v_1 \longrightarrow v_2 \longrightarrow v_1 \longrightarrow v_2$$

and

$$v_1 \longrightarrow v_2 \longrightarrow v_3 \longrightarrow v_2 .$$

We can naturally replace each positive entry with a 1 to convert a non-negative matrix to the 0–1 matrix that is the adjacency matrix for the graph $G(M)$. As in the above lemma, matrix multiplication using natural-number arithmetic counts the number of paths. However, in order to study the existence or nonexistence of paths between pairs of vertices it is easier to use **Boolean OR** as addition and **Boolean AND** as multiplication. For the remainder of this section our matrices will be **Boolean**; that is, 0–1 matrices with Boolean operations.

Other matrix properties also correspond to graph properties. A graph is **strongly connected** if for all pairs of vertices v_i, v_j there is a path from v_i to v_j. Since this statement specifies a property for *all* vertices, it implies that there is also a path from v_j to v_i, but there is no assurance that the path from v_i to v_j has the same length as the path from v_j to v_i. For example, in the strongly connected graph

$$
\begin{array}{ccc}
v_1 & \longrightarrow & v_2 \\
\uparrow & & \downarrow \\
v_4 & \longleftarrow & v_3
\end{array}
$$

there is a path of length 1 from v_1 to v_2, but the length of any path from v_2 to v_1 is at least 3.

What matrix property corresponds to strong connectedness in the associated graph? We have already seen that the entry $(M^L)_{ij}$ is positive iff there is a path of length L from v_j to v_i. So, the existence of a path of length at most K between every pair of vertices corresponds to

$$\sum_{L=0}^{K} M^L \gg 0.$$

(Recall that $A \gg 0$ means that every entry of the matrix A is positive.) This formula suggests that an infinite amount of work is involved in using matrices to check for strong connectedness. However, we are lucky, because whenever there is a path from v_j to v_i, there is a relatively short path, of length at most $(n-1)$, from v_j to v_i, and our condition can be amended to

$$G(M) \text{ is strongly connected} \iff \sum_{L=0}^{k} M^L \gg 0, \text{ for some } k \leq n-1.$$

This form suggests a calculation for $k = n - 1$ using $n - 2$ matrix multiplications and $n - 1$ matrix additions, but the calculation can actually be carried out with fewer matrix operations. For this we need to look at the calculation slightly differently. One way to recast the calculation is to notice that for any positive integers $c_0, c_1, \ldots, c_{n-1}$,

$$(7.5) \qquad \sum_{L=0}^{k} M^L \gg 0 \quad \text{iff} \quad \sum_{L=0}^{k} c_L M^L \gg 0.$$

Which c_L's should we choose to make our calculation easier? Because the matrices M and I commute, a type of Binomial Theorem holds:

$$(M + I)^k = \sum_{L=0}^{k} \binom{k}{L} M^L I^{k-L} = \sum_{L=0}^{k} \binom{k}{L} M^L.$$

Therefore, choosing $c_L = \binom{k}{L}$ in (7.5) gives

$$\sum_{L=0}^{k} M^L \gg 0 \quad \text{iff} \quad (M + I)^k \gg 0,$$

where $(M + I)^k$ can be computed relatively easily by fast exponentiation. This calculation can be further simplified by using $k = 2^r$ for $r = \lceil \log_2(n - 1) \rceil$, and r matrix multiplications (squarings) suffice. Thus, whether a graph is strongly connected can be decided in time $O(T(n) \log n)$, where $T(n)$ is the time to compute the product of two $n \times n$ matrices. (We'll discuss matrix multiplication algorithms more fully in Chapter 9.) The best currently known value for $T(n)$ is $O(n^\alpha)$, where $\alpha \approx 2.38$ (see [29, 28] for details), and this matrix method therefore decides strong connectedness in $O(n^\alpha \log n)$.

7.3.1   Next node representation

A matrix may have many zero entries, which can be considered to correspond to non-edges, edges that do not exist in the graph. A more compact representation might avoid representing these non-edges and represent only edges that actually exist. Concentrating on one vertex v, the edges from v tell us which vertices can be reached from v in one step. So, we could represent a graph by a collection of sets with one set for each vertex v; namely, the set that contains the vertices that can be reached in one step from v. The adjacency matrix of the graph actually *is* such a representation, because the column corresponding to the vertex w is a **bit vector representation** of the set of vertices that can be reached in one step from w. (A bit vector represents a subset of $\{1, \ldots, n\}$ by a 0–1 vector with n components, where j is in the subset iff the jth bit in the vector is 1.) The matrix representation uses n^2 bits. Is a more compact representation possible? Yes, when there are not too many edges. This new representation is called the **next node representation**. For each vertex v there is a list, and each item in the list is a vertex that can be reached from v in one step. There is a separate next node list for each vertex. Since the vertices have labels from $\{1, \ldots, n\}$, the name of each vertex can be represented in $O(\log n)$ bits. Because there is one item in this array of lists for each edge in the graph, the whole structure can be represented in $O(|E| \log n)$ bits, where $|E|$ is the number of edges in the graph. When the implied constants are ignored, the next node representation is smaller than the matrix representation, provided $|E|$ is less than $n^2 / \log n$.

We'd like to use this next node representation to determine strong connectedness more quickly. When a graph is strongly connected, for any vertex v there is a path from v to every vertex and also a path from every vertex to v. Conversely, if there is a vertex v such that there is a path from v to every vertex and also a path from every vertex to v, then the graph is strongly connected. The reason for this is that for any pair of vertices, say w and z, there is a path from w to v and a path from v to z, and following these two paths in order gives a path from w to z. These observations suggest a strong connectedness algorithm. Pick an arbitrary vertex v and do a search to find all the vertices that can be reached (with a path of any length) from v, and then do a "backward" search to find all the vertices that can reach v. By "reach" we mean that there is a path in the graph that can be followed from the starting vertex to the vertex to be "reached".

With the next node representation these searches are easy. One starts at any vertex v and puts all of the next nodes of v into a structure. Then one takes a vertex, say w, from the structure and puts into the structure all of w's next nodes that have not previously been put into the structure. This search halts when the structure is empty. The search is a success if each vertex has been put into the structure. It is easy to keep track of which vertices have been in the structure by using a bit array.

This sort of search can be done in $O(|E|)$ time if we assume that checking and putting a vertex into the structure are unit operations. For the forward search, the next node representation suffices, and we can find the next node to place in the structure by following a pointer and then checking the bit array to determine whether this node has already been put in the structure. For the backward search, a previous node representation is needed. Of course, this previous node representation is just the next node representation for the graph in which every edge $v_i \longrightarrow v_j$ has been turned around into the edge $v_i \longleftarrow v_j$. This can be computed quickly by running through the next node list for v say, and putting v on the previous node list for each w that is on the next node list of v. At the end of these two searches one has to check that all vertices have been found, and a bit vector representation for found vertices makes these checks easy. Overall, each search costs $O(|E|)$, and there are two $O(n)$ checks. So, assuming that $|E| > n$, strong connectedness can be determined in $O(|E|)$ time.

In Section 7.4 we will show that this sort of "graph thinking" leads to an efficient algorithm to determine whether a matrix is primitive.

7.3.2   Comments on imprimitivity

We are interested in *primitive* matrices, because by the Perron–Frobenius Theorem the difference equations associated with such matrices asymptotically have one–dimensional behavior for nonnegative initial conditions. This is the same result we found for Leslie matrices in Theorem 6.6.1. We also saw that if the characteristic polynomial of a Leslie matrix is not primitive, then it is asymptotically periodic with period $g = \gcd\{i | f_i > 0\}$. For example, if A is a 6×6 Leslie matrix in which all survival rates equal 1 and whose only positive fertility rates are f_2 and f_6, then $ch_A(x) = x^6 - f_2 x^2 - f_6 = 0$ and $g = 2$. The graph corresponding to A is

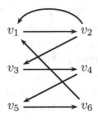

and the graph corresponding to A^2 is

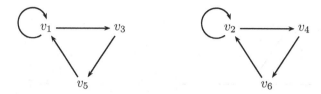

These graphs tell us that the population decomposes into two separate populations, one consisting of the odd-numbered components and one consisting of the even-numbered components. If we use the original matrix, these two populations change places at each step: the odds become evens and the evens become odds. On the other hand, if we use the square of the original matrix, the populations remain separate: the odds stay as odds and the evens stay as evens. Using the squared matrix corresponds to looking at the populations at every two time units instead of looking at them at every time step. For our example, squaring the original 6×6 matrix and taking only the components in odd rows and odd columns gives the 3×3 matrix

$$\begin{bmatrix} f_2 & 0 & f_6 \\ 1 & 0 & 0 \\ 0 & 1 & 0 \end{bmatrix}.$$

The same 3×3 matrix is obtained by taking only the even rows and columns. These matrices are identical because the survival rates are all 1. If unequal survival rates are used, these matrices are

$$\begin{bmatrix} s_1 f_2 & 0 & s_5 f_6 \\ s_1 s_2 & 0 & 0 \\ 0 & s_3 s_4 & 0 \end{bmatrix} \quad \text{and} \quad \begin{bmatrix} s_1 f_2 & 0 & s_1 f_6 \\ s_2 s_3 & 0 & 0 \\ 0 & s_4 s_5 & 0 \end{bmatrix},$$

which look slightly different, but they have the same characteristic polynomial, $x^2(s_1 f_2 - x)$.

In what follows, we call a matrix **strongly connected** if its graph is strongly connected. In general, if a strongly connected matrix M is not primitive, then the matrix M^g decomposes into g primitive matrices, where g is the greatest common divisor (gcd) of the cycle lengths in the graph $G(M)$. This decomposition is not as regular as the decomposition for Leslie matrices. In particular, the number of vertices in each component need not be the same, and the decomposition does not have to be periodic across the indices of the matrix. For example, the graph

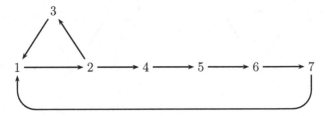

is strongly connected with $g = 3$. The third power of the corresponding matrix decomposes using the three sets of indices $\{1, 5\}, \{2, 6\}, \{3, 4, 7\}$. The graphs for this decomposition are

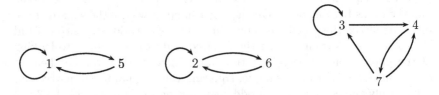

and the corresponding decomposed matrices are

$$\begin{bmatrix} + & + \\ + & 0 \end{bmatrix}, \begin{bmatrix} + & + \\ + & 0 \end{bmatrix}, \begin{bmatrix} + & 0 & + \\ + & 0 & + \\ 0 & + & 0 \end{bmatrix},$$

where the $+$'s indicate the positive entries.

Nonnegative matrices can also fail to be primitive because they are not strongly connected. For Leslie matrices this is no great problem. If an $n \times n$ Leslie matrix has f_k with $k < n$ for its last positive fertility rate, then one can analyze the $k \times k$ submatrix with the first k rows and the first k columns. If this $k \times k$ matrix is primitive, then its powers converge as specified by the theorems for Leslie matrices in Chapter 6. The last $n - k$ columns of the $n \times n$ matrix converge in $n - k$ steps to columns consisting solely of 0's. The first k components in the solution vector X converge as usual for a $k \times k$ Leslie system. The last $n - k$ components of X converge like the first k components, but these last components are also multiplied by appropriate products of survival rates. The periodic Leslie case is similar. If g is the index of imprimitivity, using the gth power of the Leslie matrix decomposes the system into g Leslie systems with $k \times k$ aperiodic submatrices. These systems behave as just described.

The case for general nonnegative matrices is, unfortunately, more complicated, because there are a myriad of ways in which primitive and periodic blocks may be connected. Let us discuss some of these possibilities in terms of the graphs associated with the matrix. Of course, this short description is insufficient to give more than a flavor of the possible complications. In a simple situation, the graph may be disconnected, with a partition of the vertex set into several subsets in which there is *no* path from a vertex in one subset to a vertex in another subset. In this situation, the disconnected components can be analyzed separately. But a strange thing can happen. One of the components could have an oscillation of period p, while another could have an oscillation of period q. Viewed together as a single system, there is an oscillation of period $\mathrm{lcm}(p, q)$. This multiplication of periods makes it possible for a system to have a period much greater than the system's dimension. This is very different from *strongly connected* systems, in which the period of any oscillation must divide the dimension.

Let us now consider the connected case, in which there is a path in at least one direction between every pair of vertices. For example, the graph $v_1 \longrightarrow v_2$ is connected, since there is a directed path from v_1 to v_2. Of course, this example is only connected and not strongly connected. To analyze connected graphs, we make use of the equivalence relation that specifies that two vertices v_i and v_j are equivalent iff there is a directed path from v_i and v_j and a directed path from v_j and v_i. The partition induced by this equivalence relation breaks the graph into **strongly connected blocks**, where the vertices in a block and the edges within a block form a strongly connected graph. We construct a new graph that has as its vertices these strongly connected blocks and that has an edge from block B_i to block B_j iff there is a vertex in B_i that has an edge to a vertex in B_j. This new graph is called a **DAG**, a **directed acyclic graph**, because it has no directed cycles (every directed cycle is inside a block) and the edges between blocks have direction. Two special kinds of blocks need to be singled out: the sources and the sinks. A **source** is a block with no in-coming edges, whereas a **sink** is a block with no out-going edges. In any system, the initial conditions determine which blocks are important. A block is **active** if the initial condition for at least one component in the block is positive. (Recall that we are dealing only with nonnegative systems with nonnegative initial conditions.) A block is **eventually active** if either the block is initially active or there is a directed path to the block from an initially active block. Of particular interest are the eventually active sinks. In analogy with fluid flow, we expect the sinks to contain all of the asymptotic behavior. If we consider the initial conditions to be an initial distribution of fluid in various containers corresponding to blocks, we expect the fluid to flow downhill and end up in the sinks. Certainly, if a block consists of a single vertex with an out-going edge, we expect the fluid to flow out of this chamber. Of course, there might be an in-coming flow that would keep fluid in the chamber, but we expect that all fluid eventually flows out. When a block has cycles, we expect the fluid to flow around these cycles. But if there is an out-going edge, we expect some fluid to flow out, and even the part of the fluid that is recirculating should eventually hit the out-going edge and flow out. So, at least asymptotically, we expect the eventual behavior of the system to be determined by the eventually active sinks. As before, this set of sinks behaves as a disconnected system (unless there is only one eventually active sink), and we can analyze the behavior as we did for disconnected systems. There is one thing wrong with this picture. It is valid only in restricted situations since the analogy to fluid flow makes sense only if fluid is neither created nor destroyed. If we assume that our system is a Markov chain, then the fluid flow analogy does make sense, because the total probability always remains equal to 1. Even in the Markov case we have passed over some difficulties. The asymptotic probability distribution within a sink depends only on the submatrix for the sink if that submatrix is primitive, but the probability of being in a particular sink depends on the initial conditions over the whole system. Further, when the sink is periodic, the maximum period is determined by the sink's submatrix, but the actual period depends on the initial conditions across the whole matrix. If the system is not a Markov chain, then blocks can have eigenvalues larger than 1, which correspond to the creation of fluid, and some vertices may lack out-edges, and self-loops, which correspond to the destruction of fluid. The analysis of such systems have to take into account both the graphical properties of the matrix and the actual numerical values in the matrix. So, we leave such systems with the comment that their analyses are complicated.

7.4   Algorithms for Primitivity

In this section we investigate algorithms for determining whether a nonnegative matrix is primitive.

7.4.1   Algorithm I

The most straightforward algorithm for primitivity is based on the observation that if a power of a nonnegative matrix is positive then all higher powers are also positive. Using graph theory, we show that the $(n-1)^2 + 1$ power of any primitive matrix is strictly positive.

Theorem 7.4.1. *If A is a primitive $n \times n$ matrix and m_0 is the least nonnegative integer m such that $A^m \gg 0$, then m_0 obeys*

$$m_0 \leq (n-2)l + n \leq (n-1)^2 + 1,$$

where l is the length of the shortest cycle in the graph $G(A)$.

Proof. If $n = 1$, then $A = [a_{11}]$ with $a_{11} > 0$, and the corresponding graph consists of a single vertex with a self-loop. For such a graph, $l = 1$. Since $A^0 = [1]$, the formulas are correct. We next note that for $n > 1$, $l = n$ implies that A is a permutation matrix and so cannot be primitive. Therefore, we may assume that $n > 1$ and $l \leq n - 1$.

Consider a cycle of length l. Since the graph is strongly connected and $l < n$, there is a vertex v on this cycle that has an edge to a vertex not on this cycle. Because A is primitive, for every large enough m there is a path of length m between every pair of vertices, and in particular, m can be taken in the form $il + 1$. Let S_i be the set of all vertices that can be reached from the vertex v in exactly $il + 1$ steps. Clearly, $|S_0| \geq 2$ and $S_0 \subseteq S_1 \subseteq \cdots \subseteq S_n$. Also, if $S_{i+1} = S_i$, then $S_{i+2} = S_{i+1} = S_i$. Therefore, either $|S_i| = n$ or

$$2 \leq |S_0| < |S_1| < \cdots < |S_i|.$$

This implies $|S_{n-2}| = n$, and every vertex is reachable from v in $(n-2)l+1$ steps. If y is any vertex, strong connectedness implies the existence of an in-coming edge from some x to y. Since x is reachable from v in $(n-2)l+1$ steps, y is reachable from v in $(n-2)l+2$ steps. From this argument we see that for every $j \geq 1$ every y is reachable from v in $(n-2)l+j$ steps. To go from any vertex x to any vertex y, one can go from x to v and then go from v to y. But there is a path of length at most $n - j \leq n - 1$ from x to v, which means that one can go from x to y in $n - j + (n-2)l + j$ steps. This gives the first upper bound. The second upper bound follows from $l \leq n - 1$.   □

So the (theoretically) simple algorithm is to compute $A^{(n-1)^2+1}$ and check to see whether all entries in this computed matrix are positive. The analysis of this algorithm is also simple. Matrix multiplications are the most time-consuming part, so let $M(n)$ be the time to compute the product of two $n \times n$ matrices. If we compute the power of A by simple powering as in

```
POWER := A
FOR I := 1 TO (n - 1)^2 DO
    POWER := A × POWER
```

then our algorithm uses $O(n^2)$ matrix multiplications, and its time complexity is $O(n^2 M(n))$. But most of these multiplications can be avoided by fast exponentiation (repeated squaring), in which POWER is multiplied by POWER rather than by the original matrix A.

```
POWER := A
I := 1
WHILE I < (n − 1)² + 1 DO
      I := 2 × I
      POWER := POWER × POWER
```

In the last procedure, the power of A is doubled on each execution of the WHILE loop. So, if the loop is executed j times, then POWER contains A^{2^j} and I contains 2^j. From the loop condition, we have $2^j \geq (n-1)^2 + 1$ and $2^j < 2(n-1)^2$. So $2\log(n-1) + 1 > j > 2\log(n-1)$, and the repeated squaring method has time complexity $O(\log n M(n))$, which is superior to $O(n^2 M(n))$.

At this point we should say more about $M(n)$, the time complexity of multiplying two $n \times n$ matrices. The classical row-times-column algorithm uses $\Theta(n^3)$ arithmetic operations (additions and multiplications). Is this the best possible? That depends on what kind of entries are in your matrices and on what kind of operations you are allowed to use. As we saw earlier, when we are trying to determine if a power of a nonnegative matrix is positive, it might be reasonable to change the matrix to a Boolean matrix by writing 1 for each positive entry and using Boolean addition and multiplication, corresponding to OR and AND. When the operations on scalars are restricted to AND and OR it is known that (refer to [112, V. 2, pp. 159–168]) any computer program for (Boolean) matrix multiplication with *no* branching has time complexity $\Omega(n^3)$ and the classical method is the best possible! (If one allows branching, the Four Russians' algorithm computes the logical product of two matrices in $O(n^3/\log n)$ time. See Aho, Hopcroft, Ullman [2].)

This may require a small explanation. So far the operations OR and AND, or addition and multiplication, are monotone in the sense that $X \geq Y \implies f(X) \geq f(Y)$ where X and Y are vectors. If $\geq$ is defined on the scalars, we can extend the ordering from scalars to vectors by requiring that the scalar ordering hold on each component of the vectors. Notice that even if the scalar ordering is a total ordering, the vector ordering may only be (and usually is) a partial ordering. For example, $\{0, 1\}$ is totally ordered by $\geq$ defined as

$$0 \geq 0, \quad 1 \geq 0, \quad \text{and } 1 \geq 1,$$

but extending this $\geq$ to vectors gives

$$\begin{matrix} 0 \\ 0 \end{matrix} \geq \begin{matrix} 0 \\ 0 \end{matrix} \quad \begin{matrix} 0 \\ 1 \end{matrix} \geq \begin{matrix} 0 \\ 0 \end{matrix} \quad \begin{matrix} 1 \\ 0 \end{matrix} \geq \begin{matrix} 0 \\ 0 \end{matrix} \quad \begin{matrix} 1 \\ 1 \end{matrix} \geq \begin{matrix} 0 \\ 1 \end{matrix} \quad \begin{matrix} 1 \\ 1 \end{matrix} \geq \begin{matrix} 1 \\ 0 \end{matrix},$$

and does not specify any relationship between

$$\begin{matrix} 0 \\ 1 \end{matrix} \quad \text{and} \quad \begin{matrix} 1 \\ 0 \end{matrix}.$$

Similarly, when the usual ordering $\geq$ on the reals is extended to vectors, we get

$$\begin{pmatrix} x_1 \\ x_2 \end{pmatrix} \geq \begin{pmatrix} y_1 \\ y_2 \end{pmatrix} \quad \text{iff} \quad x_1 \geq y_1 \text{ and } x_2 \geq y_2,$$

and there no relationship for $x_1 > y_1$ and $y_2 > x_2$. When we discuss the usual sort of operations for f, X and Y are vectors with two components and $f(X)$ and $f(Y)$ are scalars, so we could rewrite the monotonicity condition as

$$x_1 \geq y_1 \text{ and } x_2 \geq y_2 \implies f(x_1, x_2) \geq f(y_1, y_2).$$

For example,

$$x_1 \geq y_1 \text{ and } x_2 \geq y_2 \implies (x_1 \text{ OR } x_2) \geq (y_1 \text{ OR } y_2).$$

Now the question is: Will using nonmonotonic operations allow matrix multiplication to be calculated more quickly? The answer is YES, as shown by Strassen [156] in 1969. (We will return to Strassen's method in Chapter 9.) He showed that by using the nonmonotonic operation of matrix subtraction he could produce a divide-and-conquer algorithm for matrix multiplication that uses only seven half-size multiplications, rather than the eight half-size multiplications in a divide-and-conquer algorithm that uses only addition and multiplication. Strassen's algorithm has time complexity $\Theta(n^{\log 7})$. A number of researchers have found very clever ways to further reduce the complexity of matrix multiplication. At present, the best algorithm has time complexity $\Theta(n^\alpha)$, where α is less than 2.4 [29, 28]. These faster matrix multiplication methods could be used to produce a faster algorithm for determining whether a matrix is primitive. Of course, to use the methods requiring subtraction, computation must be carried out in the integers. While this adds some complexity because the size of the integers could increase, it can be shown that the increase is not very significant.

Are these "faster" methods practical? Since complexity analysis is only asymptotic, it is very possible that a "faster" method is slower than standard methods for all reasonably sized problems. For Strassen's method, $n > 1000$ is needed to be competitive with the classical method. For other nonmonotonic methods, much larger values of n seem to be needed for there to be any speedup over the classical method. On the other hand, the Four Russians' method may be competitive for reasonable values of n (say, $n \approx 100$) if some of the operations are implemented as bit vector operations.

7.4.2   Algorithm II

After all this talk about matrix multiplication, are there other methods for testing primitivity that don't use matrix multiplication? Yes, there are methods that are based on the calculation of graph properties. The basic technique used in these methods is **depth-first search**. [2] In this technique, one explores all edges in a graph by following each edge leading to an unused vertex and backing up when there are no edges to unused vertices. The complexity of depth-first search is $O(|E|)$, where $|E|$ is the number of edges in the graph. In our matrix application, $|E|$ is the number of positive entries in the matrix, and so $O(|E|)$ is $O(n^2)$.

To determine whether a directed graph is strongly connected, two depth-first searches are performed: one on the graph and one on the reversed graph. (Recall that in the reversed graph each directed edge is turned around so that the tail of the original edge becomes the head of the new edge, and the head of the original edge becomes the tail of the new edge.) The graph is strongly connected iff both these searches find all the vertices, which means that strong connectedness can be tested in $O(|E|)$.

If M is nonnegative and a power M^k is strictly positive, the corresponding graph has at least one path of length k between every pair of vertices. In particular, for every vertex there is a cycle of length k from that vertex back to itself. Further, since M^{k+1} must also be strictly positive, there is a cycle of length $k + 1$ around each vertex. These observations lead to the following theorem.

Theorem 7.4.2. *A nonnegative matrix M is primitive iff the corresponding graph $G(M)$ is strongly connected and has two relatively prime cycle lengths.*

Proof. Since k and $k + 1$ are relatively prime, the above observations show the *only if* part. For the *if* part, if there are two relatively prime cycles, every positive integer greater than some integer B can be represented as a positive linear combination of the lengths of these two cycles. (See Exercise 5.5.) Since the graph is strongly connected, there are paths between every pair of vertices. We claim that there exists a path of length $k = 3(n-1)+B$ (where the graph has n vertices) between any pair of vertices. This will show that M^k is strictly positive. A path from any v to any w can be constructed by going from v to a vertex x_1 on the first cycle, then around this cycle as many times as you like, then from x_1 to some vertex x_2 on the second cycle, then around the second cycle as often as you want, and finally from x_2 to w. The paths from v to x_1, x_1 to x_2, and x_2 to w have total length at most $3(n - 1)$. Remember that any number greater than or equal to B can be created by sums of the two given relatively prime cycle lengths, and

---

[2] A breadth-first search could be used here in place of the depth-first search.

so going around the two cycles the appropriate number of times results in a path of length $3(n-1) + B = k$. This proves $M^k \gg 0$. $\qquad\square$

Perhaps it is easy to misread this theorem. It does not say that there are two *simple* cycles whose lengths are relatively prime. For example, you can construct (refer to Exercise 7.2) a strongly connected graph on 20 vertices that has only three simple cycles, of lengths $6, 15$, and 20. Although this graph does *not* have two relatively prime simple cycles, the corresponding matrix is still primitive, because you can produce a cycle of length 21 by going around the 6-cycle and the 15-cycle. The correct corollary is the following:

Corollary 7.4.3. *A nonnegative matrix is primitive iff the corresponding graph is strongly connected and the gcd of the lengths of its simple cycles is 1. (We call such a graph a **primitive graph**.)*

The relatively prime cycle lengths may be relatively large and hard to find. Because of this, our strategy is to find some small simple cycles and then to check whether the gcd of their lengths is 1. Ideally, we would find the lengths of all simple cycles, but this seems computationally difficult. In particular, determining whether an n-vertex graph has a simple cycle of length n is the famous **Hamiltonian circuit problem** (refer to [68]), which is known to be NP–complete. So, our procedure looks for any short cycles, not just simple ones.

We start by picking an arbitrary vertex v, and then doing a backward search and a forward search from v. As seen above, this pair of searches checks whether the graph is strongly connected because each vertex appears in both the forward search and in the backward search iff the graph is strongly connected.

Once we know that the graph is strongly connected, we need a procedure for finding cycle lengths. Consider our search to be breadth-first, so for each vertex we can talk of a successor set of vertices and a predecessor set of vertices. For each vertex w, let $P(w)$ be the length of the shortest path from w to a fixed vertex v. Since there is a path of length 0 from v to v, we initialize with $P(v) = 0$, and $P(w)$ is calculated by doing a backward breadth-first search from v, which can be done in $O(|E|)$. For the forward search, define $S_0 = \{v\}$ and inductively calculate

$$S_{i+1} = \text{Succ}(S_i) - \textbf{Already Visited Vertices}\;.$$

Of course, here we are using $\text{Succ}(X)$ to mean the set of vertices that can be reached by a path of length 1 (an edge) from a vertex in the set X. We remove the **Already Visited Vertices** so that edges are not repeated and also so that this forward search can be accomplished in $O(|E|)$ time. For each $w \in \text{Succ}(S_i)$ put $P(w) + i + 1$ into the set of cycle lengths C. We now will argue that the gcd of the final (finite) set C is the gcd of all cycle lengths. For this we use the $\pm$ **closure** of a set C of natural numbers.

This closure, denoted by $C^{\pm}$, is the smallest set that both contains C and is closed under additions and differences. This means that if $a \in C^{\pm}$ and $b \in C^{\pm}$, then both $a + b$ and $|a - b|$ are in $C^{\pm}$.

Lemma 7.4.4. *If C is a finite set of natural numbers, then $\gcd(C) = \gcd(C^{\pm})$ holds. Also, if g is this common value, then $C^{\pm} = g\mathbb{N}$; that is, $C^{\pm}$ consists of all nonnegative multiples of g.*

Proof. The elements of $C^{\pm}$ can be assigned types indicating the least number of operations needed to create the element from the original elements in C. Elements of C are type-0 elements of $C^{\pm}$, and x is an element of type K if $x = a + b$ or $x = |a - b|$, where one of a and b has type $K - 1$ and the other has type at most $K - 1$. By hypothesis, g is the gcd of type-0 elements. Assume that g is the gcd of elements of type at most $K - 1$. If g divides both a and b, then g also divides their sum and their difference, and hence g divides all elements of type at most K. Because adding elements to a set cannot increase the gcd, g must be the gcd of elements of type at most K. Since every element of $C^{\pm}$ has finite type, then $\gcd(C^{\pm}) = g$. Why is $C^{\pm} = g\mathbb{N}$? First of all, since $g = \gcd(C^{\pm})$, all elements of $C^{\pm}$ are multiples of g. On the other hand, by the Euclidean Algorithm we know that $g \in C^{\pm}$, and by closure under addition all positive multiples of g must then be in $C^{\pm}$. (Closure under subtraction guarantees that 0 is in $C^{\pm}$.) $\square$

We will prove that all cycle lengths are in $C^{\pm}$, where C is the output of our primitivity algorithm. Then this will allow us to find the greatest common divisor of all cycle lengths by calculating $\gcd(C)$.

Lemma 7.4.5. *If C is the set of numbers found by the primitivity algorithm then every cycle length is in $C^{\pm}$, and therefore $\gcd(C)$ is the greatest common divisor of all cycle lengths.*

Proof. Consider any cycle, simple or not simple, and let $v_1, v_2, \ldots, v_L$ be the ordered list of the vertices in this cycle of length L. For every v_i, there is an edge $v_i \longrightarrow v_{i+1 \bmod L}$. If $P(v_i)$ is the position at which v_i is found in the backward search, and $Q(v_i)$ is the position at which v_i is first found in the forward search, then both

$$P(v_i) + Q(v_i) \quad \text{and} \quad P(v_{i+1}) + Q(v_i) + 1$$

have been put into C, because from the edge $v_i \longrightarrow v_{i+1 \bmod L}$, the vertex v_{i+1} is found on the step after v_i. Since $C^{\pm}$ is closed under addition and differences, the following natural number must be in $C^{\pm}$:

$$\left| \sum_{i=1}^{L} [P(v_{i+1}) + Q(v_i) + 1] \ - \ \sum_{i=1}^{L} [P(v_i) + Q(v_i)] \right|.$$

But this sum is L, the cycle length, and we've proved that every cycle length must be in $C^{\pm}$. $\square$

Theorem 7.4.6. *If C is the set of numbers found by the primitivity algorithm, then a strongly connected graph is primitive iff* $\gcd(C) = 1$.

Proof. From the discussion before the last result, it is enough to recall that from Theorem 7.4.2 we know that a strongly connected graph is primitive iff the gcd of its cycle lengths is 1. □

To complete an analysis of this algorithm we need to look at the complexity of calculating the gcd of a set of numbers. A naive way to compute the gcd is to begin with the two smallest numbers, use the Euclidean Algorithm to find their gcd, and replace these two elements with their gcd. This process can be applied recursively to the remaining set, and there are $O(n)$ calls to the Euclidean Algorithm when n is the size of the set. If all elements are less than $2n$ (as they are in our set C), each call to the Euclidean Algorithm could use $O(\log n)$ divisions, and in total this procedure is $O(n \log n)$. This is an overestimate. On each division, either the division is exact (and the larger number is eliminated) or one of the numbers is decreased. In fact, if the division is not exact, then the smaller number is at least halved in every two divisions. Since the smallest number has only $\log n$ bits, it is reduced to 1 once there are $2 \log n$ divisions that are not exact. This means that the gcd of an n-subset from $\{1, \ldots, 2n\}$ can be computed in $O(n)$ time. Coupled with the above this gives the following two theorems.

Theorem 7.4.7. *Whether an n vertex graph is primitive can be determined in* $\Theta(|E| + n)$*, where $|E|$ is the number of edges in the graph.*

Theorem 7.4.8. *Whether an $n \times n$ nonnegative matrix is primitive can be determined in* $\Theta(n^2)$*. Except for an initial scan of the matrix, the algorithm runs in* $\Theta(|E|)$*, where $|E|$ is the number of non-zero entries in the matrix.*

We finish this section with a couple of comments. It seems that if a graph has many edges, then the graph is most likely primitive. For instance, with some effort one could calculate a constant γ and then show that the graph must be primitive if $|E| > \gamma n^2$. It can also be shown [10] that there is a constant α such that if each vertex has at least $\alpha \log n$ edges, then the graph is almost surely primitive when the edges are assigned at random. Finally, our algorithm lets us pick v. Which v should be chosen? A reasonable choice is to choose the v with the largest number of edges. In graphs that arise in modeling, the structure of the graph may allow a fast proof of primitivity without even using our algorithm. The point of all these comments is that our complexity estimates may be overestimates.

7.5   Matrix Difference Equations with Input

In the previous sections of this chapter we have considered matrix difference equations of the form
$$X_{t+1} = MX_t.$$

These are called **homogeneous** because multiplying each component of X_t by α results in each component of X_{t+1} being multiplied by α. The solutions to such equations also satisfy the *additive* property in that if $X_{t+1} = MX_t$ and $W_{t+1} = MW_t$, then $Z_t = X_t + W_t$ also satisfies $Z_{t+1} = MZ_t$. (Notice that while the recurrence stays the same under addition, the initial conditions change, because if $X_0 = C$ and $W_0 = C$, then $Z_0 = X_0 + W_0 = 2C \neq C$ if C is non-zero.) Homogeneous difference equations are used to model systems with no input, where the current state of the system depends only on the previous state of the system. In contrast, other systems are better described by nonhomogeneous equations, because the state of the system depends on an input as well as on the internal workings. For example, the states of a sewage system depend on what is fed into the system as well as on the internal processing within the system.

We describe nonhomogeneous systems by a matrix difference equation of the form

(7.6) $X_{t+1} = MX_t + Y_t,$

but in many fields more complicated models are used. For example, control engineers often use a pair of equations,

$$X_{t+1} = MX_t + BY_t$$
$$Z_{t+1} = CX_t + DY_t,$$

where X_t is the internal state, Y_t is the input, and Z_{t+1} is the output. They then ask such questions as what input sequence produces a desired output sequence. Of course, even more complicated models (including non-linear models) are used in a variety of fields. Here we restrict ourselves to equations in the form of (7.6).

Our first result is a trivial observation, which may seem surprising.

Theorem 7.5.1. *If $X_{t+1} = MX_t + Y_t$, then the input sequence Y_t can be chosen so that the solution is any desired sequence.*

Proof. If X_0 is the initial condition and X_1 is the next desired value, taking $Y_0 = -MX_0 + X_1$ gives $MX_0 + Y_0 = MX_0 - MX_0 + X_1 = X_1$. Similarly, if X_t was the last value and X_{t+1} is the next desired value, taking $Y_t = -MX_t + X_{t+1}$ gives X_{t+1} the desired value.   □

This result may seem surprising because linear systems are so simple that they should have very limited possibilities, and therefore it's surprising that every desired sequence can be produced from a linear system. On the other hand, the result is trivial because it says that if you can produce any desired input sequence (any Y_t), then you can pass your constructed sequence (which contains all the complexity) through a linear system and obtain whatever you like. The linear system does *not* add any complexity or flexibility; rather, you had that when you were constructing the input sequence.

A major virtue of linear systems is that we can write down their solution. The solution to (7.6) with initial condition X_0 is

$$X_t = M^t X_0 + \sum_{i=0}^{t-1} M^{t-1-i} Y_i.$$

Notice that this is formally the same as the solution to the first–order one–dimensional difference equation

$$x_{t+1} = a x_t + y_t,$$

but the symbols are now vectors and matrices. This change is significant because the order of operands is now important. For example, a $k \times k$ matrix times a $k \times 1$ vector must be multiplied in the order matrix times vector. Of course, this form of solution is purely formal and doesn't really tell us much about how the solution behaves. After we give the solution, we must add some hypotheses about the structure of the matrices and the growth of the input sequence in order to get some bounds on the behavior of the solution.

7.5.1 Reduction to one dimension

General multi–dimensional systems are difficult to analyze, but linear systems have such nice additive properties that we can hope that multi–dimensional linear systems are not much more complicated than one–dimensional linear systems. We will show that a multi–dimensional linear system can in fact be decomposed into a set of one–dimensional linear systems. We start with a special case that often arises in practice.

Theorem 7.5.2. *The k–dimensional linear difference equation*

$$X_{t+1} = M X_t + Y_t$$

can be decomposed into the set of one–dimensional difference equations

$$\hat{x}_1(t+1) = \lambda_1 \hat{x}_1(t) + \hat{y}_1(t),$$
$$\hat{x}_2(t+1) = \lambda_2 \hat{x}_2(t) + \hat{y}_2(t),$$

$$\vdots$$

$$\hat{x}_k(t+1) = \lambda_k \hat{x}_k(t) + \hat{y}_k(t),$$

if the matrix M has a basis of eigenvectors. Here, the $\lambda_1, \ldots, \lambda_k$ are the eigenvalues and the $\hat{x}$'s and $\hat{y}$'s are linear combinations of the components of X and Y respectively.

Proof. If Z is a (column) eigenvector of M, then $MZ = \lambda Z$, where λ is the corresponding eigenvalue. When M has k linearly independent eigenvectors $Z_1, Z_2, \cdots, Z_k$ with corresponding eigenvalues $\lambda_1, \lambda_2, \cdots, \lambda_k$, then

$$M[Z_1 Z_2 \cdots Z_k] = [Z_1 Z_2 \cdots Z_k] \begin{bmatrix} \lambda_1 & 0 & \cdots & 0 \\ 0 & \lambda_2 & \cdots & 0 \\ & & \ddots & \\ 0 & 0 & \cdots & \lambda_k \end{bmatrix}.$$

(Here $[Z_1 Z_2 \cdots Z_k]$ is the $k \times k$ matrix whose i^{th} column is the ith eigenvector.) The equation holds because postmultiplying a matrix by a diagonal matrix results in multiplying each component of the ith column by the ith entry on the diagonal. Putting this in matrix form, we have

$$M\mathcal{Z} = \mathcal{Z}\Lambda,$$

where $\mathcal{Z}$ is the matrix of eigenvectors and Λ is the diagonal matrix of eigenvalues. Since the eigenvectors are linearly independent, $\mathcal{Z}$ is an invertible matrix and

$$\mathcal{Z}^{-1}M\mathcal{Z} = \Lambda \quad \text{and} \quad M = \mathcal{Z}\Lambda\mathcal{Z}^{-1}.$$

Replacing M in the matrix equation gives

$$X_{t+1} = \mathcal{Z}\Lambda\mathcal{Z}^{-1}X_t + Y_t,$$

and multiplying by $\mathcal{Z}^{-1}$ gives

$$(\mathcal{Z}^{-1}X_{t+1}) = (\Lambda\mathcal{Z}^{-1}X_t) + (\mathcal{Z}^{-1}Y_t).$$

So setting $\hat{X} = (\mathcal{Z}^{-1}X)$ and $\hat{Y}_t = (\mathcal{Z}^{-1}Y_t)$ (which are linear combinations of X and Y) gives

$$\hat{X}_{t+1} = \Lambda\hat{X}_t + \hat{Y}_t.$$

Because Λ is diagonal, each component of this equation is independent of the other components, and so we can write the k–dimensional difference equation as a set of k one–dimensional difference equations. $\qquad\square$

Once a matrix difference equation has been reduced to its one–dimensional form, the methods in earlier chapters can be used to find the solutions to these equations. Re-assembling these component solutions into a vector and multiplying by the matrix $\mathcal{Z}$ gives the solution X_t to the matrix difference equation.

This reduction can also be described in terms of the eigenvectors and generalized eigenvectors of M. As we observed in Section 2.4, a matrix also has row eigenvectors, by which we mean a row vector R such that $RM = \lambda R$, where λ is the corresponding eigenvalue. Starting with the equation

$$(7.7) \qquad X_{t+1} = MX_t + Y_t,$$

we can multiply by a row eigenvector R to obtain

$$RX_{t+1} = RMX_t + RY_t.$$

Since $r(t) = RX_t$ is one–dimensional, we obtain the one–dimensional recurrence

$$r(t+1) = \lambda\, r(t) + RY_t$$

with solution

$$r(t) = \lambda^t\, r(0) + \sum_{j=0}^{t-1} \lambda^{t-1-j} RY_t.$$

If M has a basis of eigenvectors, the solution X_t to (7.7) can be written as a linear combination of the solutions to a set of one–dimensional linear equations. In fact, if $V_1, \ldots, V_k$ are the column eigenvectors of M, then

$$X_t = \sum_{i=1}^{k} r_i(t)\, V_i,$$

where each $r_i(t)$ is the solution to one of the one–dimensional equations (with associated row eigenvector R_i) and the column vector V_i is normalized so that $R_i V_i = 1$.

This special situation often arises in modeling physical systems because the laws of physics often lead to real symmetric matrices and such matrices always have a basis of eigenvectors. Recall that such matrices are diagonalizable, which means that they are similar to a diagonal matrix. When the matrix is not diagonalizable, the above method fails, but we can fall back on using the Jordan Canonical Form for the matrix. (See Section 7.1 and Appendix C.) Performing the above construction with the Jordan form leads to a set of difference equations of the form

$$\hat{X}_{t+1} = \begin{bmatrix} \lambda & 1 & 0 & \cdots & 0 \\ 0 & \lambda & 1 & \cdots & 0 \\ & & \ddots & \ddots & \\ & & & & 1 \\ 0 & 0 & 0 & \cdots & \lambda \end{bmatrix} \hat{X}_t + \hat{Y}_t.$$

This means that every matrix difference equation decomposes into a set of matrix difference equations in which the matrices are Jordan blocks. Each Jordan block is similar to a companion matrix whose characteristic polynomial is $(x - \lambda)^r$, where r is the size of the block (see Exercise 7.13), which means that any matrix (even with entries in a general field) is similar to a blockwise companion matrix. (We've seen this before in Section 2.3.2, and the basis in Corollary 2.3.5 records the similarity transformation.) A blockwise companion matrix that is similar to M is called a **Rational Canonical Form** for M. The reason we've used the article "a" rather than "the" here is that Rational Canonical Form is not unique, but rather, a matrix can be similar to a variety of rational forms. (See Exercise 7.12.) The next theorem uses a divisibility condition on the characteristic polynomials of the companion matrices to identify one type of Rational Canonical Form. Of course, one can choose other relations on the characteristic polynomials to obtain a different Rational Canonical Form. (Refer to [78, Sections 7.1–7.2].)

Theorem 7.5.3 (Rational Canonical Form). *A square matrix M is similar to a block companion matrix*

$$\begin{bmatrix} C_1 & & & \\ & C_2 & & \\ & & \ddots & \\ & & & C_k \end{bmatrix}$$

*in which the sequence of characteristic polynomials is a **divisor sequence**, which means that each element divides the next element of the sequence. Recall that a companion matrix C has the form*

$$\begin{bmatrix} c_1 & \cdots & & c_r \\ 1 & & & 0 \\ & 1 & & \\ & & \ddots & \\ & & 1 & 0 \end{bmatrix},$$

with characteristic polynomial $ch_C(x) = x^r - c_1 x^{r-1} - \cdots - c_r.$

This is a very powerful theorem, because it applies to matrices with entries in any field and because all operations used to convert a matrix to Rational Canonical Form are rational operations in the field. This is in sharp contrast to the Jordan Canonical Form which requires one to find the roots of polynomials. This is difficult to do for many reasons, including the fact that some polynomials do *not* have roots in the field. For instance, one may start with a rational or real matrix and have to go to complex matrices in order to obtain the Jordan form, and even then, some roots may not be expressible by an algebraic formula. Although Rational Canonical Form always exists, it may not be easy to find. Algorithms for this problem are known (for example, refer to Harrison [75]), but none is straightforward enough to be included here. However, knowledge of the *existence* of Rational Canonical Form can be used to obtain the following result.

Theorem 7.5.4. *Any k–dimensional matrix difference equation can be reduced to a one-dimensional difference equation whose order is at most k.*

We do not prove this result in general, but instead give an idea of how Rational Canonical Form could be used to prove it. Under the similarity that transforms the matrix into its rational decomposition, the original difference equation is changed into a set of matrix difference equations in which all matrices are in companion form. To see how a companion matrix difference equation can be decomposed into one–dimensional difference equations, let us simply consider a 3×3 example, the companion matrix for the polynomial $x^3 - c_1 x^2 - c_2 x - c_3$. The component equations for the matrix difference equation are:

$$\begin{aligned}
x_1(t+1) &= c_1 x_1(t) + c_2 x_2(t) + c_3 x_3(t) + y_1(t)\,, \\
x_2(t+1) &= x_1(t) + y_2(t)\,, \\
x_3(t+1) &= x_2(t) + y_3(t)\,.
\end{aligned}$$

From this,

$$x_3(t) = x_2(t-1) + y_3(t-1)\,, \quad \text{where} \quad x_2(t-1) = x_1(t-2) + y_2(t-2)$$

and so

$$x_3(t) = x_1(t-2) + y_2(t-2) + y_3(t-1)\,,$$

which gives

$$\begin{aligned}
x_1(t+1) &= c_1 x_1(t) + c_2[x_1(t-1) + y_2(t-1)] \\
&\quad + c_3[x_1(t-2) + y_2(t-2) + y_3(t-1)] + y_1(t) \\
&= c_1 x_1(t) + c_2 x_1(t-1) + c_3 x_1(t-2) \\
&\quad + c_2 y_2(t-1) + c_3 y_2(t-2) + c_3 y_3(t-1) + y_1(t)\,,
\end{aligned}$$

a third–order one–dimensional difference equation. The other components are just shifted versions of the sequence $\langle x_1(t) \rangle$ with some of the input sequence added. For example,

$$x_3(t) = x_1(t-2) + y_2(t-2) + y_3(t-1)\,,$$

which is a trivial difference equation in which its output, $x_3(t)$, is equal to its input because x_3 does not appear on the right side of the equation.

7.5.2 Reduction to homogeneous form

The matrix difference equation (7.7) can be reduced to homogeneous form if the input is well-behaved. Specifically, if there is a homogeneous matrix difference equation $Z_{t+1} = A Z_t$ and $Y_t = P Z_t$, then (7.7) can be rewritten as $\mathcal{X}_{t+1} = \mathcal{M} \mathcal{X}_t$, where

$$\mathcal{X}_{t+1} = \begin{pmatrix} X_{t+1} \\ Z_{t+1} \end{pmatrix} = \begin{bmatrix} M & P \\ 0 & A \end{bmatrix} \begin{pmatrix} X_t \\ Z_t \end{pmatrix}\,.$$

This allows one to write the solution to (7.7) as

$$X_t = Q \, \mathcal{M}^t \begin{pmatrix} X_0 \\ Z_0 \end{pmatrix},$$

where Q is the k–dimensional projection matrix that returns the first k components of its input vector. This reduction replaces a nonhomogeneous equation by a homogeneous equation at the cost of increasing the size of the matrices in the homogeneous equation.

Since the new matrix $\mathcal{M}$ has a special form, we can hope that computing powers of $\mathcal{M}$ may be easier than computing powers of a general matrix. If the matrices M and A do not have any eigenvalues in common, then an easier computation is possible. If V_i is an eigenvector or generalized eigenvector of M, then $(V_i^T, 0^T)$ is a corresponding eigenvector or generalized eigenvector of $\mathcal{M}$. Similarly, if U_j is an eigenvector or generalized eigenvector of A corresponding to the eigenvalue λ_j, and if λ_j is not an eigenvalue of M, then there exists a vector $\widehat{V}_i$ such that $(\widehat{V}_i^T, U_j^T)$ is a corresponding eigenvector or generalized eigenvector of $\mathcal{M}$. The assumption that λ_j is not an eigenvalue of M is needed to ensure that $(M - \lambda_j I)^m$ is a nonsingular matrix, and hence that the linear equation for $\widehat{V}_i$ has a solution. These considerations give the following result.

Theorem 7.5.5. *If M and A have no eigenvalues in common then a solution to*

$$X_{t+1} = M X_t + Y_t$$

can be written in the form

$$X_t = \sum_{i=1}^{k} \alpha_i t^{m_i} \lambda_i^t V_i + \sum_{j=1}^{k_2} \beta_j t^{m_j} \gamma_j^t U_j.$$

7.6   Exercises

Ex 7.1. Let

$$M = \begin{bmatrix} -4 & 9 \\ -4 & 8 \end{bmatrix}.$$

Use the two solutions $x_1(n) = n2^{n-1}$ and $x_2(n) = 2^n$ to find an expression for M^n. Compare your result to that found in (7.3).

Ex 7.2. Construct a graph on 20 vertices that has only three simple cycles, of lengths 6, 15, and 20. Find the least k such that there is a path of length k between every pair of vertices. Construct the corresponding matrix M and show that k is the least positive exponent such that $M^k \gg 0$.

Ex 7.3. Let A be a 10×10 Leslie matrix in which f_6 is the last positive fertility rate. Assume that the initial population is $x_1 = x_2 = x_3 = x_4 = x_5 = x_6 = 0$, $x_7 = 100$, $x_8 = 100$, $x_9 = 90$, $x_{10} = 10$. Find the asymptotic population vector for $X_{t+1} = AX_t$.

Ex 7.4. Let $A = \begin{bmatrix} 0 & 0 & 2 \\ 1 & 0 & 0 \\ 0 & 1 & 0 \end{bmatrix}$ and $X_0 = \begin{pmatrix} 1 \\ 0 \\ 0 \end{pmatrix}$. Calculate the asymptotic population vector for $X_{t+1} = AX_t$. How does this calculation reflect the remarks about imprimitive matrices?

Ex 7.5. Let g be the gcd of the cycle lengths for a nonnegative strongly connected matrix. Is M^g primitive? Does M^g decompose into a set of primitive submatrices?

Ex 7.6. Given the following graph, determine the gcd g of the cycle lengths, and find the decomposition of the graph into g disconnected subgraphs.

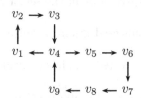

Ex 7.7. Let $X_{t+1} = MX_t$ with $M = \begin{bmatrix} 0 & 0 & 0 \\ 1/3 & 0 & 1 \\ 2/3 & 1 & 0 \end{bmatrix}$. Find the graph for this system. Find the DAG for this graph. Is the system a Markov chain? Is it periodic? Find the solution for $X_0 = (x_1(0), x_2(0), x_3(0))^T$ and give initial conditions that lead to an asymptotic fixed point.

Ex 7.8. Let $X_{t+1} = MX_t$ with $M = \begin{bmatrix} 0 & 0 & + & 0 \\ + & 0 & 0 & 0 \\ 0 & + & 0 & + \\ + & 0 & 0 & 0 \end{bmatrix}$. Draw the graph and DAG for this system. Find the period g of the system and show how M^g decomposes into subsystems.

Ex 7.9. Let $X_{t+1} = MX_t$ with $M = \begin{bmatrix} 0 & 2 & 0 \\ 1 & 0 & 0 \\ 1 & 0 & 1 \end{bmatrix}$. Find the graph and DAG for this system. Let $X_0 = (x_1(0), x_2(0), x_3(0))^T$ and find

$$\lim_{t \to \infty} \frac{X_{2t}}{2^t}.$$

Compare this to the predicted asymptotic behavior based on a graphical analysis.

Ex 7.10. Let $X_{t+1} = MX_t$ where $M = \begin{bmatrix} 0 & 1 & .1 & 0 & 0 \\ 1 & 0 & 0 & 0 & 0 \\ 0 & 0 & 0 & 0 & 1 \\ 0 & 0 & .9 & 0 & 0 \\ 0 & 0 & 0 & 1 & 0 \end{bmatrix}$. Draw the

graph and DAG for this system. Is this a Markov chain? Predict the asymptotic behavior by a graphical analysis. Use a computer to follow the evolution of X_t starting from various nonnegative initial conditions. Do the computed results for X_t agree with your predictions?

Ex 7.11. A graph is in **Leslie normal form** iff it has n vertices $v_1, v_2, \ldots, v_n$ and for all i, $v_i \to v_{i+1 \bmod n}$ and all other edges have the form $v_j \to v_1$; that is, all other edges go to v_1. (Multiple edges are not allowed, but a self-loop $v_1 \to v_1$ is allowed.) Two strongly connected graphs are **g-equivalent** iff the gcd of the set of cycle lengths is the same for each graph. Two strongly connected graphs are **c-equivalent** iff the set of cycle lengths for the two graphs is the same.

(a) Is every strongly connected graph c-equivalent to a graph in Leslie normal form?

(b) Is every strongly connected graph g-equivalent to a graph in Leslie normal form?

(c) If every strongly connected graph is equivalent to a graph in Leslie normal form, is the Leslie graph unique? If not, how many Leslie graphs are equivalent to a given strongly connected graph?

(d) Is there a *minimum* Leslie graph equivalent to a given strongly connected graph?

(e) If there is such a minimum Leslie graph, does it have a special form?

Ex 7.12. Show that a matrix may be similar to more than one blockwise companion matrix.

Hint: Consider the effect of various vectors on the matrix $\begin{bmatrix} 1 & 1 \\ 0 & 2 \end{bmatrix}$.

Ex 7.13. Show that a $k \times k$ Jordan block is similar to a companion matrix whose characteristic polynomial is $(x - \lambda)^k$ where λ is the eigenvalue of the Jordan block.

Hint: Consider the effect of this companion matrix on the eigenvectors and generalized eigenvectors.

Ex 7.14. When λ_i is a simple eigenvalue of the matrix M, show that the solution to

$$X_{t+1} = MX_t$$

can be written as

$$X_t = a\lambda_i^t C + Y_t,$$

where C is the column eigenvector for λ_i, R is the row eigenvector for λ_i, $a = (RX_0)/(RC)$, and Y_t is a non-zero vector such that $RY_t = 0$. (Compare this with Theorem 2.4.2.)

Ex 7.15. Let $X_{t+1} = MX_t$ with $M = \begin{bmatrix} 3 & -3/2 & -1/2 \\ 1 & 0 & 0 \\ 3 & -2 & 0 \end{bmatrix}$. Find the eigen-

values of the matrix and use them to predict the asymptotic behavior of solutions to the difference equation. Find the solution for the initial value vector $(2, 0, 0)^T$. Compare this solution to the solution with initial value vector $(1, 1, 1)^T$. Find the companion form and the Jordan form for the matrix and use them to explain the differences between the two solutions.

Ex 7.16. Write the following coupled pair of difference equations as a matrix difference equation

$$s_n = 2t_{n-1} + s_{n-2}, \quad t_n = -s_{n-1} + t_{n-2}.$$

(What size matrix do you need?) Show that all solutions to these equations have periods that divide 8. Solve the initial value problem with initial values

$$s_0 = 1, \, s_1 = 0, \, t_0 = 0, \, t_1 = 1.$$

Compare your solution to the solution found in Exercise 4.20.

Ex 7.17. Let $X_{t+1} = HX_t$ where $H = \begin{bmatrix} 1 & 1 & 1 & 1 \\ 1 & -1 & 1 & -1 \\ 1 & 1 & -1 & -1 \\ 1 & -1 & -1 & 1 \end{bmatrix}$. Show that H

has only two eigenvalues but four linearly independent eigenvectors. Show that the solution obeys

$$X_t = \begin{cases} 2^t X_0 & \text{if } t \text{ even,} \\ 2^{t-1}X_1 & \text{if } t \text{ odd,} \end{cases}$$

where $X_1 = HX_0$. Give some initial conditions such that X_1 is very different from X_0.

Ex 7.18. Let $X_{t+1} = MX_t$ with $M = \begin{bmatrix} 5 & 3 & 2 \\ 3 & 2 & 1 \\ 2 & 1 & 1 \end{bmatrix}$. Find the eigenvalues

and eigenvectors of the matrix M and use them to predict the asymptotic behavior of solutions to the difference equation. Show that if $X_0 \gg 0$, then $X_t = \Theta(\lambda_0^t)$, where λ_0 is the unique positive eigenvalue of M. More strongly, show that if $X_0 \gg 0$, then $\lim_{t\to\infty}(X_t - \lambda_0^t E_1) = 0$, where $ME_1 = \lambda_0 E_1$. Show that for all X_0, $\lim_{t\to\infty}(X_t - \lambda_0^t E_1) = 0$, where $ME_1 = \lambda_0 E_1$, but for some X_0's one must take $E_1 = 0$.

Ex 7.19. Let $X_{t+1} = MX_t$ with $M = \begin{bmatrix} 2 & -1 & 0 \\ -1 & 2 & -1 \\ 0 & -1 & 2 \end{bmatrix}$. Find conditions under which $X_t = \Theta((2 + \sqrt{2})^t)$ and conditions under which $X_t = \Theta(2^t)$.

Ex 7.20. Show that exactly one of the following two statements holds for a strongly connected graph on n vertices:
  (a) The graph is primitive and there is a vertex v such that the lengths of all cycles of length $\leq n$ that contain v have gcd $= 1$.
  (b) The graph is *not* primitive and for all vertices v the lengths of all cycles of length $\leq 2n - 1$ that contain v have gcd > 1.
Use this result to create an algorithm that takes a strongly connected graph as input and decides whether the graph is primitive. Show that your algorithm has complexity $O(n|E|)$ for a graph with n vertices and $|E|$ edges.

Ex 7.21. Let $X_{t+1} = MX_t + Y_t$ with $Y_{t+1} = MY_t$. If M has a basis of eigenvectors $V_1, \ldots, V_k$, show that every solution X_t can be written as

$$X_t = \sum_{i=1}^{k} (\alpha_i t + \beta_i) V_i,$$

where the α_i's depend only on the initial conditions for Y_t.

8

Modular Recurrences

In this chapter we consider recurrences modulo a fixed positive integer. For any positive integer $m \geq 2$, the output of the operation of reducing an integer modulo m (which we usually refer to as mod m) is the remainder after division by m, where the remainder is chosen to lie in the set $\{0, 1, \ldots, m - 1\}$. For instance, computing a few terms of the Fibonacci sequence mod 6 gives

$$(8.1) \qquad 1, 1, 2, 3, 5, 2, 1, 3, 4, 1, 5, 0, 5, \ldots,$$

where each term is the "mod 6 sum" of the two previous terms. We are interested in answers to the following types of questions:

- Is this sequence periodic or eventually periodic mod 6?

- What is its period?

- What is the largest period of a sequence that satisfies the Fibonacci recurrence mod 6?

- How many different sequences satisfy this recurrence mod 6?

The first question has a fairly quick answer, which we give in the next section. If we have only basic properties of modular arithmetic at our disposal, answers to the other questions can be quite complicated. Instead of using only these basic properties, we give a more sophisticated yet accessible point of view, which will shed light on the general structure of recurrences mod m and provide straightforward answers to these questions. We end the chapter with applications of modular recurrences to pseudorandom number generation and to factorization of integers.

8.1   Periodicity

There's no apparent pattern in the first few terms of the Fibonacci numbers mod 6 as listed in (8.1). Continuing further into the sequence,

$$(8.2) \quad 1, 1, 2, 3, 5, 2, 1, 3, 4, 1, 5, 0, 5, 5, 4, 3, 1, 4, 5, 3, 2, 5, 1, 0, 1, 1, 2, 3, \ldots,$$

we see that the initial pair 1, 1 recurs. Because it is obtained from a second–order recurrence, the sequence will repeat once the initial pair occurs again. Even without listing any elements of the sequence, we know that because there are only 36 different pairs of elements mod 6, at least one repetition of a pair must occur within the first 37 terms of the sequence. A rewording of this is helpful. Let X be the set of all ordered pairs of integers mod 6, and let f be the Fibonacci function mod 6 defined on X by

$$(8.3) \qquad\qquad f(x, y) = (y, x + y \bmod 6),$$

where the second coordinate is specified to be the least nonnegative remainder mod 6, and so $f(x, y)$ is a function on X. Since there are only 6^2 different pairs of integers mod 6, any list of 37 consecutive pairs in a sequence defined by the second–order recurrence mod 6 must contain a repeated pair. This argument holds for any second–order recurrence mod 6 with any pair of initial values.

Putting this in a general context: For a function f defined on a set X, we will use $f^{(n)}$ to denote the nth **iterate of** f,

$$f^{(n)} = \overbrace{f \circ f \circ f \circ \cdots \circ f}^{n \text{ times}},$$

and the **orbit** of $x \in X$ is the sequence

$$x, f(x), f^{(2)}(x), f^{(3)}(x), f^{(4)}(x), \ldots,$$

the sequence formed by starting with the value x and applying f again and again. If there is some $n > 0$ such that $f^{(n)}(x) = x$, the orbit of x is called a **periodic orbit**, x is called a **periodic point**, and its **period** is the least positive t such that $f^{(t)}(x) = x$. When $t = 1$ holds, x is called a **fixed point**. From (8.2) we see that $(1, 1)$ is a periodic point of the Fibonacci function mod 6, and its period is 24.

The orbit of a periodic point with period t can be visualized as forming a closed loop that returns to its starting point after t iterations of the function f. Sometimes an orbit might get into a loop without returning to its initial value. (Refer to Figure 8.1.) We will call x an **eventually**

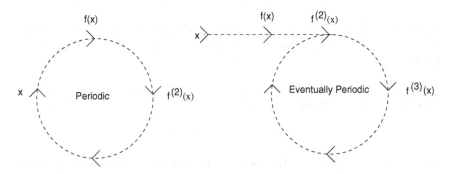

FIGURE 8.1. Periodic and Eventually Periodic Orbits.

periodic point if there exists an integer t such that $f^{(n+t)}(x) = f^{(n)}(x)$ holds for all sufficiently large n, and the smallest such t is its **period**. Notice that under this definition every periodic point is also eventually periodic.

For example, for the function $f(x) = -x^2 + 2x + 1$ on the set $X = \mathbb{Z}$ the orbit of $x = 0$ is the sequence $0, 1, 2, 1, 2, 1, 2, \ldots$, and $x = 0$ is an eventually periodic point of f that is not periodic, and both $x = 1, 2$ are periodic points of f. For each of these three values of x the period is two.

Theorem 8.1.1. *If f is a function on a finite set X, then every element of X is an eventually periodic point of f, and its period is at most the number of elements in X. If f is a one-to-one function on any (not necessarily finite) set, then every eventually periodic point of f is periodic. In particular, if f is a one-to-one function on a finite set, then every element of X is a periodic point of f.*

Proof. If X has n elements, the set $\{x, f(x), \ldots, f^{(n)}(x)\}$ contains $n + 1$ elements from X and so must contain a duplicate, say $f^{(i)}(x) = f^{(j)}(x)$ for some $0 \leq i < j \leq n$. Then

$$f^{(i+1)}(x) = f(f^{(i)}(x)) = f(f^{(j)}(x)) = f^{(j+1)}(x),$$

$$\vdots$$

$$f^{(i+k)}(x) = f^{(j+k)}(x) \text{ for all } k \geq 0.$$

Therefore, for $m \geq i$,

$$f^{(m+(j-i))}(x) = f^{(j+(m-i))}(x) = f^{(i+(m-i))}(x) = f^{(m)}(x),$$

and we see that x is an eventually periodic point whose period is at most $j - i \leq n$.

Suppose f is one-to-one and the period of $x \in X$ is t. Then there exists a minimal N such that $f^{(n+t)}(x) = f^{(n)}(x)$ for all $n \geq N$. To show that x is periodic, we prove that $N = 0$. If N were non-zero, then we would have

$$f(f^{(N-1)}(x)) = f^{(N)}(x) = f^{(N+t)}(x) = f(f^{(N+t-1)}(x)),$$

and the fact that f is one-to-one would give $f^{(N-1)}(x) = f^{(N-1+t)}(x)$. This is a contradiction to the assumed minimality of N, and $N = 0$ must be true.   □

For general integer $m \geq 2$, the Fibonacci function mod m (also called a **modular Fibonacci function**) is

$$F(x, y) = (y, x + y \bmod m).$$

Since this is a one-to-one function on the ordered pairs of integers mod m, and the set of integers mod m is finite, the previous result gives the following corollary.

Corollary 8.1.2. *Each modular Fibonacci function has a periodic orbit for every pair of initial values.*

Let's compute the period of the Fibonacci sequence mod m for four more values of m:

$$m = 2: \quad 1, 1, 0, 1, 1, \ldots, \text{ whose period is 3;}$$
$$m = 3: \quad 1, 1, 2, 0, 2, 2, 1, 0, 1, 1, \ldots, \text{ whose period is 8;}$$
$$m = 4: \quad 1, 1, 2, 3, 1, 0, 1, 1, \ldots, \text{ whose period is 6;}$$
$$m = 12: \quad 1, 1, 2, 3, 5, 8, 1, 9, 10, 7, 5, 0, 5, 5, 10, 3, 1, 4, 5,$$
$$9, 2, 11, 1, 0, 1, 1, \ldots, \text{ whose period is 24.}$$

We have already computed the period mod 6 to be 24, which is the product of the periods mod 2 and mod 3. Because of this, you might conjecture that this holds in general, but the period mod 12 is 24 which does not equal $8 \cdot 6$. After some deliberation you can see that 24 does equal $\mathrm{lcm}(8, 6)$, the least common multiple (lcm) of 8 and 6. Why the lcm occurs is explained in the next theorem.

Theorem 8.1.3. *Let $X = \mathbb{Z}^k$ for some k and let f be any function defined on X. For $x \in X$ and positive integers m_1, m_2, let t_i be the period of x under f modulo m_i for each of $i = 1, 2$. Then $\mathrm{lcm}(t_1, t_2)$ is the period of x under f modulo $\mathrm{lcm}(m_1, m_2)$.*

Proof. Let $m = \mathrm{lcm}(m_1, m_2)$ and $L = \mathrm{lcm}(t_1, t_2)$. If t is the period of x under f mod m, there exists N such that for all $n \geq N$,

$$f^{(n+t)}(x) \equiv f^{(n)}(x) \,(\bmod m),$$

and since each of m_1, m_2 divides m, then

$$f^{(n+t)}(x) \equiv f^{(n)}(x) \,(\text{mod } m_i) \quad \text{for all } n \geq N.$$

Therefore, t must be a common multiple of t_1 and t_2, and L divides t.

On the other hand, for each of $i = 1, 2$, there exists N_i such that $f^{(n+t_i)}(x) \equiv f^{(n)}(x) \,(\text{mod } m_i)$ for all $n \geq N_i$. Using $N = \max\{N_1, N_2\}$ and the fact that L is a multiple of both t_1 and t_2, we have that for all $n \geq N$,

$$f^{(n+L)}(x) \equiv f^{(n)}(x) \,(\text{mod } m_i).$$

This means that the difference $f^{(n+L)}(x) - f^{(n)}(x)$ is divisible by each m_i and so also by their least common multiple m, giving

$$f^{(n+L)}(x) \equiv f^{(n)}(x) \,(\text{mod } m),$$

and x has a period L that is a multiple of t. Since we've already shown that L divides t (and they're both positive), then $L = t$. □

How do we find the period of the Fibonacci function for a general modulus? If we happened to know (or could easily find) a factorization of the modulus into a product of two **relatively prime** integers m_1, m_2 (that is, $\gcd(m_1, m_2) = 1$), then the theorem could be applied to find the period mod m. That's exactly what we noted for $m = 12$. The following result follows directly from this observation.

Corollary 8.1.4. *If $m = p_1^{\alpha_1} \cdots p_s^{\alpha_s}$ is the prime factorization of m and t_i is the period of x under f modulo $p_i^{\alpha_i}$, then the period of x under f modulo m is $\text{lcm}(t_1, \ldots, t_s)$.*

8.1.1 Periodicity of linear modular recurrences

We've been working with functions defined on $\mathbb{Z}^k$ that are then reduced mod m, but we could equally well have considered f to be defined on the set $X = \mathbb{Z}_m^k$. (Here $\mathbb{Z}_m$ means the set $\{0, 1, \ldots, m-1\}$ under the operations of addition and multiplication mod m.)

As we saw with the Fibonacci recurrence, a kth order linear recurrence
(8.4)
$$s_{j+k} \equiv c_1 s_{j+k-1} + \cdots + c_k s_j + c_{k+1} \,(\text{mod } m), \text{ where } c_k \not\equiv 0 \,(\text{mod } m),$$

defines the function

(8.5) $\quad S(x_1, \ldots, x_k) = (x_2, \ldots, x_k, \ c_k x_1 + \cdots + c_1 x_k + c_{k+1} \bmod m)$

on $\mathbb{Z}_m^k$, where the last component is chosen to be the least nonnegative value of its congruence class mod m. (Notice that we are allowing nonhomogeneous equations with the constant forcing term c_{k+1}.)

Let's talk a bit more about notation. Normally, when 26 (mod 25) is written we think of 1, but 26 (mod 25) is of course an infinite set of integers, all integers that are congruent to 26 modulo 25. We've been using the notation $t \bmod m$ (as contrasted with $t \pmod m$) to denote the least non-negative integer that is congruent to t modulo m. For example, $3 \cdot 9 \bmod 5$ equals 2. Another way of saying this is that $3 \cdot 9$ equals 2 in $\mathbb{Z}_5$.

The following lemma follows from (8.5) because

$$S(x_1, \ldots, x_k) = S(y_1, \ldots, y_k) \iff x_2 = y_2, \ldots, x_k = y_k, c_k x_1 = c_k y_1.$$

Lemma 8.1.5. *S is a one-to-one function on $\mathbb{Z}_m^k$ iff c_k has a multiplicative inverse modulo m.*

Orbits of S correspond to choices of initial values $s_0, \ldots, s_{k-1}$ in the recurrence (8.4). Since S is a function defined on a finite set with m^k elements, from this lemma and Theorem 8.1.1 we obtain the following result.

Theorem 8.1.6. *Each $(s_0, \ldots, s_{k-1})$ is an eventually periodic point of S. Moreover, if c_k has a multiplicative inverse modulo m, then every orbit under S is periodic.*

Because every element of $\mathbb{Z}_m$ has an additive inverse, an element that has a multiplicative inverse can be referred to as an **invertible element** without confusion as to which operation is meant. For the Fibonacci sequence, $c_k = 1$ is of course invertible for every modulus. This means that (regardless of the initial pair) every modular sequence generated by the Fibonacci recurrence is periodic.

It can be checked that the orbit of $(1, 5)$ under the recurrence $s_{j+2} \equiv s_{j+1} + 3s_j \pmod{18}$ is

$$1 , 5 , 8 , 5 , 11 , 8 , 5 , 11, \ldots,$$

an eventually periodic sequence that is not periodic. Note that this is consistent with the last result, since 3 is not invertible mod 18 and $S(5, 8) = (8, 5) = S(11, 8)$ shows that S is not one-to-one.

The periodicity of modular recurrences therefore depends only on the *algebraic* property of the invertibility of c_k. We next describe an efficient procedure for determining whether an element is invertible, which then becomes a method for deciding whether a specific modular linear recurrence is periodic. The procedure is the famous **Euclidean Algorithm**, which we've already seen several times.[1] We used this algorithm in Chapter 6 (and for polynomials in Chapter 4) to compute the greatest common divisor (gcd) of n-bit numbers in $O(n)$ operations. Because we're interested in

---

[1] The Euclidean Algorithm can be found in the seventh volume of Euclid's *Elements*. A translation of this can be found at
aleph0.clarku.edu/~djoyce/java/elements/bookVII/propVII2.html .

using the algorithm to compute more than the gcd, it's helpful to review the process here. Recall that the gcd of two non-zero integers a, m (which we denote by $\gcd(a, m)$) is a positive integer that divides both a and m and is defined by the property that $\gcd(a, m)$ is divisible by all divisors of both a and m.

Theorem 8.1.7 (Euclidean Algorithm). *Let m be any positive integer and a be any non-zero integer. Then the sequence of divisions*

$$m = q_1 a + r_1; \quad a = q_2 r_1 + r_2; \quad r_1 = q_3 r_2 + r_3; \quad \cdots$$

(where $0 \leq r_1 < a$ and $0 \leq r_{i+1} < r_i$ for all $i > 1$) ends after finitely many steps, and the last non-zero remainder r_K is $\gcd(a, m)$.

Proof. Since the sequence of remainders $\langle r_i \rangle$ is a strictly decreasing sequence of nonnegative integers, it must be a finite sequence, and its last remainder is zero,

$$m = q_1 a + r_1; a = q_2 r_1 + r_2; \ldots ; r_{K-2} = q_K r_{K-1} + r_K; r_{K-1} = q_{K+1} r_K.$$

From the definition of gcd, $\gcd(a, m) = \gcd(m - ka, a)$ holds for all integers k, and we successively obtain

$$\gcd(a, m) = \gcd(m - q_1 a, a) = \gcd(r_1, a) = \gcd(r_1, a - q_2 r_1) = \cdots$$
$$= \gcd(r_{K-1}, r_{K-2} - q_2 r_{K-1}) = \gcd(q_{K+1} r_K, r_K) = r_K .$$

$$\square$$

For instance, to compute $\gcd(437, 12)$:

$$437 = 36 \cdot 12 + 5, \quad 12 = 2 \cdot 5 + 2, \quad 5 = 2 \cdot 2 + 1, \quad 2 = 2 \cdot 1 + 0,$$

and $\gcd(437, 12) = 1$ is obtained. Whenever $\gcd(a, m) = 1$ holds, the steps of the Euclidean Algorithm can be reversed to obtain $a^{-1} \pmod{m}$. For instance, successively solving backwards for each remainder in this example gives

$$1 = 5 - 2 \cdot 2 = 5 - 2 \cdot (12 - 2 \cdot 5) = 5 \cdot 5 - 2 \cdot 12$$
$$= 5 \cdot (437 - 36 \cdot 12) - 2 \cdot 12 = 5 \cdot 437 - 182 \cdot 12$$
$$= 5 \cdot 437 - 182 \cdot 12 \equiv -182 \cdot 12 \equiv 255 \cdot 12 \pmod{437},$$

and we see that 255 is the inverse of 12 in $\mathbb{Z}_{437}$. This method works in general and is the basis for the following result.

Theorem 8.1.8. *Let m be any positive integer and a be any integer. Then a is invertible modulo m iff $\gcd(a, m) = 1$.[2]*

---

[2]The Euler phi function, $\phi(m)$, counts the number of invertible elements modulo m.

Proof. If a is invertible, there exists x such that $ax \equiv 1 \pmod m$, which means that $ax + my = 1$ for some y. Therefore, the gcd must divide 1, and so equals 1. This proves the *only if* direction. On the other hand, when $\gcd(a, m) = 1$ we can "solve the Euclidean Algorithm backwards" to obtain x, y such that $1 = ax + my$, and $ax \equiv 1 \pmod m$. □

Combining these results gives a general answer to the first question at the beginning of this chapter.

Theorem 8.1.9. *If S is the function in (8.5), then every orbit of S is periodic iff $\gcd(c_k, m) = 1$. In particular, when the modulus m is prime, every solution to a linear recurrence modulo m is periodic.*

8.1.2 Fast modular computations

In this section we describe Montgomery multiplication, a quick way to compute the product $a * b \pmod m$. The method was first suggested by Peter Montgomery [115] in 1985. The technique is helpful for implementing modular exponentiation used in many cryptosystems, for example, in RSA [137] and in any RSA-type key exchange in other cryptographic methods.

The technique relies on the fact that there are moduli for which arithmetic is quick, for instance a computer's machine word size. For any such modulus r and any m with $\gcd(m, r) = 1$, Montgomery multiplication translates operations mod m to the faster operations mod r. Articles [166, 58, 167] can be consulted for a discussion of an efficient hardware implementation of Montgomery multiplication. In particular, [58] claims that their implementation is twice as fast as the methods previously used for modular arithmetic. (Refer also to [14].)

We may assume that our factors a and b are greater than 0 and less than m. We write our modulus m and each of our factors in their base-r representation, where the digits are indicated by subscripts, for example, $a = a_0 + a_1 r + \cdots + a_k r^k$, and each a_i satisfies $0 \le a_i < r$. Since we are assuming that $1 = \gcd(m, r) = \gcd(-m_0, r)$, then $-m_0$ is an invertible element of $\mathbb{Z}_r$ and $-m_0 n \equiv 1 \pmod r$ for some $1 \le n < r$. Computing n is a one-time calculation.

Montgomery multiplication (as given in [58]) involves the calculation of a sequence $R_0, R_1, \ldots, R_k$ of integers for which $P = r^{k+1} R_k$ equals $ab \bmod m$ and $0 < P < 2m$. Then either P or $P - m$ is the required product $ab \bmod m$. (Notice that if r is a power of 2, then P can be calculated easily from R_k by shifting. More general r's are considered in [167].)

The sequence $\langle R_i \rangle$ is inductively calculated in tandem with another sequence, which we call $\langle Q_i \rangle$. For the choice of $Q_0 = a_0 b_0 n \bmod r$, the natural number $a_0 b + Q_0 m$ is divisible by r and is congruent to $a_0 b \pmod m$. We set $R_0 = (a_0 b + Q_0 m)/r$. Continuing, for $Q_1 = R_0 + a_1 b_0 n \bmod r$, it can be checked that $R_0 + a_1 b + Q_1 m$ is divisible by r and is congruent to

$a_1 b \,(\mathrm{mod}\ m)$. Setting $R_1 = (R_0 + a_1 b + Q_1 m)/r$, we have

$$r^2 R_1 \;=\; r R_0 + a_1 r b + Q_1 m r \;\equiv\; a_0 b + a_1 r b \;=\; (a_0 + a_1 r) b \,(\mathrm{mod}\ m).$$

Here's the algorithm:

> PROCEDURE   MULT(a, b)
> $R := 0$
> FOR $i := 0$ TO k DO
> $\qquad Q := (R + a_i b_0) * (-m^{-1}) \bmod r$
> $\qquad R := (R + a_i b + Q m)/r.$

Note that only the least-significant digit of b is needed. Refer to Exercise 8.5 for a justification of the algorithm. Because of the initial investment of time for the computation of $-m^{-1} \bmod r$ and the final adjustment of multiplying by r^{k+1}, the algorithm is not particularly helpful for calculating just one product.

8.2   Finite Fields

In this section we begin the study of finite fields, generalizations of the integers modulo a prime. They provide a more sophisticated context for investigating modular recurrences. Évariste Galois [66] was the first to use some properties of finite fields, and the first systematic theory was written by Leonard E. Dickson in [53]. Because they're *finite*, recurrences in finite fields satisfy the periodicity results proved in Section 8.1.

Let $\mathbb{F}$ be a finite set with two operations, $+$ and $*$. Then $\mathbb{F}$ **is a finite field** under these operations if the following hold:

1. $\mathbb{F}$ is an **abelian group** under $+$, which means that $+$ is an associative and commutative operation on $\mathbb{F}$; $+$ has a special identity element denoted by 0 such that $a + 0 = a$ holds for all $a \in \mathbb{F}$ and each element a has an (additive) inverse $b \in \mathbb{F}$ satisfying $a + b = 0$.

2. The non-zero elements of $\mathbb{F}$ form an **abelian group** under $*$, which as above means that $*$ is an associative and commutative operation and has a special identity element denoted by 1 such that $a * 1 = a$ holds for all $a \in \mathbb{F}$ and each non-zero element a has a (multiplicative) inverse $b \in \mathbb{F}$ satisfying $a * b = 1$.

3. The operations of $+$ and $*$ are connected by the **distributive law**, which means that $a_1 * (a_2 + a_3) = a_1 * a_2 + a_1 * a_3$ for all $a_1, a_2, a_3 \in \mathbb{F}$.

For example, $\mathbb{Z}_5 = \{0, 1, 2, 3, 4\}$ is a field under the operations of addition and multiplication mod 5, where the additive and multiplicative identities are respectively 0 and 1; 0 is its own additive inverse; and $1, 2, 3, 4$ have additive inverses $4, 3, 2, 1$ and multiplicative inverses $1, 3, 2, 4$. The associative, commutative, and distributive properties are inherited from the set of integers. Although $\mathbb{Z}_4 = \{0, 1, 2, 3\}$ is an abelian group under the operation of addition mod 4, the fact that $\gcd(2, 4) = 2$ implies that 2 is not invertible under multiplication mod 4, and $\mathbb{Z}_4$ is therefore not a field.

Theorem 8.2.1. $\mathbb{Z}_m$ *is a finite field under the operations of addition mod* m *and multiplication mod* m *iff* m *is prime.*

Proof. As in the two examples above, we observe that the associative, commutative, and distributive properties are all inherited from the integers. Also, 0 is the additive identity and $\mathbb{Z}_m$ is an abelian group under addition mod m. By Theorem 8.1.8, a is invertible mod m iff $\gcd(a, m) = 1$. Ensuring that $\gcd(a, m) = 1$ for all $1 \leq a < m$ is equivalent to requiring that m be prime. □

The number of elements in a finite field must be a prime power. (Refer to Exercise 8.6.)

Let $\mathbb{F}_4$ be the set of polynomials $\{0, 1, x, x + 1\}$. If we define addition on $\mathbb{F}_4$ to be the usual polynomial addition followed by the reduction of the coefficients mod 2, it can be checked that $\mathbb{F}_4$ is an abelian group under addition. If we define multiplication as in Table 8.1,

TABLE 8.1. The multiplication table for the field with four elements

$*$	0	1	x	$x + 1$
0	0	0	0	0
1	0	1	x	$x + 1$
x	0	x	$x + 1$	1
$x + 1$	0	$x + 1$	1	x

then 1 is the multiplicative identity, and each of the non-zero elements $1, x, x + 1$ has a multiplicative inverse (respectively $1, x + 1, x$). The associative law for multiplication and the distributive law are tedious to check, but $\mathbb{F}_4$ satisfies these laws.

It's helpful to examine this example further. In earlier chapters the notation $\mathbb{R}[x]$ meant the set of polynomials with coefficients from $\mathbb{R}$. Likewise, $\mathbb{F}[x]$ will be the set of polynomials with coefficients from a field $\mathbb{F}$. For instance, $\mathbb{Z}_2[x]$ is the set of all polynomials whose coefficients are either 0 or 1. We're usually interested not only in a *set* of polynomials, but also in its algebraic properties. In $\mathbb{F}[x]$ we can define the polynomial operations of addition and multiplication similarly to the operations in $\mathbb{R}[x]$ with the coefficients calculated using the operations in $\mathbb{F}$. For example, you can check that when $f(x) = x^2 + x$ and $g(x) = x^3 + x + 1$ are considered as elements of $\mathbb{Z}_2[x]$, their sum and product are

$$f(x) + g(x) = x^3 + x^2 + 1 \quad \text{and} \quad f(x)g(x) = x^5 + x^4 + x^3 + x.$$

This discussion allows us to describe the four-element field $\mathbb{F}_4$ above in another way. Let S be the set of all constant and linear polynomials in $\mathbb{Z}_2[x]$. Since S is an abelian group under addition in $\mathbb{Z}_2[x]$, we have a suitable addition for S. What about multiplication? The set S is not closed under the usual polynomial multiplication in $\mathbb{Z}_2$, because for instance $(x+1)^2 = x^2+1$ is not an element of S. But a slight modification of the usual multiplication using the polynomial $f(x) = x^2 + x + 1$ (which is not in S) works. Namely, the product of two elements in S will be defined by first obtaining the product $p(x)$ of the two polynomials as elements of $\mathbb{Z}_2[x]$ and then "reducing $p(x)$ mod $f(x)$," in other words, using the Division Algorithm in $\mathbb{Z}_2[x]$ to find the unique remainder when $p(x)$ is divided by $f(x)$. Since we're dividing by a quadratic polynomial, the Division Algorithm always yields a remainder in the set S. For example, $x(x+1) = x^2 + x = f(x) + 1$ gives the product $x * (x+1) = 1$ in S. (In practice, $p(x) \mod f(x)$ is often computed by repeated subtraction of multiples of $f(x)$ rather than by using long division.)

This procedure can be generalized to obtain a finite field with p^m elements for any prime p and positive exponent m. The construction relies on the fact that for every p and m there is an **irreducible polynomial** in $\mathbb{Z}_p[x]$ with $\deg(f) = m$, where irreducible means that $f(x)$ cannot be factored into polynomials of smaller degree in $\mathbb{Z}_p[x]$. The construction and more information about finite fields can be found in [98, p. 91 ff]. (Also, refer to Exercises 8.8–8.12 at the end of this chapter.) In particular, for each choice of prime p and exponent m there is essentially only one finite field with p^m elements. These fields are frequently denoted by $\mathrm{GF}(p^m)$ and are the finite **Galois fields**, named in honor of É. Galois. In what follows we drop the $*$ symbol for multiplication, and instead use juxtaposition of elements to denote multiplication. Our proof of Theorem 8.1.9 generalizes to finite fields.

Theorem 8.2.2. *In a finite field every linear recurrence is periodic.*

8.3   Periods of First–Order Modular Recurrences

In what follows we will allow R to be either the set of integers mod m or a finite field and investigate the period of first–order linear recurrences $s_{j+1} = as_j + b$ in R, where $a \neq 0$. We first note that when $a = 1$, then $s_n = s_0 + nb$ and the period of $S(x) = x + b$ is therefore the least integer $t \geq 1$ such that $tb = 0$ in R. We next consider the case in which $a - 1$ is invertible.

Lemma 8.3.1. *Let a be an element of R such that a−1 has a multiplicative inverse c in R. If $\langle s_j \rangle$ is a solution to the recurrence $s_{j+1} = as_j + b$ in R, then*

$$(8.6) \qquad\qquad s_{n+j} = a^j s_n + bc(a^j - 1) \quad \text{for all } n, j.$$

Proof. For fixed n we prove (8.6) by induction on $j \geq 1$. Since $c(a-1) = 1$ holds in R,

$$s_{n+1} = as_n + b = as_n + bc(a-1).$$

Assuming that (8.6) holds for all $1 \leq j < J$,

$$s_{n+J} = as_{n+J-1} + b = a(a^{J-1}s_n + bc(a^{J-1} - 1)) + bc(a-1)$$
$$= a^J s_n + bc(a^J - a + a - 1) = a^J s_n + bc(a^J - 1),$$

completing the proof. $\qquad\qquad\qquad\qquad\qquad\qquad\qquad\qquad\qquad\square$

Theorem 8.3.2. *Let S be the function defined on R by $S(x) = ax + b$, where b is arbitrary and $a - 1$ is any element of R that has a multiplicative inverse c in R. Then $s_0 = -bc$ is the only fixed point of S, and every $s_0 \in R$ is an eventually periodic point of S whose period equals the least integer t such that $(a^t - 1)(S^{(n+1)}(s_0) - S^{(n)}(s_0)) = 0$ holds for all sufficiently large n.*

Proof. We see that s_0 is a fixed point iff $as_0 + b = s_0$, which can be uniquely solved for $s_0 = -bc$. From Theorem 8.1.6, every orbit of S is eventually periodic, and if t is any multiple of the period of s_0, then $s_{n+t} = s_n$ for sufficiently large n. By (8.6) this becomes

$$a^t s_n + bc(a^t - 1) = s_n,$$
$$(a^t - 1)(s_n + bc) = 0.$$

Multiplying the last equation by $a - 1$ gives

$$0 = (a^t - 1)((a - 1)s_n + b) = (a^t - 1)(s_{n+1} - s_n).$$

Because $a - 1$ is invertible in R, each of these steps is reversible, and we obtain

$$s_{n+t} = s_n \iff (a^t - 1)(s_{n+1} - s_n) = 0,$$

which completes the proof. $\qquad\qquad\qquad\qquad\qquad\qquad\qquad\qquad\square$

Let us consider orbits of $S(x) = 12x + 4$ on $R = \mathbb{Z}_{21}$. Since $11^{-1} = 2$, then $-bc = 13$ is the only fixed point of S. For instance, the orbits of 1 and 2 are

$$1, 16, 7, 4, 10, 19, 1, \ldots \quad \text{and} \quad 2, 7, 4, 10, 19, 1, 16, 7, \ldots,$$

where 1 is a periodic point, while 2 is *eventually* periodic. The periods are the same, but is that only because 7 is a common term of both orbits? In

Exercise 8.15 you're asked to find the periods of other orbits under this map.

When R is a finite field, the situation is quite simple. Every orbit under $S(x) = ax + b$ is periodic, *and* the period is the same for every initial value.

Theorem 8.3.3. *Consider the iteration of $S(x) = ax + b$ in the finite field $\mathbb{F}$. When $a = 1$, the period of every orbit is the* **characteristic** *of the field $\mathbb{F}$, the least positive integer n such that the sum of n copies of any element in $\mathbb{F}$ is zero. (Refer to Exercise 8.4.) When $a \neq 1$, let c be the multiplicative inverse of $a - 1$. Then $x = -bc$ is the only fixed point of S, and for all other $x \neq -bc$ the period equals the period of 1 under the linear function $S_0(x) = ax$. This number is called the* **order** *of a in $\mathbb{F}$ and will be denoted by $\mathrm{ord}(a)$.*

Proof. From our remark before Lemma 8.3.1, the period of $S(x) = x + b$ is the characteristic of $\mathbb{F}$. For any $S(x) = ax + b$ with $a \neq 1$, the orbit of any $s_0 \neq -bc$ is periodic with period $t \geq 2$, which all $s_{m+1} \neq s_m$ for all m, and $s_{m+1} - s_m$ is an invertible element of $\mathbb{F}$. Therefore, the condition in the last theorem becomes $a^t = 1$, and the period is $t = \mathrm{ord}(a)$. $\square$

We can rewrite the fact that $\mathrm{ord}(a)$ is the period of the sequence $\langle a^j \rangle$ as

$$(8.7) \qquad a^i = a^j \iff i \equiv j \,(\mathrm{mod}\,\mathrm{ord}(a)),$$

a useful observation in what follows. The next algebraic result is a computational aid for computing $\mathrm{ord}(a)$. Its statement appeared in a letter written by Pierre de Fermat to Frénicle de Bessy in 1640, and was later proved by Gottfried Leibniz. It's usually called Fermat's *Little* Theorem, to differentiate it from Fermat's *Last* Theorem, which was proved in 1994 by Andrew Wiles with the assistance of Richard Taylor [168, 158]. Alf van der Poorten [162] has published a fairly accessible account of both the history and the proof of Fermat's Last Theorem, the culmination of the work of many mathematicians.

Theorem 8.3.4 (Fermat's Little Theorem). *If $\mathbb{F}$ is a finite field with q elements, then every non-zero $a \in \mathbb{F}$ satisfies $a^{q-1} = 1$, and so $\mathrm{ord}(a)$ divides $q - 1$.*

Proof. Let a be a fixed non-zero element of $\mathbb{F}$, and let $a_1, \ldots, a_{q-1}$ be any listing of all non-zero elements in $\mathbb{F}$. If a^{-1} is the multiplicative inverse of a, then for any i, j,

$$aa_i = aa_j \iff a^{-1}aa_i = a^{-1}aa_j \iff a_i = a_j,$$

which means that the sets $\{aa_1, \ldots, aa_{q-1}\}$ and $\{a_1, \ldots, a_{q-1}\}$ are equal. Because multiplication is commutative, the product of the elements in the second set equals the product of the elements in the first set,

$$a_1 \cdots a_{q-1} = (aa_1) \cdots (aa_{q-1}) = a^{q-1}(a_1 \cdots a_{q-1}).$$

The product $a_1 \cdots a_{q-1}$ is a non-zero element of the field $\mathbb{F}$, and multiplying the above equation by this element's inverse yields $a^{q-1} = 1$. □

Corollary 8.3.5. *The period of any first–order linear recurrence with $a \neq 1$ in a finite field with q elements divides $q - 1$.*

We now use this result to determine the periods of $S(x) = 215x + 3$ in $\mathbb{Z}_m$ for the prime $m = 12323$ without actually calculating any orbit. Since m is a prime, $\mathbb{F} = \mathbb{Z}_m$ is a finite field, and all orbits are periodic. From Exercise 8.3, 5816 is the multiplicative inverse of $a - 1 = 214$ in $\mathbb{F}$ and $s_0 = -3 * 5816 = 7198$ is the only fixed point of S. The period of any $s_0 \neq 7198$ equals the order of 215 in $\mathbb{F}$ and it remains to find $t = \mathrm{ord}(215)$. Using Fermat's Little Theorem, t divides $m-1 = 12322 = (2)(61)(101)$, and there are eight possible values for t. (What are these eight values?) For each divisor k of 12322, calculating a^k can be done by **fast exponentiation**, the same process that we have already used several times to find powers of matrices. For instance, $101 = 2^6 + 2^5 + 2^2 + 1$ gives

$$a^{101} = a^{2^6} a^{2^5} a^{2^2} a \equiv 4797 * 1822 * 3809 * 215 \equiv 1 \, (\mathrm{mod}\, m).$$

Since 101 is prime, $\mathrm{ord}(215) = 101$, and S has one fixed point and 122 orbits whose period is 101.

We close this section with one more algebraic result. Every odd element in $\mathbb{Z}_8$ (there are four of them) is a root of the *quadratic* equation $x^2 - 1 = 0$. This overabundance of roots (there are more than $\deg(x^2 - 1)$ of them) can never happen when the coefficient set is a *field*. (Refer to Exercise 8.23.) This simple fact can have some surprising consequences. For instance, if $\mathbb{F}$ is a field with $q = p^m$ elements, then the order of $s \in \mathbb{F}$ divides $p - 1$ iff s is a root of $x^{p-1} - 1 = 0$; there are at most $p - 1$ elements of $\mathbb{F}$ whose order divides $p-1$. Fermat's Little Theorem gives the following result, which will be useful later.

Theorem 8.3.6. *Let $\mathbb{F}$ be a finite field with $q = p^m$ elements. If s is a non-zero element of $\mathbb{F}$, then $\mathrm{ord}(s)$ divides $p - 1$ iff $s \in \mathbb{Z}_p$.*

8.3.1   First–order modular recurrences with maximal period

In this section we prove that any finite field with q elements has an element whose order is $q - 1$, which we know is the largest possible order. These elements are called **primitive** elements of $\mathbb{F}$. The orbits of $S(x) = ax+b$ are easily described when a is a primitive element: there are only two orbits, the fixed point and everything else! The reason for this is that Theorem 8.3.3 can be applied, since a cannot be either 0 or 1 and so both a and $a - 1$ are invertible elements of $\mathbb{F}$. The next lemma is used in our proof of the existence of primitive elements.

Lemma 8.3.7. *Let a, b be non-zero elements of $\mathbb{F}$ with $\mathrm{ord}(a) = r$ and $\mathrm{ord}(b) = s$. If $\gcd(r, s) = 1$, then $\mathrm{ord}(ab) = rs$.*

Proof. Let $n = \text{ord}(ab)$. Then $(ab)^n = 1$. Also,

$$1 = ((ab)^n)^r = (ab)^{rn} = (a^r)^n b^{rn} = b^{rn},$$

and rn must be a multiple of $\text{ord}(b) = s$. Since $\gcd(r, s) = 1$, this implies that s divides n. Interchanging the roles of a and b we obtain that n is a common multiple of r and s. Again using the fact that $\gcd(r, s) = 1$, we have $\text{lcm}(r, s) = rs$, and so rs divides n. We complete the proof by showing that n divides rs. (Since rs and n are both positive, this gives $n = rs$.) For this,

$$(ab)^{rs} = (a^r)^s (b^s)^r = 1,$$

implying that rs is a multiple of $\text{ord}(ab) = n$. □

Theorem 8.3.8 (Primitive Element Theorem). *Every finite field with q elements has at least one element whose order equals $q - 1$.*

Proof. Let N be the maximum element in the finite set $\{\text{ord}(b) : b \in \mathbb{F}\}$, and let d be some element of $\mathbb{F}$ with $\text{ord}(d) = N$. Fermat's Little Theorem implies that $N \leq q - 1$. If we can prove that the order of every element divides N, then each of the $q - 1$ non-zero elements of $\mathbb{F}$ would satisfy $x^N - 1 = 0$, and by Theorem 8.3.6, $N \geq q - 1$ also would hold.

By way of contradiction, we assume that there exists an element $c \in F$ whose order does not divide N, from which it follows that there exists a prime p such that p^γ divides $\text{ord}(c)$ and p^γ does not divide N. Let $\text{ord}(c) = p^\gamma N_1$, where $\gcd(N_1, p) = 1$, and let $\beta < \gamma$ such $N = p^\beta N_2$ with $\gcd(N_2, p) = 1$. For $a = c^{N_1}$ and $b = d^{p^\beta}$ we have $\text{ord}(a) = p^\gamma$ and $\text{ord}(b) = N_2$. Since these orders are relatively prime, the last lemma implies that $\text{ord}(ab) = p^\gamma N_2 > p^\beta N_2 = N$, contrary to the maximality of N, and so the order of every element of $\mathbb{F}$ does in fact divide N, giving $N = q - 1$. □

In Sections 69–75 of his famous *Disquisitiones Arithmeticae* [70], Carl Friedrich Gauss considered the question of finding primitive elements for a prime modulus. Gauss wrote (in Latin) "Euler admits that it is extremely difficult to pick out these numbers (primitive elements) and that their nature is one of the deepest mysteries of numbers." Finding a primitive element is still considered a hard problem. It is related to the discrete log problem, the basis for the (assumed) security of several cryptosystems.

In 1927 Emil Artin first conjectured that every nonsquare positive integer is a primitive element for infinitely many prime moduli. Some specific examples of this conjecture had already been constructed in the years between Gauss' work and 1927. For instance, 2 was already known to be a primitive element for all primes p of the form $p = 2q + 1$, where q is prime (these primes are now often called "safe primes".) In [80] Christopher Hooley proved that Artin's conjecture would follow from the Generalized Riemann Hypothesis, a result that is thought to be true but is considered to be very difficult to prove.

Although it is hard to find primitive elements, our proof of the Primitive Element Theorem can be combined with trial and error to construct primitive elements. For example, to find a primitive element in the finite field $\mathbb{F} = \mathbb{Z}_{31}$, we could first try $a = 2$. Since the powers of 2 modulo 31 are $2, 4, 8, 16, 1$, then $\operatorname{ord}(2) = 5 < 30$, and 2 is not a primitive element. We notice that 5 does not occur in this sequence and compute the powers of 5 mod 31. These are: $5, 25, 1$, which gives $\operatorname{ord}(5) = 3$ in $\mathbb{F}$. Since $\gcd(\operatorname{ord}(2), \operatorname{ord}(5)) = 1$, from the lemma, $\operatorname{ord}(2*5) = 15$. Also, $\operatorname{ord}(-1) = 2$ implies $\operatorname{ord}(21) = \operatorname{ord}(-10) = 30$, and 21 is a primitive element in $\mathbb{F}$.

We can use the existence of primitive elements to derive the following somewhat surprising result.

Corollary 8.3.9. *If $\mathbb{F}$ is a finite field with q elements, then for any divisor d of $q - 1$ there exists a first–order recurrence in $\mathbb{F}$ whose period is d.*

Proof. Let a be any primitive element in $\mathbb{F}$. For any divisor d of $q-1$, there exists k such that $kd = q - 1$. The period of the recurrence $s_{n+1} = a^k s_n + b$ (for any b) is $\operatorname{ord}(a^k) = d$.   □

8.4   Periodic Second–Order Modular Recurrences

In the last section we proved that when the modulus is prime, the period of all non-fixed points under a first–order modular recurrence $s_{n+1} = as_n + b$ is the order of the coefficient a. For higher order sequences the situation is less clear and seems to be more complicated.

Our analysis for second–order recurrences involves the matrix form of the recurrence. Let R be either a finite field or $\mathbb{Z}_m$ for some $m \geq 2$ and let $\langle s_j \rangle$ be a solution to a second–order homogeneous recurrence in R,

$$s_{n+2} = c_1 s_{n+1} + c_2 s_n, \quad c_2 \neq 0.$$

Although linear algebra cannot be used here unless R is a field, matrices are still a convenient way to represent higher order recurrences. Recall that the **companion matrix** of the recurrence is

(8.8)
$$A = \begin{bmatrix} c_1 & c_2 \\ 1 & 0 \end{bmatrix},$$

and consecutive pairs $S_i = (s_{i+1}, s_i)^T$ are connected by the matrix equation $S_n = AS_{n-1}$, implying

$$S_n = A^n S_0.$$

When c_2 is invertible in R, it can be checked that

$$A^{-1} = \begin{bmatrix} 0 & 1 \\ c_2^{-1} & -c_1 c_2^{-1} \end{bmatrix},$$

and A is an element of the finite set S of invertible 2×2 matrices with entries in R. In Exercise 8.14 you prove that this implies there exists a minimal positive integer k (called the **order of the matrix**) such that $A^k = I$, from which we get $S_0 = S_k = A^k S_0$, proving that k is a multiple of the period of the sequence. Although the order of the matrix can be larger than the period (see Exercise 8.25), we will next obtain conditions on the initial values that ensure that the period *equals* the order of the matrix.

Let T be the map $T(S) = AS$ defined on R^2. Then T is a linear map since it satisfies $T(cS + X) = cT(S) + T(X)$ even when R is not a field. If the first two state vectors, $S_0 = (s_1, s_0)^T$ and $S_1 = (s_2, s_1)^T$, form a spanning set for R^2, then each (linear map) T^n is determined by its effect on the set $\{S_0, S_1\}$. In particular, $A^n = I$ iff both $T^n(S_0) = S_0$ and $T^n(S_1) = S_1$, which implies that the order of the matrix equals the period of the solution. We've proved the following theorem.

Theorem 8.4.1. *Let R be either a finite field or $\mathbb{Z}_m$ for some $m \geq 2$. If $s_{n+2} = c_1 s_{n+1} + c_2 s_n$ is a second–order recurrence in R and c_2 is invertible, then every solution is periodic and the period always divides the order of the companion matrix of the recurrence. Moreover, if $\{S_0, S_1\}$ is a spanning set for R^2, then the period equals the order of the companion matrix.*

Since our proof relied only on the invertibility of the companion matrix, an analogue of this result holds for k^{th} order recurrences. (Refer to Exercise 8.27.)

When R is a field and the eigenvalues of A are distinct, it can be checked that any non-trivial S_0, S_1 are linearly independent, and we have the following theorem.

Theorem 8.4.2. *Let $\mathbb{F}$ be a finite field and let $s_{n+2} = c_1 s_{n+1} + c_2 s_n$ be a second–order recurrence in $\mathbb{F}$ with two distinct eigenvalues $\lambda_1 \neq \lambda_2$ in $\mathbb{F}$. Then every non-zero solution is periodic and its period equals the order of the companion matrix unless it has the form $s_n = \gamma \lambda_i^n$, in which case the period is $\operatorname{ord}(\lambda_i)$.*

8.4.1 Periods of modular Fibonacci sequences

We now specialize to the Fibonacci recurrence, and first answer the questions posed at the beginning of this chapter. As already noted, every Fibonacci orbit is periodic, and the period mod 6 is $\operatorname{lcm}(3,8) = 24$. If $\langle f_j \rangle$ is the Fibonacci sequence, we noted in Chapter 2 that the orbit of any $(a,b) \in \mathbb{Z}^2$ under $F(x,y) = (y, x+y)$ is the sequence $\langle b f_{j+1} + a f_j \rangle$ (where $f_{-1} = 1$). Therefore, the period of any orbit under F modulo 6 must divide the period of the orbit of $(0,1)$, and so 24 is the largest period obtained from the Fibonacci recurrence mod 6.

Our final question was how many *different* orbits are generated by the Fibonacci recurrence mod 6, where two orbits are considered the same when they're translates of each other. First note that $(0,0)$ is the only fixed point. Among the $6^2 - 1 = 35$ non-zero elements of $\mathbb{Z}_6^2$, the usual Fibonacci sequence (which is the orbit of $(0,1)$) generates 24 different pairs of consecutive elements, and this leaves $35 - 24 = 11$ non-zero pairs to be accounted for. In the orbit of $(2,2)$, every element of the orbit is even, and so its period equals the period of $(1,1)$ mod 3, which we've already calculated to be 8. (Because its period is not 24, this orbit is completely disjoint from the orbit of $(0,1)$.) Similarly, since the period of $(1,1)$ mod 2 is 3, the orbit of $(3,3)$ mod 6 is 3. These two periods account for the other 11 elements and give a total of three different non-zero orbits under the Fibonacci recurrence. (Also refer to [163].)

What about the period of the Fibonacci sequence for other moduli? A paper by D.D. Wall [165] was the first systematic approach to this problem, and although many of his results have been generalized, his 1960 paper contains most of what is currently known about modular Fibonacci periods.

Recalling that the period mod $m = p_1^{a_1} \cdots p_r^{a_r}$ is the least common multiple of the periods mod $p_i^{a_i}$, it is enough to consider moduli that are powers of a prime number. Also, if $t = t(p^{j+1})$ is the period of the Fibonacci sequence mod p^{j+1} for any $j \geq 1$, then

$$f_t \equiv 0 \,(\mathrm{mod}\, p^{j+1}) \ \text{ and } f_{t+1} \equiv 1 \,(\mathrm{mod}\, p^{j+1})$$

are congruences that also hold modulo all divisors of p^{j+1}. Therefore, for all $i \leq j$ it is also true that $(f_t, f_{t+1}) \equiv (0,1)\,(\mathrm{mod}\, p^i)$, and we have

$$t(p^{j+1}) \text{ is divisible by } t(p^j)\,.$$

Since the first two state vectors, $S_0 = (1,0)^T, S_1 = (1,1)^T$, form a spanning set for every $\mathbb{Z}_m^2$, the period is always the order of the companion matrix, $F = \begin{bmatrix} 1 & 1 \\ 1 & 0 \end{bmatrix}$. Using the fact that $F^3 = I + 2F$, it can be shown (refer to Exercise 8.26) that for all $n > 0$, $t(2^n) = 3 \cdot 2^{n-1}$. The periods $t(5^n) = 4 \cdot 5^n$ can be found in a similar manner.

What about powers of other primes $p \neq 2, 5$? It has been conjectured that $t(p^j) = p^{j-1}t(p)$ holds for all prime powers, and this was verified by Wall for all $p \leq 10,000$. His paper would be very long indeed if for each prime less than $10,000$ he performed an induction similar to what we have just outlined for $p = 2$. Rather, he proved the surprising fact that if p is a prime for which $t(p^2) \neq t(p)$, then it is true that $t(p^{j+1}) = p^j t(p)$ for all $j \geq 1$. We don't give his argument here but simply comment that he used several combinatorial identities to derive the result. In January 2003 Jonathan Goff, a graduate student in mathematics at Oregon State University, extended Wall's calculations by verifying that $t(p^2) \neq t(p)$ for all primes less than one million and for all primes $p \equiv \pm 1\,(\mathrm{mod}\, 10)$ that are less than twenty million.

Let's look again at the Fibonacci sequence mod 5. Rather than computing all terms until the second occurrence of $0, 1$, we'll consider what happens algebraically. In $\mathbb{Z}_5[x]$ the characteristic polynomial $ch(x) = x^2 - x - 1$ factors as $ch(x) = (x - 3)^2$. Since $\lambda = 3$ is a repeated eigenvalue and $\mathbb{Z}_5$ is a field, [3] there exists a polynomial p with $\deg(p) \leq 1$ such that

$$f_n = p(n)3^n .$$

(Note that this is written as an equation in $\mathbb{Z}_5$, and is actually the same as saying that $f_n \equiv p(n)3^n \pmod 5$.) From $f_0 = 0$ and $f_1 = 1$ we obtain $p(n) = 3^{-1}n$, which gives

$$f_n = n3^{n-1} \text{ for all } n \geq 0 .$$

If $t = t(5)$ is the period, in $\mathbb{Z}_5$ we have

$$t3^{t-1} = f_t = f_0 = 0 \quad \text{and} \quad (t+1) \cdot 3^t = f_{t+1} = f_1 = 1 ,$$

which can be solved to find that t is the least positive integer that simultaneously satisfies $t \equiv 0 \pmod 5$ and $3^t \equiv 1 \pmod 5$. Since $\text{ord}(3) = 4$ in $\mathbb{Z}_5$, the period is $t = \text{lcm}(4, 5) = 20$.

For this argument to work for other primes $p \neq 2, 5$, we'd like to find the eigenvalues of the recurrence modulo p, and in order to do so, we must factor the characteristic polynomial. But $ch(x)$ might be an irreducible quadratic in $\mathbb{Z}_p[x]$. When is $ch(x)$ irreducible? Since a *quadratic* polynomial is irreducible in $\mathbb{Z}_p[x]$ iff it has no roots in $\mathbb{Z}_p$, we'll use the technique of completing the square to determine whether $ch(x)$ has roots. The oddness of p means that 2 is invertible, say $2c = 1$ in $\mathbb{Z}_p$, and

$$ch(x) = x^2 - x - 1 = x^2 - 2cx - 4c^2 = (x - c)^2 - 5c^2,$$

giving $ch(x+c) = x^2 - 5c^2$. Since $ch(x)$ is irreducible exactly when $ch(x+c)$ is irreducible, we obtain

$ch(x)$ is reducible in $\mathbb{Z}_p[x]$ $\iff$ $p \neq 2$ and $x^2 - 5c^2$ has roots in $\mathbb{Z}_p$.

The last restriction is the same as requiring that $5c^2$, and so 5, is a square in $\mathbb{Z}_p$ (which is often called a **quadratic residue**.) For what odd primes does this happen? For this type of question, number theorists invoke the famous **Law of Quadratic Reciprocity**, which was first stated by Legendre and proved by Gauss [70, Section 123] [4]. Using quadratic reciprocity, it's possible to prove that $p \equiv \pm 1 \pmod{10}$ are exactly the odd primes for which 5 is a non-zero square in $\mathbb{Z}_p[x]$. These are therefore the primes p for which the Fibonacci recurrence mod p has eigenvalues in $\mathbb{Z}_p[x]$. What about the other odd primes, $p \equiv \pm 3 \pmod{10}$? For those primes $ch(x)$ is

---

[3] Note that this argument cannot be used for $m = 25$, since $\mathbb{Z}_{25}$ is not a field.

[4] Consult the website http://www.rzuser.uni-heidelberg.de/~hb3/rchrono.html for a chronology of the many proofs of the Law of Quadratic Reciprocity.

irreducible in $\mathbb{Z}_p[x]$. But (refer to Exercise 8.12) $\mathbb{K} = \mathbb{Z}_p[x]/(ch(x))$ is a field with p^2 elements, and $ch(x)$ has a root in $\mathbb{K}$.

Setting $\mathbb{K} = \mathbb{Z}_p[x]/(ch(x))$ for primes $p \equiv \pm 3 \,(\mathrm{mod}\, 10)$, and setting $\mathbb{K} = \mathbb{Z}_p$ for primes $p \equiv \pm 1 \,(\mathrm{mod}\, 10)$, then for each prime $p \neq 2, 5$ we have defined a field $\mathbb{K}$ in which $ch(x)$ has roots in $\mathbb{K}$. We next show that these eigenvalues are different. If $\lambda_1 = \lambda_2$ were to hold, then $(x - \lambda_1)^2 = x^2 - x - 1$, and $\lambda_1^2 = -1$. Since $ch(\lambda_1) = 0$, then $\lambda_1^2 = \lambda_1 + 1$, from which we get $\lambda_1 = -2$. Therefore, $5 = ch(-2) = 0$ in $\mathbb{K}$, contradicting $p \neq 5$. So, for $p \neq 2, 5$, the modular Fibonacci matrix F has distinct eigenvalues and so is diagonalizable. The change of basis matrix is the invertible matrix $B = \begin{bmatrix} \lambda_1 & \lambda_2 \\ 1 & 1 \end{bmatrix}$ with

$$BFB^{-1} = \begin{bmatrix} \lambda_1 & 0 \\ 0 & \lambda_2 \end{bmatrix} \quad \text{and} \quad BF^j B^{-1} = \begin{bmatrix} \lambda_1^j & 0 \\ 0 & \lambda_2^j \end{bmatrix}.$$

This means that $\mathrm{lcm}(\mathrm{ord}(\lambda_1), \mathrm{ord}(\lambda_2))$ is the order of F, which equals $t(p)$ (recall Theorem 8.4.2).

You might rightfully question the helpfulness of relating the period to the order of certain elements because we have already said that calculating the order of an element is a hard problem. (In particular, refer to our discussion of Artin's conjecture on primitive elements.) The answer to this objection is that this connection allows us to prove other results about the period, and the next theorem is an example of this. (A different proof of this result can be found in [165, Theorems 6, 7].)

Theorem 8.4.3. *For prime $p \neq 2, 5$, let t be the period of the Fibonacci sequence modulo p.*
  (a) If $p \equiv \pm 1 \,(\mathrm{mod}\, 10)$, then t divides $p - 1$.
  (b) If $p \equiv \pm 3 \,(\mathrm{mod}\, 10)$, then t divides $2(p + 1)$, and also $(2p + 2)/t$ is odd.

Proof. From $ch(x) = x^2 - x - 1 = (x - \lambda_1)(x - \lambda_2)$ we have $\lambda_1 = -\lambda_2^{-1}$. Since $\mathrm{ord}(\lambda) = \mathrm{ord}(\lambda^{-1})$ and $\mathrm{ord}(-1) = 2$, then

$$\mathrm{lcm}(\mathrm{ord}(\lambda_1), \mathrm{ord}(\lambda_2)) = \mathrm{lcm}(\mathrm{ord}(\lambda_1), 2),$$

and the above discussion gives $t = \mathrm{lcm}(\mathrm{ord}(\lambda_1), 2)$.

First considering the case in which $p \equiv \pm 1 \,(\mathrm{mod}\, 10)$, recall that $\mathbb{K} = \mathbb{Z}_p$, and by Fermat's Little Theorem we know that the order of the non-zero element $\lambda_1 \in \mathbb{K}$ divides $p - 1$, an even number, and so $\mathrm{lcm}(\mathrm{ord}(\lambda_1), 2)$ does divide $p - 1$.

In the other case in which $p \equiv \pm 3 \,(\mathrm{mod}\, 10)$, the field $\mathbb{K}$ contains an element α such that $\alpha^2 = 5$. Since 5 is an element of $\mathbb{Z}_p$, then $\alpha^{2(p-1)} = 5^{p-1} = 1$. This means that $x = \alpha^{p-1}$ is a root of the polynomial $x^2 - 1 = 0$, which has only the two roots $x = \pm 1$ in any field. Since $\alpha \notin \mathbb{Z}_p$, we know that $\alpha^{p-1} \neq 1$ (refer to Theorem 8.3.6), and so $\alpha^{p-1} = -1$ follows. What does this have to do with $\text{ord}(\lambda_1)$? In the "completing the square" argument above we had $ch(x) = (x - c)^2 - 5c^2$, where $2c = 1$. Because $ch(\lambda_1) = 0$, $(\lambda_1 - c)^2 = 5c^2 = \alpha^2 c^2$, implying $\lambda_1 = c(1 \pm \alpha)$, and we can assume the positive sign occurs. Since $\mathbb{K}$ has p^2 elements, then (refer again to Exercise 8.7)

$$(1 + \alpha)^p = 1 + \alpha^p = 1 + \alpha\alpha^{p-1} = 1 - \alpha,$$

which when combined with $c^{p-1} = 1$ gives

$$\lambda_1^{p+1} = c^{p+1}(1 - \alpha)(1 + \alpha) = c^2(1 - \alpha^2) = c^2(1 - 5) = -(2c)^2 = -1.$$

Then $\lambda_1^{2(p+1)} = 1$, and so $\text{ord}(\lambda_1)$ divides $2(p + 1)$. Further, writing $2(p + 1) = 2^k n$ where n is odd, we know that $\text{ord}(\lambda_1)$ divides $2^k n$ and doesn't divide $2^{k-1}n$ (since $\lambda_1^{p+1} \neq 1$). This means that $\text{ord}(\lambda_1)$ is even, and $t = \text{lcm}(\text{ord}(\lambda_1), 2) = \text{ord}(\lambda_1)$ with $(2p + 2)/t$ odd. $\qquad\square$

We calculated $t(p)$ for all primes $p \leq 1000$ and found that the upper bound in the theorem is attained quite often for these primes. Of the seventy-eight primes that are congruent to $\pm 1 \,(\text{mod } 10)$, the period equals the upper bound for forty of them. The situation for the "irreducible" primes is even stronger. Among all primes less than 1000 there are eighty-eight primes congruent to $\pm 3 \,(\text{mod } 10)$ and for all but fifteen of these the period equals the upper bound. Table 8.2 lists the irreducible primes with smaller periods, and the second column gives the quotient $(2p + 2)/t$.

TABLE 8.2. The exceptional primes congruent to $\pm 3 \,(\text{mod } 10)$

p	$(2p + 2)/t$
$47, 107, 113, 233, 353, 563, 677, 743, 977$	3
$307, 797$	7
$263, 557, 953$	9
967	11

We close this discussion by computing the Fibonacci period $t(2^3\, 3^2\, 17\, 47^2)$. Corollary 8.1.4 implies that the period is $\text{lcm}(t(8), t(9), t(17), t(47^2))$, where by Wall's calculations we know that

$$t(8) = 2^2\, t(2), \quad t(9) = 3\, t(3), \quad \text{and} \quad t(47^2) = 47\, t(47).$$

From the earlier calculations given in Section 8.1, we have $t(2) = 3$ and $t(3) = 8$. The remaining two primes $p = 17, 47$ are congruent to $-3 \,(\text{mod } 10)$, and from Table 8.2 we see that 17 is not exceptional, which means that $t(17) = 2(18) = 36$ and that $t(47) = 2(48)/3 = 32$. Combining these facts gives the period

$$\text{lcm}(12, 24, 36, (32)(47)) = (32)(9)(47) = 13536.$$

8.5   Applications

8.5.1   Application 1: Pseudorandom number generation

Random numbers are often needed in scientific computation, especially simulations and probabilistic algorithms that have some stochastic components. They are also used in cryptology and even computer games. The term "random sequence" means a sequence that passes certain statistical tests. For instance, one simple randomness requirement for sequences from $\mathbb{Z}_{10}$ might be the property that each digit should occur about the same number of times, so that in suitably long subsequences each decimal digit occurs close to one-tenth of the time. In contrast to sequences generated mechanically or by a physical phenomenon, sequences generated by a mathematical iterative process (such as a recurrence) can be *proved* to have good statistical properties. Another advantage of such deterministic procedures is that they will dependably generate the same sequence when the initial conditions are unchanged. This allows for the exact reproduction of data for numerical experiments. Sequences that are generated deterministically and have good statistical properties are called sequences of **pseudorandom numbers (PRNs)**. Because a PRN generator cannot satisfy all possible statistical properties (which is another reason for calling them *pseudo*random), the practitioner should know which statistical properties are required for the application before choosing the generator. Chapter 3 of Knuth [88] and Chapter 7 in Niederreiter [119] can be consulted for information on the common statistical tests used to test for pseudorandomness. In this section we concentrate on two *structural* properties, the period length and the lattice structure.

A linear modular recurrence is the simplest example of a deterministic process that can be used for generating PRNs. These are frequently called **linear congruential generators**. From the perspective of PRN generation, the Primitive Element Theorem guarantees that every prime p has at least one linear PRN generator mod p whose period is $p - 1$, the longest possible period. Knuth [88, Chapter 3] has an extensive discussion of the periods for general moduli. For many years after their introduction by D.H. Lehmer [94] in 1949, the linear congruential generator was the preferred method of PRN generation. However, an article by G. Marsaglia [106] raised some serious questions about their use by proving that every linear PRN generator has an inherent lattice structure. As he said in the paper, "for the past 20 years such regularity might have produced bad, but unrecognized, results in Monte Carlo studies ... ." In a later article [107] he developed the **Lattice Test for PRNs**, which can be stated as follows. For any deterministic sequence $\langle s_n \rangle$ (with period $N \geq 2$) generated in a finite field $\mathbb{F}$ and each $d \geq 1$, define the d-dimensional points $U_1, \ldots, U_{N-1}$ by

$$U_i = (s_i - s_0, s_{i+1} - s_1, \ldots, s_{i-1+d} - s_{d-1}) \in \mathbb{F}^d,$$

and let L_d be the subspace of $\mathbb{F}^d$ spanned by these points. For instance,

$$L_1 = \mathrm{Span}\{U_1\} = \mathrm{Span}\{s_1 - s_0\} = \mathbb{F}$$

(since $s_1 \neq s_0$) and

$$L_2 = \mathrm{Span}\{U_1, U_2\} = \mathrm{Span}\{(s_1 - s_0, s_2 - s_1), (s_2 - s_0, s_3 - s_1)\},$$

which may have dimension one or two over $\mathbb{F}$. The **lattice dimension** for the sequence $\langle s_n \rangle$ is defined to be the largest integer $D \geq 1$ for which $L_D = \mathbb{F}^D$. When the lattice dimension is small, the points display a high degree of regularity, and so the generation is predictable and not random. A linear generator fares poorly under this test, since an inductive argument shows that $s_n = as_{n-1} + b$ satisfies

$$s_{n+i} - s_i = a^i(s_n - s_0) \quad \text{for all } i,$$

and the points U_i therefore have the form $U_i = (s_i - s_0)(1, a, a^2, \ldots, a^{d-1})$. This means that its lattice dimension is always 1.

The remainder of this discussion on PRNs is devoted to an analysis of a commonly used *nonlinear* generator whose lattice dimension mod p has been proved to be at least $(p+1)/2$, much better than linear generators. The generator is called an **inversive generator** on the finite field $\mathbb{F}$ and has the form $f(x) = ax^{-1} + b$ (where we set $0^{-1} = 0$, but otherwise the arithmetic is performed in $\mathbb{F}$.) Because f is a one-to-one function on $\mathbb{F}$, any sequence generated by the first–order recurrence $x_{n+1} = f(x_n)$ is periodic, and the number of elements in $\mathbb{F}$ is an upper bound on the period. Below we show that any finite field has inversive generators with this maximal period.

Although the Euclidean Algorithm can be used to compute inverses quickly, finding inverses does take longer than performing "polynomial" operations, and so in this respect inversive generators are slower. On the other hand, research has shown that inversive generators do usually have better statistical properties than polynomial generators. (Refer to [119, Chapter 8].) For instance, extensive computations by Poul Petersen [127] show that for all primes $p \leq 10^5$ the lattice dimension of inversive generators mod p with maximal period is either $p - 6$, $p - 4$, or $p - 2$. (Table 8.3 gives more details. Also, in [64] it's shown that the lattice dimension of a maximal period inversive generator is always odd.)

As said above, we will show that every finite field has inversive generators of maximal period. The polynomial $ch_I(x) = x^2 - bx - a$ associated with the generator $f(x) = ax^{-1} + b$ is useful in this analysis. For example, the generator $f(x) = 3x^{-1} + 2$ in $\mathbb{F} = \mathbb{Z}_{11}$ has $ch_I(x) = x^2 - 2x - 3$, which factors as $ch_I(x) = (x - 3)(x - 10)$ in $\mathbb{F}[x]$. Its roots $x = 3, 10$ are the only fixed points of f, and every other element of $\mathbb{F}$ is in the orbit of $x = 2$, which is $2, 9, 6, 8, 1, 5, 7, 4, 0$. This example illustrates a general property of

TABLE 8.3. Lattice dimensions for maximal period inversive generators mod p for $5 \le p < 10^5$

Number of primes less than 10^5	9590
Number of maximal period inversive generators	$5,579,945,320,208$
Number whose dimension is $p - 6$	1
Number whose dimension is $p - 4$	1829
Number whose dimension is $p - 2$	$5,579,945,318,378$

an inversive generator: its fixed points are precisely the roots of $ch_I(x)$ in the field $\mathbb{F}$. To see this, we first note the fact that a is non-zero implies that 0 is not a fixed point of f and is not a root of $ch_I(x)$. Therefore, any root c of $ch_I(x)$ is invertible, and multiplying by c^{-1} yields

$$ch_I(c) = c^2 - bc - a = 0 \quad \Longleftrightarrow \quad c^2 = bc + a \quad \Longleftrightarrow \quad c = b + ac^{-1} = f(c).$$

Because our goal is to find inversive generators with maximal period, we will consider generators that have no fixed points, which is equivalent to requiring that $ch_I(x)$ be irreducible in $\mathbb{F}[x]$. As we used earlier with the Fibonacci recurrence, $\mathbb{K} = \mathbb{F}[x]/(ch_I(x))$ is then a field in which $ch_I(x)$ factors into linear factors, $ch_I(x) = (x - \alpha)(x - \beta)$ in $\mathbb{K}[x]$. We ensure that the roots are distinct by requiring that the discriminant $b^2 + 4a$ be non-zero. Although the recurrence is not linear, there is a relationship between the orbit of 0 and the roots of $ch_I(x)$; namely, we will show that for every $j \ge 0$ with $f^{(j-1)}(0) \ne 0$,

$$(8.9) \qquad\qquad f^{(j)}(0) = \frac{\alpha^{j+1} - \beta^{j+1}}{\alpha^j - \beta^j}.$$

From $ch_I(x) = (x - \alpha)(x - \beta) = x^2 - bx - a$, we have $a = -\alpha\beta$ and $b = \alpha + \beta$, and this gives

$$f(0) = b = \alpha + \beta = \frac{\alpha^2 - \beta^2}{\alpha - \beta},$$

which is (8.9) for $j = 1$. Further, if for $j \ge 1$ we have

$$f^{(j)}(0) = \frac{\alpha^{j+1} - \beta^{j+1}}{\alpha^j - \beta^j} \ne 0,$$

then

$$f^{(j+1)}(0) = -\alpha\beta \frac{\alpha^j - \beta^j}{\alpha^{j+1} - \beta^{j+1}} + \alpha + \beta = \frac{\alpha^{j+2} - \beta^{j+2}}{\alpha^{j+1} - \beta^{j+1}},$$

and the $(j + 1)^{\text{st}}$ element of the orbit of 0 is as given by (8.9).

Theorem 8.5.1. *Let $\mathbb{F}$ be a finite field with q elements and let f be an inversive generator on $\mathbb{F}$ such that $ch_I(x)$ is irreducible in $\mathbb{F}[x]$ and has non-zero discriminant. If t is the period of $x = 0$ under f, then the order of β^{q-1} in $\mathbb{K} = \mathbb{F}[x]/(ch_I(x))$ is $t + 1$.*

Proof. Since f is a one-to-one function, every orbit of f is periodic, and the period t is the least positive integer such that $f^{(t)}(0) = 0$. As above, the irreducibility of $ch_I(x)$ implies that $\mathbb{K} = \mathbb{F}[x]/(ch_I(x))$ is a field that contains elements α, β such that (8.9) holds. Therefore,

$$f^{(t)}(0) = 0 \iff \alpha^{t+1} - \beta^{t+1} = 0 \iff (\alpha\beta^{-1})^{t+1} = 1.$$

Since t is the minimal positive integer with this property, $t + 1$ must be the order of $\alpha\beta^{-1}$ in $\mathbb{K}$. We complete the proof by showing that $\alpha = \beta^q$. From Exercise 8.7 we know that $ch_I(\beta^q) = (ch_I(\beta))^q$, which means that β^q is also a root of $ch_I(x)$. The fact that $\beta \notin \mathbb{F}$ implies $\beta^q \neq \beta$ (refer to Exercise 8.16), and β^q is forced to equal α, the other root of $ch_I(x)$. $\qquad\square$

There is a similarity between this last result and what happens with first–order linear recurrences, since the period of 0 is related to the order of an element in an associated algebraic structure. Because calculating the order of an element can be quite lengthy, this result is more useful for theoretical rather than computational purposes. For instance, it's used in [64] to show that the lattice dimension of an inversive generator is always odd.

For the sake of concreteness, let us consider the orbit of 0 in $\mathbb{F} = \mathbb{Z}_{11}$ under $f(x) = 7x^{-1} + 10$. Then $ch_I(x) = x^2 + x + 4$, which can be checked to be irreducible in $\mathbb{F}[x]$ with non-zero discriminant. We won't explicitly determine the field $\mathbb{K} = \mathbb{F}[x]/(ch_I(x))$, but rather recall that it's a finite field with $(11)^2 = 121$ elements in which $ch_I(x)$ factors into two linear factors. We want to calculate or otherwise determine the order of $\gamma = \beta^{q-1} = \beta^{10}$, where $\beta \in \mathbb{K}$ is a root of $x^2 + x + 4$ and its order divides $q^2 - 1 = 120$. This means that the order of $\gamma = \beta^{10}$ divides 12. Calculating $\gamma = \beta^{10}$ using fast exponentiation, from $\beta^2 = -\beta - 4$ we have

$$\beta^4 = (\beta + 4)^2 = 7\beta + 1 \,; \quad \beta^8 = (7\beta + 1)^2 = -2\beta + 3 \,,$$

and

$$\gamma = \beta^8 \cdot \beta^2 = (-2\beta + 3)(-\beta - 4) = 3\beta + 2 \,.$$

Since

$$\gamma^2 = (3\beta + 2)^2 = 3\beta + 1 \quad \text{and} \quad \gamma^3 = (3\beta + 1)(3\beta + 2) = -1 \,,$$

the order of $\gamma = \beta^{q-1}$ is 6, which means that the period of 0 is five. This can also be verified by direct calculation, tracing the orbit of 0 as $0, 10, 3, 5, 7$.

Theorem 8.5.2. *Every finite field has inversive generators with one orbit. This orbit contains every element of the field and so has the maximal period among all sequences generated by an inversive generator.*

Proof. Let $\mathbb{F}$ be a finite field with q elements, and let g be any irreducible quadratic polynomial in $\mathbb{F}[x]$. (In Exercise 8.34 you show that such polynomials exist for all q.) From the Primitive Element Theorem, there exists

$\gamma \in \mathbb{F}[x]/(g)$ whose order is q^2-1. Therefore, $\gamma \notin \mathbb{F}$, and the quadratic polynomial $G(x) = (x-\gamma)(x-\gamma^q)$ is irreducible in $\mathbb{F}[x]$. For $G(x) = x^2 - Bx - A$, we consider the associated inversive generator $f(x) = Ax^{-1} + B$, which has $ch_I(x) = G(x)$. Since $\text{ord}(\gamma) = q^2 - 1$, then $\text{ord}(\gamma^{q-1}) = q + 1$, and the period of $x = 0$ under f does equal q.   □

8.5.2   Application 2: Integer factorization

The security of the RSA cryptosystem [137] relies on the difficulty of factoring integers that are a product of two large primes. Because of this, there was an increase of interest in factorization algorithms after the RSA cryptosystem was introduced in the late 1970s.

Factoring a natural number m can be viewed as a search problem, searching the set $\mathcal{S} = \{2, \ldots, \sqrt{m}\}$ for divisors of m, but for large m searching by trial division is not practical.[5] Instead of searching all of $\mathcal{S}$, most modern factoring techniques generate a subset $\mathcal{T}$ whose elements are likely to have a factor in common with m, and search for $t \in \mathcal{T}$ such that $d = \gcd(t, m) \neq 1, m$. Once such $t \in \mathcal{T}$ has been found, the factorization $m = d \cdot \frac{m}{d}$ is obtained. (A complete factorization of m is then found by continuing to factor each of $d, \frac{m}{d}$.) The key insight was that searching for elements that have a factor in common with m can be used to get a divisor of m, and such a search is more reasonable than searching for exact divisors. For instance, in the extreme (RSA) case in which m is a product of two primes p, q, each of size $O(\sqrt{m})$, the integers

$$p, 2p, \ldots, (q-1)p \quad \text{and} \quad q, 2q, \ldots, (p-1)q$$

are all divisible by p or q, and we are more likely to locate one of these $p + q - 2 = O(\sqrt{m})$ integers than one of the two integers p, q.

A Certificate of Compositeness. Most modern factorization methods are *probabilistic*, in the sense that the method is likely to return a factorization for composite m, but no specific run is guaranteed to produce a factorization of m. Because of this, these probabilistic methods are inconclusive when applied to a prime, and the methods are used only after m has obtained a "certificate of compositeness." Since the most obvious way of showing that a number is composite is to demonstrate a factorization, at first this may seem like a strong requirement. But there are some relatively quick tests for compositeness. We mention a few that are based on material already developed in this chapter, and refer the interested reader to [30, Chapter 3] and `cr.yp.to/primetests.html` for more information.

---

[5]The authors of a recent text [30, p. 111] have calculated that "in one day of current workstation time, perhaps (the primality of) a 19-digit number could be resolved" using trial division. This translates to the factorization of a (very small) 38-digit RSA-composite requiring a full day of computation by trial division.

In the summer of 2002, Agrawal, Kayal, and Saxena found a deterministic primality test that has polynomial time, specifically $O(\log^{12+\epsilon}(m))$. [6] The number theory community was amazed and delighted by the simple elegance of the test and its proof. As of this writing, the test is not yet of practical use.

Let m be the number to be tested for compositeness. For a fixed natural number a, $1 < a < m$, we use the Euclidean Algorithm to calculate $d = \gcd(a, m)$. If $d \neq 1$, we know for certain that m is composite, and have obtained its certificate of compositeness as well as the divisor d. We may therefore assume that $\gcd(a, m) = 1$. The contrapositive of Fermat's Little Theorem yields a **Fermat Test** for compositeness; namely, if there exists a, $1 < a < m$, such that $a^{m-1} \not\equiv 1 \,(\mathrm{mod}\, m)$, then m is not prime. Notice that since, for example, the composite number $m = 21$ satisfies $a^{m-1} \equiv 1 \,(\mathrm{mod}\, m)$ for $a = 8$, the condition is only sufficient and not necessary for compositeness.

A composite integer m that satisfies $a^{m-1} \equiv 1 \,(\mathrm{mod}\, m)$ for $1 < a < m$ is called a **pseudoprime to the base** a, where the term "pseudoprime" is used because for the base a the composite number m behaves as if it were prime. In 1950 Paul Erdős [59] proved that pseudoprimes are relatively rare when compared to primes, which means that performing a battery of Fermat Tests might be a useful strategy for verifying compositeness. However, in 1994 W.R. Alford, Andrew Granville, and Carl Pomerance [3] proved the existence of infinitely many composite m, that are pseudoprimes to every base which is relatively prime to m. These numbers are called **Carmichael numbers**, named in honor of R.D. Carmichael's 1912 work [24].

All is not lost, because we can invoke our theory and obtain a variant, called the **Strong Fermat Test**, that doesn't have this problem. The idea behind this is the following. If m were prime, then by Fermat's Little Theorem, $b = a^{\frac{m-1}{2}}$ would be a solution to $x^2 - 1 = 0$, a quadratic equation that can have only the two solutions $x = \pm 1$ in the field $\mathbb{Z}_m$. Therefore, if we can find a non-zero base a such that $a^{\frac{m-1}{2}}$ is something other than $\pm 1 \,(\mathrm{mod}\, m)$, we are assured that m is not prime. Because the exponent is slightly smaller, this test is a bit easier to perform than a Fermat Test, but a more important property is that there are no "strong Carmichael" numbers. In fact, in 1980 Monier [114] and Rabin [132] independently proved that every odd composite $m > 9$ is a **strong pseudoprime** for at most one-quarter of the bases mod m. (If the widely believed but still-unproved Generalized Riemann Hypothesis is true, we'd actually be guaranteed that every odd composite m passes a Strong Fermat Test for at least one positive base $a < 2 \log^2(m)$.) Therefore, m must be composite if there is a non-zero base a for which $a^{\frac{m-1}{2}} \not\equiv \pm 1 \,(\mathrm{mod}\, m)$. For example, the Strong Fermat

---

[6] www.cse.ittk.ac.in/primality.pdf

Test with $a = 2$ proves that $m = 18923$ is composite, since $\frac{m-1}{2} = 9461$ and $2^{9461} = 8144$ in $\mathbb{Z}_m$.

The Pollard Rho Method. We will now discuss one of the earliest modern methods for factoring an integer m. The method involves a clever use of orbits of a polynomial modulo m, which we know must be eventually periodic and so can be drawn in the shape of the Greek letter ρ, where the tail indicates the pre-period of the orbit. (Refer to Figure 8.1.). The method is now usually called the **Pollard Rho Method** although in his original article [128] Pollard referred to the algorithm as the Monte Carlo Method because it is probabilistic and relies on a polynomial that has "sufficiently random" orbits. In practice, a quadratic polynomial is usually used unless some special knowledge of the divisors of m indicates that a polynomial of higher degree would work better.

The method relies on the iterates of $f(x)$ being sufficiently random that a generic orbit of $f(x)$ modulo m has the property that there is a proper divisor d of m such that the orbit gets into a cycle modulo d well before it begins to repeat modulo m. Such an orbit yields a nontrivial divisor of m, since $a \equiv b \pmod{d}$ implies $g = \gcd(a-b, m)$ is a multiple of d, while $a \not\equiv b \pmod{d}$ ensures that g is a *proper* divisor of m. So, a successful orbit is sufficiently random that it becomes periodic modulo some divisor d before it becomes periodic modulo m. In [74] Richard Guy stated that for prime p the orbits of $f(x) = x^2 + 1$ modulo p seem to cycle quite quickly, and he conjectured that the number of iterates needed to cycle is $O(\sqrt{p}\ln(p))$.

Writing the conditions for successful orbits in terms of iteration of the function $f(x)$, we want the sequence of iterates to contain some $s \in \mathbb{Z}_m$ such that there are integers n, k (say $n < k$) for which

$$f^{(n)}(s) \equiv f^{(k)}(s) \pmod{d}$$

but

$$f^{(i)}(s) \not\equiv f^{(j)}(s) \pmod{m} \text{ for all } 0 \le i, j \le k.$$

Then for $A = f^{(n)}(s) - f^{(k)}(s) \pmod{m}$, $g = \gcd(A, m)$ is a proper divisor of m. Popular implementations of the method recognize that it's not necessary to find the *smallest* values of n and k for which $f^{(n)}(s) \equiv f^{(k)}(s) \pmod{d}$. For example, Floyd's cycle-detecting method uses the idea of subtracting the "fast" sequence $\langle f^{(2n)}(s)\rangle$ from the "slow" sequence $\langle f^{(n)}(s)\rangle$ and then using the sequence

$$\gcd(f^{(n)}(s) - f^{(2n)}(s), m).$$

(Refer to the work of R. Brent and J. Pollard [13] for other refinements.) Here's an algorithm:

PROCEDURE   POLLARD$(m,f(x))$
Randomly generate s.
$t := f(f(s))$; $d := \gcd(s - t, m)$
WHILE $d = 1$ DO
    $s := f(s)$
    $t := f(f(t))$
    $d := \gcd(s - t, m)$
ENDWHILE
RETURN(d)

If POLLARD$(m, f(x)) = m$ is returned, the procedure can be repeated using another choice of either the polynomial f or the seed s.

We have already commented that $m = 18923$ can be shown to be composite by the Strong Fermat Test with $a = 2$. Let's now use the Pollard Rho Method with $f(x) = x^2 + 1$ to factor m. For instance, the first few elements of the orbit of 2 under $f(x)$ mod m are

$$2, 5, 26, 677, 4178, 8679, 11502, 5312, 3152, 530, 15979, 403, 11026, 11325 \,.$$

Applying the Pollard Rho Method gives $d_n = 1$ for all $n \leq 8$ and then the divisor $d_9 = 127$. From this example you see that the gcd sequence can begin with a rather long string of ones. Because of this, implementations of the method often take giant steps through the gcd sequence: Calculating several $A_i = f^{(n_i)}(s) - f^{(2n_i)}(s)$, finding their product A mod m, and then computing $g = \gcd(A, m)$. Since $A < m$, g is a proper divisor provided at least one A_i has a nontrivial common divisor with m.

It is useful to modify the condition in the WHILE statement so that the do-loop ends when it seems likely that the run will not be successful. This translates to having a good estimate of when the first duplication is likely to occur for a given modulus; that is, we want an estimate for the length of the letter ρ in Figure 8.1. In mathematical terms, the probability that there is a repeated element among k elements chosen from a set with n elements is $1 - P(k, n)$, where

$$P(k,n) = \frac{n(n-1)\cdots(n-(k-1))}{n^k} = \left(1 - \frac{1}{n}\right)\cdots\left(1 - \frac{k-1}{n}\right)$$

is the probability that the k choices are distinct. Estimating this probability is often referred to as The Birthday Problem, because it can be used to guess the likelihood that at least two people in a crowd of M people have the same birthday. (Refer to Exercise 8.38.) In Exercise 8.37 you find fairly tight bounds on this probability, which yields the following heuristic for terminating the loop. Remember that we don't want a duplication mod

m, and from Exercise 8.37(c) we see that a duplication mod m is unlikely before $0.8\sqrt{m}$ iterations. On the other hand, since m is composite, it has a divisor d within the interval $(1, \sqrt{m})$, and so from Exercise 8.37(b) we expect a duplication mod d within $1.39\sqrt{d}$ steps, that is, within $1.39\sqrt[4]{m}$ steps, which for large m is quicker than $0.8\sqrt{m}$ steps. Because of this, it is reasonable to stop the run if it doesn't yield a divisor within $1.39\sqrt[4]{m}$ iterations.

8.6 Exercises

Ex 8.1. For any positive integer m, show that there are infinitely many Fibonacci numbers that are divisible by m.
Hint: Consider the Fibonacci sequence mod m.

Ex 8.2. Find all periodic and eventually periodic orbits of $f(x) = -x^2 + 2x + 1$.

Ex 8.3.   (a) Show that $\gcd(19, 157) = 1$. Find an integer solution to $19x + 157y = 1$ and show that there exist no positive integer solutions to $19x + 157y = 1$.
  (b) Find the multiplicative inverse of $214 \pmod{12323}$.
  (c) Show that there are integer points (x, y) on the line $179x + 2351y = 1$. Find the integer point on the line that is closest to the origin.

Ex 8.4. For the finite field $\mathbb{F}$, define $n \cdot 1$ to be the sum of n copies of the multiplicative identity 1 in $\mathbb{F}$.
  (a) Show that there exists a least positive integer n_0 such that $n_0 \cdot 1 = 0$, the additive identity in $\mathbb{F}$.
  (b) Show that n_0 must be prime. This is called the **field characteristic** of $\mathbb{F}$.

Ex 8.5. This problem contains a justification of Montgomery multiplication. (Refer to the procedure in Section 8.1.2.)
  (a) Write the recursions for the sequences $\langle Q_i \rangle$ and $\langle R_i \rangle$.
  (b) Use congruence modulo r to show that each R_i is an integer.
  (c) Use induction to show that for each $0 \le i \le k$,

$$r^{i+1} R_i \equiv a_0 + a_1 r + \cdots + a_i r^i \pmod{m}.$$

  (d) Complete the justification by proving that $r^{k+1} R_k < 2m$.
  (e) What is the output from $\text{MULT}(r^{k+1}a, r^{k+1}b)$?

Ex 8.6. Let $\mathbb{F}$ be a finite field of characteristic p. Show that $\mathbb{F}$ can be regarded as a finite-dimensional vector space over $\mathbb{Z}_p$, and therefore $\mathbb{F}$ has p^n elements for some $n \ge 1$.

Ex 8.7. Let $\mathbb{F}$ be a finite field with $q = p^n$ elements. Show that for any $f \in \mathbb{F}[x]$, $(f(x))^p = f(x^p)$ and that $(f(x))^q = f(x^q)$ follows by induction. Hint: Use the Binomial Theorem.

Ex 8.8. Let $\mathbb{F}$ be any finite field and n be a positive integer. Let V be the set of all polynomials in $\mathbb{F}[x]$ of degree less than n.
  (a) Show that V is an abelian group under the operation of polynomial addition in $\mathbb{F}[x]$.
  (b) Show that V is a vector space of dimension n over $\mathbb{F}$. Therefore, the number of elements in V equals q^n, where q is the number of elements in $\mathbb{F}$.

Ex 8.9. Let $g(x) \in \mathbb{F}[x]$ be a fixed polynomial, with $n = \deg(g)$. Let V be the set of all polynomials in $\mathbb{F}[x]$ whose degree is less than n, which was shown to be a vector space in the last exercise. Impose still more structure on V by defining an additional operation $\star$ as follows. For any $a(x), b(x) \in V$ we use the Division Algorithm for polynomials to obtain $q(x), r(x) \in \mathbb{F}[x]$ such that

$$a(x)b(x) = q(x)g(x) + r(x) \text{ where } \deg(r) < n,$$

and define the operation $\star$ on V by $a(x) \star b(x) = r(x) \in V$. Show that $\star$ is an associative commutative operation on V that is distributive over the usual addition defined as in Exercise 8.8. Is V always a field under the operations of $+$ and $\star$? In what follows we use the notation $\mathbb{F}[x]/(g(x))$ for V.

Ex 8.10. Let $\mathbb{F} = \mathbb{Z}_7$ and $g(x) = (x^2 + 1)(x^3 + x + 1)$. Show that neither $x^2 + 1$ nor $x^3 + x + 1$ has a multiplicative inverse in $F[x]/(g(x))$.

Ex 8.11. Let $\mathbb{F}$ be any finite field and let $g(x)$ be a **reducible polynomial** in $\mathbb{F}[x]$; that is, there exist polynomials $a(x), b(x) \in \mathbb{F}[x]$ such that $g(x) = a(x)b(x)$ with $1 \leq \deg(a), \deg(b) < \deg(g)$. Show that neither $a(x)$ nor $b(x)$ has an inverse under $\star$ in $\mathbb{F}[x]/(g(x))$.

Ex 8.12. Let $\mathbb{F}$ be any finite field. Let $g(x)$ be a polynomial that is irreducible in $\mathbb{F}[x]$. Verify that $L = \mathbb{F}[x]/(g(x))$ is a finite field under the operations of addition and $\star$.

Ex 8.13. Verify that $x^2 + 1$ is irreducible in $\mathbb{Z}_3[x]$ and use it to construct a field with nine elements.

Ex 8.14. In Section 8.2 we defined an *abelian* group. When the requirement of commutativity is removed, the structure is called a **group**. Let R be either a finite field or $\mathbb{Z}_m$ for some $m \geq 2$.
  (a) Show that the set S of all 2×2 invertible matrices with entries in R is a finite group under multiplication.
  (b) Show that for any $A \in S$ there exists a positive integer k such that $A^k = I$, which is called the **order** of A in the group S. (For this, you

might mimic the argument we used to prove that every invertible element of $\mathbb{Z}_m$ has an order.)

Ex 8.15. Find the orbits of $x = 3$ and $x = 4$ under $S(x) = 12x + 4$ on $R = \mathbb{Z}_{21}$. Formulate a conjecture about the periods of all orbits under this map. Check your conjecture by computing the orbits.

Ex 8.16. Suppose $\mathbb{F}, \mathbb{K}$ are finite fields with $\mathbb{F} \subseteq \mathbb{K}$, and let q be the number of elements in $\mathbb{F}$. For $\alpha \in \mathbb{K}$, show that $\alpha \in \mathbb{F}$ iff $\alpha^q = \alpha$. This is a generalization of Theorem 8.3.6.

Ex 8.17. Let S_n denote the set of all permutations of n elements, where a **permutation** means an ordering (or bijection) of the integers $1, 2, \ldots, n$. Show that S_n is a group under composition of functions. What is the identity element in this group?

Ex 8.18. Now consider the action of shuffling a deck of n cards. Every shuffle of n cards can be viewed as an element of S_n, and so every shuffle has an order in the group S_n. What is the practical meaning of the order of a shuffle?

Ex 8.19. For this problem consider S_{12}, the group of shuffles of a deck with twelve cards. We'll call a shuffle **perfect** if the shuffle begins by splitting the deck into two equal piles and then alternates between the two piles with the card on the bottom of the original second pile becoming the bottom card. (This is also called a **riffle shuffle** or **riffling**.)For instance, an example of a perfect shuffle of six cards changes $1, 2, 3, 4, 5, 6$ into $4, 1, 5, 2, 6, 3$.
  (a) Show that after i perfect shuffles, the original first card is in the 2^i position mod 13. After how many shuffles is the first card returned to its original place?
  (b) What is the order of a perfect shuffle in S_{12}?

Ex 8.20. What is the order of a perfect shuffle in S_{52}?

Ex 8.21. (This problem is based on the work of Joseph Keller in [83].) Consider a riffle shuffle of a deck with k cards. By this we mean cutting the deck once at random and then riffling together the two parts formed by the cut. For this we assume that cutting satisfies a uniform distribution, that is, the probability of any cut is $1/(k-1)$. Let p_n be the probability that the original bottom card is on the bottom of the deck after n riffle shuffles.
  (a) Show that $p_0 = 1$ and for all $n \geq 0$,
$$p_{n+1} = \frac{1}{2}\left(p_n + (1 - p_n)\frac{1}{k-1}\right).$$
  (b) Use the theory of linear recurrences to show that
$$p_n = \frac{1}{k} + \frac{k-1}{k}\left(\frac{k-2}{2(k-1)}\right)^n.$$

Ex 8.22. For this exercise use $\mathbb{F} = \mathbb{Z}_p$ where p equals the prime 5009. Without calculating any orbits, find the period of $x = 0$ under each of $f(x) = 3x + 1$; $f(x) = 4x + 7$.

Ex 8.23. Use induction to show that the number of solutions to any polynomial equation with coefficients in a field is bounded by the degree of the polynomial.

Ex 8.24. For this exercise, use $\mathbb{F} = \mathbb{Z}_p$, where p equals the prime 1361. Check that 3 is a primitive element in $\mathbb{F}$. Find a first–order linear recurrence mod p whose period is 85.

Ex 8.25. Let A be the companion matrix for the Fibonacci recurrence mod 6.
(a) What is the order of the matrix A?
(b) Let $\langle s_n \rangle$ be the sequence that satisfies the Fibonacci recurrence mod 6 and has initial state vector $(3, 3)$. Find the period of this sequence by considering the sequence mod each of the primes 2 and 3.

Ex 8.26.   (a) Use induction to show that the Fibonacci sequence $\langle f_j \rangle$ satisfies

$$f_{2n} = f_n(2f_{n-1} + f_n) \text{ and } f_{2n+1} = f_n^2 + f_{n+1}^2.$$

Hint: Verify the two identities in tandem.
(b) Use part (a) to show that the Fibonacci period mod 2^{n+1} is $3 \cdot 2^n$.
(c) For this problem let F be the companion matrix for the Fibonacci recurrence. Show that $F^3 = I + 2F$ and use this to find the period of the Fibonacci sequence mod 2^n.

Ex 8.27. Let R be either $\mathbb{Z}_m$ or a finite field. Let $s_{n+k} = c_1 s_{n+k-1} + \cdots + c_k s_n$ be a linear recurrence in R with c_k an invertible element of R.
(a) Show that the companion matrix is invertible.
(b) If the first k state vectors form a spanning set for R^k, show that the period of any non-zero solution to the recurrence equals the order of the matrix.

Ex 8.28. Let $\mathbb{F}$ be a finite field and $\langle s_n \rangle$ a non-zero sequence that satisfies a homogeneous second–order recurrence in $\mathbb{F}$ with $s_0 \neq 0$.
(a) Show that the period of $\langle s_n \rangle$ is less than the order of the companion matrix of the recurrence iff $s_1 s_0^{-1}$ is a root of the characteristic polynomial of the recurrence.
(b) If $\mathbb{F} = \mathbb{Z}_p$ for some prime $p \equiv \pm 3 \pmod{10}$, show that the period of any non-zero solution to the Fibonacci recurrence is the order of the companion matrix.

Ex 8.29. For any $m \geq 2$, let $X = \mathbb{Z}_m^2$ and $f(x_1, x_2) = (x_2, x_1 + x_2 \bmod m)$.

(a) If t is the period of the usual Fibonacci sequence $\langle f_j \rangle$, show that

$$f_{t-j} = (-1)^j f_j \text{ for all } 0 \le j \le t.$$

(b) If the period of $(a, b) \in X$ under f is odd, show that $m = 2$ must hold.

(c) Show that $(0, 0)$ is the only fixed point of f.

(d) Show that no orbit of f has period equal to 2.

Ex 8.30. (a) What is the largest period for a sequence that satisfies the Fibonacci recurrence mod 5?

(b) How many *different* sequences satisfy the Fibonacci recurrence mod 5?

(c) Show that there is only one non-zero Fibonacci orbit mod 3.

(d) Without calculating any sequences, show that every Fibonacci period mod 7 must divide 16.

(e) Find a Fibonacci period mod 7 that equals 16.

Ex 8.31. (a) Factor $x^2 - x - 1$ in $\mathbb{Z}_{11}[x]$.

(b) Without calculating the actual orbit, find the period of the orbit of $(0, 1)$ under the Fibonacci recurrence mod 11. Check your answer by calculating the orbit.

Ex 8.32. This is a problem from the *American Mathematical Monthly*, March 1992, page 278.

(a) Given a positive integer m, show that the modular Fibonacci period $t(m)$ satisfies $t(m) \le 6m$ for all m and that equality holds for infinitely many m.

(b) Show that an analogous result holds for the Lucas sequence with the upper bound of $6m$ replaced by $4m$.

Ex 8.33. Find the period of 0 under $f(x) = 3x^{-1} - 2$ in $\mathbb{Z}_{11}$ (where 0^{-1} is defined to be 0). Find an inversive generator in $\mathbb{Z}_{11}$ whose period is 11.

Ex 8.34. Count the number of reducible quadratic polynomials in a field with q elements, and use that information to show that every finite field has at least one irreducible quadratic polynomial.

Ex 8.35. (a) Use a Fermat Test to show that $2047 = 2^{11} - 1$ is a composite number.

(b) Show that $m = 561$ is a Carmichael number.

(c) Prove: If m fails the Fermat Test for $a = 2$, then $N = 2^m - 1$ fails the Strong Fermat Test for $a = 2$.

Ex 8.36. (a) Show that the graph of $H(x) = -\ln(1 - x)$ lies above the line $y = x$ by proving that $H(x)$ is an increasing function on $[0, 1)$, which is concave upward and satisfies $H(0) = 0$ and $H'(0) = 1$.

(b) Show that

$$H\left(\frac{1}{m}\right) + \ldots + H\left(\frac{k-1}{m}\right) > \frac{1}{m} + \cdots + \frac{k-1}{m} = \frac{k(k-1)}{2m}.$$

(c) Use the monotonicity of $H(x)$ and integration by parts to show that

$$H(0) + H\left(\frac{1}{m}\right) + \cdots + H\left(\frac{k-1}{m}\right) < \frac{k^2}{m}.$$

Ex 8.37. For any fixed integer $m \geq 2$ and any $1 \leq k < m$, let $P(k, m)$ be the probability that k different elements are chosen from a set with m elements.

(a) Use the last problem to show that

$$(8.10) \qquad \exp\left(-\frac{k^2}{m}\right) < P(k, m) < \exp\left(-\frac{k(k-1)}{2m}\right),$$

where $\exp(x) = e^x$, the usual exponential function.

(b) Prove: If $k > 1.39\sqrt{m}$, then $P(k, m) < \frac{1}{2}$.

(c) Prove: If $k < 0.8\sqrt{m}$, then $P(k, m) > \frac{1}{2}$.

Ex 8.38 (The Birthday Problem). Estimate the probability that at least two people in a crowd of M people have the same birthday. Estimate the number of people needed to ensure that this probability is greater than $1/2$. What size crowd ensures that the probability is greater than $2/3$?

Ex 8.39. Using an analysis similar to Exercise 8.36, show there exists a constant c such that for sufficiently large n, the partial sum $\sum_{i=1}^{n} \frac{1}{i}$ is approximated by $\ln(n) + c$.

Ex 8.40. Let $m \geq 2$ be an integer and f a function defined on $\mathbb{Z}^k$ for some $k \geq 1$ and $s \in \mathbb{Z}^k$. Then $\langle f^{(k)}(s) \rangle$ is periodic mod m iff there exists a positive integer n such that

$$\langle f^{(2n)}(s) \rangle \equiv \langle f^{(n)}(s) \rangle \pmod{m}.$$

Ex 8.41. Factor $m = 7031$ using the Pollard Rho Method with the orbit of 3 under $f(x) = x^2 + 1$.

9

Computational Complexity

Analysis of algorithms is intimately related to recurrences. In this chapter we present many algorithms that are *recursive* in the sense that they call themselves. We'll see that the analysis of each algorithm quickly leads to a recurrence that we can solve using the techniques of the previous chapters. The solution of the recurrence then provides information about the amount of resources used by the algorithm. We will also see that many easily stated unsolved problems are close to the edge of standard material. The analysis and improvement of basic algorithms provides a treasure chest of research problems that are fun (and maybe even profitable) to solve and are accessible to students.

An **algorithm** is a procedure that solves a problem *and* is suitable for implementation as a program on a digital computer. This informal definition makes two important points. First, an algorithm solves a problem. There are computer programs that never terminate, and it would be very difficult to say whether such a program does anything, let alone solves a problem. Second, each step of the algorithm should be well-defined and should be representable, at least in principle, by a program. For example, $s := p/q$ is not well-defined if q is allowed to be zero. Also, "Find the smallest $x \in X$ for which the statement $P(x)$ is true" is not necessarily well-defined, since it depends on the truth value of the statement $P(x)$ and the set X. For instance, if $P(x)$ were true for all negative integers, then "Find the smallest $x \in X$ for which the statement $P(x)$ is true" is not well-defined for $X = \mathbb{Z}$ but is well-defined for $X = \mathbb{N}$.

For our purposes and for many purposes, the above somewhat informal definition of algorithm is sufficient. Formal definitions of the term "algorithm" were created by Turing, Markoff, and others (refer to [138]). It's very satisfying that all these formal definitions of algorithm define the same class of algorithms: When A is an algorithm in one of these senses, then in every one of the other senses there exists an algorithm that is equivalent to A.

9.1   Analysis of Algorithms

The **analysis of an algorithm** attempts to assay the amount of resources used by the algorithm. For any *solvable* problem there are an infinite number of algorithms that solve the problem, so how do we decide which is the *best* algorithm? An obvious idea is that *best* means uses the fewest resources. Typical resources are time and space, and in this chapter our analyses concentrate on time.

An abstract world of abstract computers and abstract programs is constructed from the real world of actual computers and actual computer programs. This construction is rarely formal, because exact definitions of abstract entities are often not stated. A number of real-world limitations disappear in the abstract world. For example, real computers have a fixed finite memory size and there's an upper bound on the size of numbers that can be represented by the computer. In the abstract world these limitations don't exist. There, computers are assumed to have finite but unbounded memories with no bound on the size of numbers that can be used. In practice, there are examples of algorithms that work relatively quickly when arbitrarily large numbers are used, but implementing them on real computers results in much slower algorithms. These algorithms make perfect sense in the abstract world, but have little or no relevance for the real world.

9.1.1   Measuring run time

We want to know how much time it takes an algorithm to perform a particular task. For a real computer program this can be done using a stopwatch to time the execution of the program. This elapsed time is often called the **wall clock time**. Another way of timing a real program is to use your computer's TIME command so that the computer types out the time used when the program is run. This time measure is often called **CPU[1] time** and may differ drastically from wall clock time.

Both wall clock time and CPU time suffer from the real-world problem of inexact repeatability. Two different runs of the same program may not take the same amount of time, although it is certainly possible to gather

---

[1]CPU means Central Processing Unit.

valid statistical data. A more serious problem is that both wall clock time and CPU time are highly dependent on the exact hardware and software implementation used, as well as on the input data. Specifically, if you change operating systems or run the program on a different computer or change the input data, you will often be unable to reliably predict how long the "same" program will take to run.

We want to call an algorithm faster (it uses less time) than another algorithm if when we run the two algorithms on a computer the faster one always finishes first. To make this a fair test some variables have to be removed. For example, we'd have to code the two algorithms in the same programming language; compile the two programs using the same compiler; run the two programs under the same operating system on the same computer and not interfere with either program while it's running. In practice, even if we could control all these conditions, to our chagrin we might find that algorithm A is faster under conditions C, while algorithm B is faster under conditions D.

To avoid this unhappy situation we calculate **unitless time**. For this we find the run time $T(n)$ as a function of n, where n is some measure of the size of the problem. For example, we could use the number of digits as the measure of problem size if the problem is addition of two integers. We could use the number of elements in a list if the problem is to sort a finite list. We could use the number of edges (or the number of vertices) in a graph if the problem is to determine whether a finite graph has a certain property. We consider two algorithms to use the same time if their run times **have the same order**. For our purposes, two run times $T_1(n)$ and $T_2(n)$ in the variable n have the same order when $T_1(n) = \Theta(T_2(n))$, where Big-Theta is the notation defined in Chapter 1; namely, the statement $T_1(n) = \Theta(T_2(n))$ means that there exist positive constants c_1, c_2 such that for all sufficiently large n,

$$c_1 T_1(n) \leq T_2(n) \leq c_2 T_1(n).$$

In particular, "having the same order" doesn't distinguish between algorithms whose run times are constant multiples of each other.

It's worthwhile to discuss Big-Theta notation further. Assume that A is an algorithm and that the size of the input data for A is represented by the variable n. A typical result might be that A has run time $\Theta(n^2)$. Since we're using unitless time, we have no idea what this means in terms of seconds or nanoseconds. Indeed, we can think of the time unit as an unknown function of many details, among them the machine and the programming language. For example, the fact that the run time is $\Theta(n^2)$ allows us to reasonably predict that when all these details are kept constant, doubling the size of the input will quadruple the run time of A.

There is even more hidden in Θ-notation. Because it's asymptotic, the above prediction might not be valid unless n is quite large. For example, the actual CPU time for $n = 2$ can be the same as for $n = 1$ when the output for these values is computed by a table lookup. Further, for some programs the squaring prediction might be valid for $n \geq 100$, whereas for others the prediction might be valid only for $n \geq 1024$ or some larger number.

So how is the measure of run time used to compare two algorithms? Assume for the moment that algorithm A has run time $\Theta(n^2)$ and that algorithm B has run time $\Theta(n \log n)$. If we program these two algorithms, will the program for B be faster than the program for A? Yes, but only for large enough n. Depending on the constants implicit in the Θ notation, it may be that B is faster than A only for $n > 10^{23}$, in which case for reasonably sized data the "slower" A might actually be faster.

To summarize, if we find that the order of the run time for algorithm A is strictly less than that for algorithm B, then we can be confident that for any large enough problem A will run faster than B. On the other hand, if the run times of A and B have the same order, then we won't be able to predict which one will be faster for any given input.

9.1.2   An example: The Towers of Hanoi puzzle

In this section we illustrate the above discussion by looking at a concrete example, the run time of an algorithm for solving the **Towers of Hanoi puzzle**.

Ball's *Mathematical Recreations and Essays* [5] contains one of the first mathematical formulations of the Towers of Hanoi puzzle. The puzzle consists of three towers or pegs (usually called A, B, C), and n disks of different sizes (numbered 1 through n) such that the ith disk is larger than the jth disk whenever $i > j$. Initially, the disks are stacked on Tower A in order of size, with the largest disk on the bottom and the smallest on top. The problem is to move the stack of disks from Tower A to Tower C, moving the disks one at a time in such a way that a larger disk is never stacked above a smaller disk. (Refer to Figure 9.1.) An extra constraint is that the sequence of moves should be as short as possible. An algorithm is therefore said to solve the Towers of Hanoi problem if when we input the number of disks and the names of the three towers, the algorithm returns a sequence of moves that conforms to the above rules.

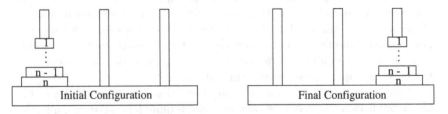

FIGURE 9.1. The Towers of Hanoi Puzzle.

Three simple observations form the key to the problem. The first is that moving the largest disk requires all other disks to be out of the way; that is, the other $n - 1$ smaller disks must be located (in the proper order) on some other tower. In order for the final result to be stacked on Tower C, Tower C must be free for the largest disk, and so the $n - 1$ disks have been moved from Tower A to Tower B, which is the Towers of Hanoi puzzle for $n - 1$ disks. After the largest disk is moved to Tower C, the other $n - 1$ disks must be moved from Tower B to Tower C, which is again the Towers of Hanoi puzzle for $n - 1$ disks. This recursive procedure leads to the following recursive algorithm.

<u>PROCEDURE</u>   HANOI(A, B, C, n)
    <u>IF</u> $n = 1$ <u>THEN</u> Move the top disk from A to C
        <u>ELSE</u> HANOI$(A, C, B, n - 1)$
            Move the top disk from A to C
            HANOI$(B, A, C, n - 1)$.

This is called a recursive procedure because it calls itself. Also, unlike the worm Ouroboros, it doesn't endlessly swallow its own tail. Each time it calls itself it decrements the size parameter n by 1, which means that eventually the sequence of calls "bottoms out" with a call to HANOI with $n = 1$. This call makes one move and then returns to the previous call. The operation of this algorithm can be seen in Figure 9.2, where we give the sequence of calls and the states of the puzzle for $n = 3$.

There's one point in the trace that might seem at least slightly strange: How did the Move from A to C in the algorithm give Move from C to B in the trace? The answer is that "A" in the algorithm refers to the first parameter in the call, and "C" refers to the third parameter in the call. So when Move from A to C is referred to within a call to HANOI(C,A,B,n), it makes a move from C to B in the execution of the solution of the puzzle.

Now that we have an algorithm for solving the Towers of Hanoi puzzle, let's analyze its run time. Because we're calculating *unitless* time, we don't need to know how long it takes to perform operations such as moving a disk or issuing a procedure call. We do have to distinguish between operations that take a constant amount of time (that is, time that is independent of the value of n) and operations whose run time depends on n. Aho, Hopcroft, and Ullman in [2] call the assumption that each operation takes the same amount of time the **uniform cost criterion**. Under this condition the run time $T(n)$ used for n disks satisfies the first–order recurrence

(9.1) $$T(n) = 2T(n - 1) + c \quad \text{for all } n > 1,$$

where c is a positive constant. This is true because within the procedure for n disks there are two calls to the procedure for $n - 1$ disks, and the

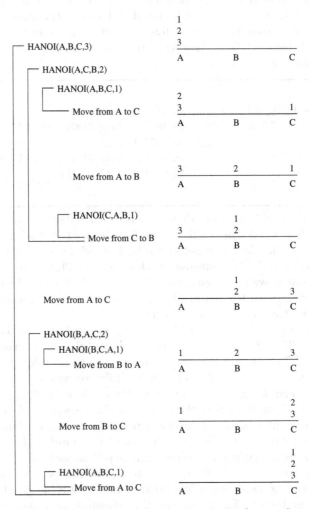

FIGURE 9.2. Trace of the program HANOI for $n = 3$.

constant c is the sum of the constant run times for the various operations. From the work in Chapter 3 we know that (9.1) has the solution

$$T(n) = T(1)2^{n-1} + c(2^{n-1} - 1),$$

and so

$$\frac{T(1)}{2} 2^n \le T(n) \le \frac{T(1) + c}{2} 2^n,$$

which gives $T(n) = \Theta(2^n)$.

Instead of assuming the uniform cost criterion, we could have assumed that the run time is proportional to the number of moves. Under that assumption, our argument could proceed as follows. Let $M(n)$ be the number of moves performed when the algorithm is called with n disks. Then

$$M(1) = 1 \quad \text{and} \quad M(n) = 2M(n-1) + 1 \text{ for all } n > 1,$$

which has the solution $M(n) = 2^n - 1$. Because the run time $T(n)$ is proportional to $M(n)$, then $T(n) = \Theta(M(n)) = \Theta(2^n)$, which is the same as we obtained above.

In our analysis of the algorithm HANOI we've made several other assumptions that bear examination. In any actual implementation on a real digital computer, the number n needs to be stored and manipulated internally. The time required for this is at the very least dependent on the number of computer words required to represent n, but our argument assumed that each operation takes a constant time, independent of n. In addition, representing a state of the puzzle in a computer's memory requires increasing memory as n increases and, presumably, increasing time to manipulate this memory. All of the operations in HANOI depend on the space needed to store n in memory, and this is proportional to the number of bits needed to represent n, which is $\lceil \log_2 n \rceil$. We can ignore the ceiling function and consider space to be proportional to $\log n$, where the constant of proportionality depends on the word size of the computer. Aho, Hopcroft, and Ullman [2] call this the **logarithmic cost criterion** and suggest that it should be used when the numbers in the algorithm don't have fixed bounds. Using this criterion, the run time for the n-disk puzzle is

$$T(n) = 2T(n-1) + c\log(n),$$

with initial condition $T(1) = t_1$. (As above, c and t_1 are unknown positive constants.) Again referring to Chapter 3, the solution to this recurrence is

$$T(n) = 2^n \left[\frac{t_1}{2} + c \sum_{i=1}^{n} \frac{\log i}{2^i} \right],$$

where each summand is positive, and the sum is therefore less than the infinite sum

$$0 < \sum_{i=1}^{n} \frac{\log i}{2^i} \le \sum_{i=1}^{\infty} \frac{\log i}{2^i},$$

which converges by the Ratio Test. The sum is bounded above and below by positive constants, and

$$c_1 2^n \leq T(n) \leq c_2 2^n,$$

where for instance c_1, c_2 can be chosen as $c_1 = \frac{t_1}{2}$ and $c_2 = \frac{t_1}{2} + c \sum_{i=1}^{\infty} \frac{\log i}{2^i}$. So again $T(n) = \Theta(2^n)$. The increase in operation time due to the size of n therefore has no effect on the order of our estimate of the run time, because any changes were absorbed into the "implied constant" in $\Theta(2^n)$.

If we consider building a physical Towers of Hanoi puzzle and the time to move disk n grows as some function $g(n)$, we would have

$$T(n) = 2T(n-1) + g(n).$$

Provided the eminently reasonable assumption that $g(n+1) < 2g(n)$ holds for all sufficiently large n, a modification of the above argument again yields $T(n) = \Theta(2^n)$. From this we see that the conclusion that HANOI has $\Theta(2^n)$ run time is **robust**: Specific details of the implementation of the algorithm have no effect on the run time. Table 9.1 gives some actual run times and compares them to the predicted run time $c\,2^n$, where c is computed as $c = T(10)/2^{10}$. Notice that the ratio $T(n)/T(n-1)$ is approximately 2, which is consistent with $T(n) = \Theta(2^n)$.

TABLE 9.1. Run times for the HANOI Algorithm showing the predicted $\Theta(2^n)$ behavior.

	Run Times for the HANOI Algorithm			
n	Number of Moves	Actual Time	Predicted Time	$T(n)/T(n-1)$
5	31	10	74	
6	63	30	147	3
7	127	230	294	7.7
8	255	250	589	1.1
9	511	1287	1178	5.1
10	1023	1810	2355	1.4
11	2047	4270	4708	2.4
12	4095	10471	9421	2.5
13	8191	19398	18842	1.9
14	16383	39447	37683	2.0
15	32767	77669	75364	2.0
16	65535	147832	150733	1.9
17	131071	301652	301466	2.0

While our recursive algorithm solves the Towers of Hanoi problem, there are many other algorithms that also solve this problem. For example, Buneman and Levy [19] give a compact iterative algorithm for Towers of Hanoi.

Among the variety of algorithms for this problem, is there a *best* algorithm? We don't want to go deeply into this question, but obviously the answer to this question depends on the definition of *best*. For Towers of Hanoi, it can be shown that every algorithm must use at least $\Omega(2^n)$ time. So, if *best* means least-time order, our recursive algorithm is *best*. But the Buneman and Levy algorithm with a reasonable data structure can also be shown to have $\Theta(2^n)$ run time and so is another *best* algorithm. So, there can be more than one *best* algorithm. Cull and Ecklund [41] consider a variety of Towers of Hanoi algorithms and show that every Towers of Hanoi algorithm must use at least $\Omega(2^n)$ time and at least $n - c$ bits of space, for some constant c. They give an algorithm that simultaneously achieves these lower bounds on time and space.

We close this section with a problem from the Advanced Problem Section of the June–July 1939 issue of the *American Mathematical Monthly* (page 363), in which B.M. Stewart proposed a generalization of the Towers of Hanoi puzzle to any number $k \geq 3$ of towers.

> Given a block in which are fixed k pegs and a set of
> n washers, no two alike in size, and arranged on one
> peg so that no washer is above a smaller washer. What
> is the minimum number of moves in which the n washers
> can be placed on another peg, if the washers must be moved
> one at a time, subject always to the condition that no
> washer be placed above a smaller washer?

Two years later, two solutions to the problem were published in the Advanced Problems Section of the March 1941 issue of the same journal (pages 216–217). These solutions yield algorithms for solving the Towers of Hanoi puzzle for k towers. Many people believe that these are optimal, but a proof of optimality is still an open question after 60 years. We close this section with a comment from a March 2002 interview with Donald Knuth [87, p. 321]. "In the case of the 4-peg 'Tower of Hanoi', there are many, many ways to achieve what we think is the minimum number of moves, but we have no good way to characterize all solutions. So that's why I personally came to the conclusion that I was never going to solve it, and I stopped working on it in 1972. But I spent a solid week working on it pretty hard."

9.2 Computer Arithmetic

Binary representation is usually used to store an integer on a digital computer, and the number of bits for the representation of N is $\lceil \log_2(N) \rceil$. This is then broken into blocks that fit into the internal words of the computer's memory and that are stored as some sort of list. Every machine has a limit on the size of integers that can be represented in this fashion. In many classical integer data types this was limited to one or two machine words, whereas many present-day applications require no pre-set bounds on the size of integers.

In any storage scheme there are details that are specific to the machine and the operating system. Among these are the word size and the manner in which the list is stored. In analysis of algorithms these details are avoided by considering the number of bit operations required to perform a calculation. A bit operation is either a **unary operation** (an operation performed on one bit, such as reversing a single bit) or a **binary operation** (an operation on two bits, such as addition). We assume that the run time is proportional to the number of bit operations. In order for this to make sense, the operations on words must introduce a constant multiple to the run time, and our run time estimates are not dominated by the overhead required for keeping track of the list structure.

9.2.1 Addition and subtraction

How long does it take to add two n-bit natural numbers M and N? That is, how many bit operations are required to add the integers? For instance, consider the algorithm we all use for adding two binary representations by hand. The maximal number of bit operations for the sum is n additions with carries. If we write the carry bits in a row above the rows containing M and N, this row always has a zero in the rightmost bit and extends at most one place to the left. Taking this into account, at most $2n$ bit operations are used to perform the addition of two n-bit integers.

On the other hand, it takes at least n bit operations to write down the number $M + N$. This is true even in the extreme case in which one of the summands is zero and we know a priori that the sum is just equal to the other integer. Because of this, n is a lower bound on the number of bit operations needed to add M and N. Since $M - N = M + (-N)$, the cost of subtracting N from M equals the cost of adding N to M plus the cost of negating N. Since negation can be accomplished in n bit operations, we conclude that the time required to add or subtract two n-bit integers is $\Theta(n)$.

9.2.2 Multiplication and division

The standard algorithm used for hand computation of the product $M * N$ in binary involves multiplying M by each bit of N (resulting either in 0 or M in binary), stacking up the results with appropriate left shifts, and then adding entries in the stack. For example, to compute $7 * 5 = 35$ the standard algorithm is

$$
\begin{array}{rccc}
 & 1 & 1 & 1 \\
* & 1 & 0 & 1 \\
\hline
 & 1 & 1 & 1 \\
0 & 0 & 0 & \\
1 & 1 & 1 & \\
\hline
1 \quad 0 \quad 0 \quad 0 & 1 & 1 \\
\end{array}
$$

When M and N are n-bit integers, this algorithm requires n additions of integers that have at most $2n$ bits, and the complexity of this standard algorithm is $\Theta(n^2)$.

This is not the best algorithm for integer multiplication, since there's an easily implemented algorithm that has complexity $\Theta(n^{\log_2(3)})$ (refer to Exercise 9.18). There's also another algorithm, due to Schönhage and Strassen [145], which has asymptotic complexity $O(n \log n \log(\log n))$ and therefore its run time has order only slightly larger than the order of addition. However, because the constant involved in the Big-Oh notation is large, the Schönhage-Strassen algorithm isn't better than the more straightforward algorithm until n is quite large.

What about the cost of division? Later in this chapter (section 9.4.4) we will use Newton's method to design an algorithm for division that has the same run time as multiplication.

9.3   An Introduction to Divide-and-Conquer

Historically, the idea of divide-and-conquer is often attributed to Julius Caesar [173]. In the context of problem solving, divide-and-conquer may reasonably be attributed to René Descartes [49, 50], the same Descartes whose Rule of Signs was used in Chapter 5 to prove that every nonnegative polynomial has only one positive root. The keystone of Descartes' analytic method is to break a complicated problem into easier constituent parts and then to solve the individual parts. An understanding of the complicated problem is then built up from the solution of its parts. In *How to Solve It* [129], George Pólya stressed that breaking down a problem into several smaller problems of the same kind is a typical step in solving a mathematical problem, and he further pointed out that this sort of analysis easily leads to an inductive proof of the correctness of the solution. In the 1930's, philosophers such as Gödel, Kleene, and Ackermann recognized the central role of this recursive technique, and their theoretical analyses led Turing to describe an abstract digital computer. Eventually, **Turing machines** were embodied as physical digital computers, and the problem of programming these computers led back to the divide-and-conquer technique. (For example, refer to [172] and [102].)

The term **divide-and-conquer algorithm** is now usually reserved for the design strategy that solves a problem of size n by solving several problems of size n/c for some constant c, so that in its current usage the "divide" means that there is an actual division of the problem size. The HANOI algorithm that we discussed in Section 9.1.2 is not a divide-and-conquer algorithm in this sense because it uses subproblems of size $n - 1$. (Perhaps a better term for this type of algorithm is *subtract*-and-conquer?) Many commonly used divide-and-conquer algorithms split the problem into two subproblems, each having half the size of the original problem. Because such an algorithm cannot divide indefinitely, we specify a size limit above which the problem is divided and below which the algorithm calls another algorithm that solves the small-size instances. The solutions are then combined to give a solution to the original problem.

The recursive structure of a divide-and-conquer algorithm leads directly to an inductive proof of correctness and also to a recurrence for the run time of the algorithm.

9.3.1 Example: Polynomial multiplication

Let's look at an example that makes this discussion more concrete. Our example is **polynomial multiplication** (or **convolution**), where the output is the one polynomial that is the product of two input polynomials. We assume an **arithmetic model of computing**, which means both that our computer can perform arithmetic operations and that we want an algorithm that multiplies the two input polynomials using only arithmetic operations on the coefficients. As above, the run time for an algorithm is the number of arithmetic operations used in the algorithm.

Assume that each input polynomial has degree $n - 1$ and hence n coefficients, which we record in a vector with n components. The output polynomial then has degree $2(n - 1)$, recorded as a vector with $2n - 1$ components. The standard algorithm for multiplying polynomials is similar to the standard algorithm for multiplying integers. If $P(x)$ and $Q(x)$ are the input polynomials, multiplication proceeds by multiplying $P(x)$ by the constant coefficient of $Q(x)$, shifting one space to the left, multiplying $P(x)$ by the linear coefficient of $Q(x)$, and so on. This results in n vectors, which are added componentwise to get the final result. Here's an example that multiplies two quadratic polynomials in this fashion:

$$
\begin{array}{rrrrrrr}
 & & & 2x^2 & + & 3x & - & 7 \\
 & & & 3x^2 & - & 2x & + & 2 \\
\hline
 & & & 4x^2 & + & 6x & - & 14 \\
 & -4x^3 & - & 6x^2 & + & 14x & & \\
6x^4 & + & 9x^3 & - & 21x^2 & & & \\
\hline
6x^4 & + & 5x^3 & - & 23x^2 & + & 20x & - & 14 \\
\end{array}
$$

This method for multiplying two polynomials has run time $\Theta(n^2)$ because each of the n^2 table entries is a product of two coefficients and each table entry occurs once as a summand.

An algorithm for polynomial multiplication

FOR $i := 0$ TO $n - 1$ DO
    FOR $j := 0$ TO $n - 1$ DO
        $c_{i+j} = c_{i+j} + a_i * b_j$

Let's construct a divide-and-conquer algorithm for this problem. The basic philosophy of a divide-and-conquer approach to polynomial multiplication is to think of the input polynomial as constructed from several smaller polynomials. In practice, we divide each polynomial "in half," and for simplicity we first consider polynomials with an even number of coefficients, say $n = 2m$. Then we can write

$$P(x) = a_{2m-1}x^{2m-1} + \cdots + a_1 x + a_0 = x^m P_1(x) + P_0(x),$$

where each of the polynomials $P_1(x) = a_{2m-1}x^{m-1} + \cdots + a_{m+1}x + a_m$ and $P_0(x) = a_{m-1}x^{m-1} + \cdots + a_1 x + a_0$ has $m = n/2$ coefficients. (For example, if $P(x) = 5x^3 + 12x^2 + 7x + 8$, then $P_1(x) = 5x + 12$ and $P_0(x) = 7x + 8$.) Splitting both input polynomials $P(x)$ and $Q(x)$ of degree $2m$ in this way and suppressing the argument variable x results in

$$PQ = (x^m P_1 + P_0)(x^m Q_1 + Q_0),$$

giving

(9.2) $$PQ = x^n P_1 Q_1 + x^m [P_1 Q_0 + P_0 Q_1] + P_0 Q_0.$$

Since multiplying a polynomial by an integer power of x simply shifts the sequence of coefficients, the original problem of finding the product PQ has been reduced to the four subproblems of finding the products $P_1 Q_1$, $P_1 Q_0$, $P_0 Q_1$, and $P_0 Q_0$.

Because m (and also the original n) might be odd, we will need to modify this procedure to get a divide-and-conquer algorithm that works for any pair of polynomials. Only a slight modification is needed. Let $n - 1$ equal the maximal degree of the factors $P(x), Q(x)$ and define $k \geq 1$ to be the least exponent such that $2^k \geq n$, that is, $k = \lceil \log_2(n) \rceil$. Filling in zero entries where necessary, we treat $P(x)$ and $Q(x)$ as polynomials with 2^k coefficients. Dividing each of $P(x)$ and $Q(x)$ into their polynomial halves, we obtain four subpolynomials having 2^{k-1} coefficients. We continue this subdivision process for k stages, or for as long as we have more than one coefficient. When we reach the stage at which each subpolynomial has one coefficient, we multiply the pairs of constants that remain. Equation (9.2) is then used to retrace the steps and arrive at the product polynomial.

The run time analysis for this divide-and-conquer algorithm is fairly simple. From the first division we obtain four subproblems, each a multiplication of polynomials of degree 2^{k-1}. The multiplications by x^{2^k} and $x^{2^{k-1}}$

are shifts in the coefficient vectors and are easily carried out before the vectors are added. Therefore, if $T(n)$ denotes the time needed to multiply two polynomials with n coefficients (where we consider $n = 2^k$), then

$$(9.3) \qquad\qquad T(n) = 4T(n/2) + bn,$$

where $4T(n/2)$ represents the time needed to multiply the four half-size polynomials, and the bn term comes from the time needed to shift and add the results. If we were doing an exact operation count we'd write down an explicit value for b. However, since we don't know the actual time required for coefficient addition or multiplication, we simply assume that b is some positive constant.

Notice that the recurrence (9.3) is not one of our usual linear recurrences, because $T(n)$ depends on $T(n/2)$ rather than on $T(n-1)$. Despite this novelty, it's possible to convert (9.3) to a linear recurrence. For this, we exploit the fact that n is assumed to be a power of 2. Introducing the new variable $t_k = T(2^k)$, (9.3) becomes

$$t_k = 4t_{k-1} + b2^k \text{ with } t_0 = T(1),$$

a nonnegative recurrence with dominant eigenvalue $\lambda_0 = 4$. It has the form

$$t_k = 4t_{k-1} + b\lambda_1^k, \quad \text{where } \lambda_1 < \lambda_0,$$

and Theorem 5.5.3 implies that $t_k = \Theta(4^k)$, giving $T(n) = \Theta(n^2)$ when n is a power of two. What about other n's? Even when n is not a power of two, we've already established that we can write the factors as polynomials with 2^k coefficients where $k = \lceil \log_2(n) \rceil$, and a polynomial with 2^{k-1} can be treated as a polynomial with n coefficients, so we have

$$T(2^{k-1}) \leq T(n) \leq T(2^k);$$

and since $T(2^k) = \Theta(n^2)$,

$$\frac{c_2 n^2}{4} \leq c_2 \left(\frac{n}{2}\right)^2 \leq c_2 2^{2(k-1)} \leq T(n) \leq c_1 2^{2k} \leq c_1(2n)^2 \leq 4c_1 n^2$$

and $T(n) = \Theta(n^2)$. What's the point of developing this algorithm when we already have an easy-to-understand iterative algorithm with run time $\Theta(n^2)$? The answer is that a further examination of this algorithm will allow us to a make a slight modification that speeds things up.

Returning to the formula in (9.2) we notice that for two of the half-size products, $P_1 Q_0$ and $P_0 Q_1$, only their sum and not the separate products is needed. This is important, because there's a way to calculate their sum without calculating both products! The trick is to use subtraction and the polynomial identity

$$(9.4) \qquad P_1 Q_0 + P_0 Q_1 = (P_1 + P_0)(Q_1 + Q_0) - P_1 Q_1 - P_0 Q_0.$$

Since each of P_1 and P_0 has $n/2$ coefficients, so does their sum $P_1 + P_0$. Similarly, $Q_1 + Q_0$ has $n/2$ coefficients, and their product can be computed with two half-size polynomial additions and one half-size polynomial multiplication. Since we also need both of $P_1 Q_1$ and $P_0 Q_0$, the product $(P_1 + P_0)(Q_1 + Q_0)$ can be computed with three multiplications, four additions, and two subtractions of half-size polynomials. So we've replaced one of the multiplications in the earlier algorithm with four additions and two subtractions. This would be a loss if these operations all had the same cost. But the operations are *polynomial* operations, and additions and subtractions cost $\Theta(n)$, while multiplications (so far) have cost $\Theta(n^2)$. Therefore, the gain is enormous. We obtain the recurrence

$$(9.5) \qquad\qquad T(n) = 3T(n/2) + Bn,$$

where the constant B in this equation is slightly larger than the constant b in the previous recurrence (9.3). It's easy to check that

$$T(n) = a\, n^{\log_2(3)} - 2Bn$$

satisfies this recurrence for some constant a that depends on the initial condition. Therefore, $T(n) = \Theta(n^{\log_2(3)})$, where $\log_2(3) \approx 1.5849$, and this new algorithm is faster than $\Theta(n^2)$ for large enough n. (In Exercise 9.18 you see that this algorithm can be used to obtain the promised $\Theta(n^{\log_2(3)})$ multiplication algorithm for integers.)

What about the implied constant in this algorithm as compared with the one in the standard algorithm? Table 9.2 records a comparison of the run times for the two algorithms for various pairs of polynomials with n coefficients. In these examples we were careful to use polynomials with small-integer coefficients so that the efficiency of either algorithm is unaffected by the size of the coefficients. Before we use the data to estimate the constants, we should ask if the data is consistent with with the predicted leading term of our run time formulas. For this, we can look at the ratio $T(2n)/T(n)$. For an $\Theta(n^2)$ algorithm this ratio should approach 4, while for an $\Theta(n^{\log_2 3})$ algorithm this ratio should approach 3. The data shows that for $n \geq 512$, these ratios are reasonably close to their predicted asymptotic values. To estimate the leading constants, we can use the ratios $T_I(n)/n^2$ and $T_R(n)/n^{\log_2 3}$. For the larger values n, these ratios settle down and so we can come up with approximate asymptotic run time formulas: $T_I(n) \approx .00047\, n^2$ and $T_R(n) \approx .03\, n^{\log_2 3}$. In conclusion, this data shows that the "faster" algorithm will be faster, but only for rather large n, and the data is sufficient to give formulas which allow us to predict the run times of these implementations, but again only for large values of n.

TABLE 9.2. Comparison of the run times for two polynomial multiplication algorithms. The iterative algorithm is the classical method with run time $\Theta(n^2)$. The recursive algorithm is the half-size method with run time $\Theta(n^{\log_2 3})$. The ratios of $T(2n)/T(n)$ are very close to the predicted values of 4 and 3. The "faster" algorithm catches up to the "slower" method for large n.

Run Times for n-coefficient Polynomial Multiplication.			
n	Iterative Time (ms)	n	Recursive Time (ms)
2	0	2	0
4	0	4	0
8	0	8	0
16	0	16	0
32	0	32	15
64	0	64	31
128	15	128	78
256	31	256	218
512	140	512	625
1024	500	1024	1875
2048	2000	2048	5578
4096	8000	4096	16812
8192	32000	8192	50438
16384	128094	16384	151578
32768	512376	32768	454734

9.4   Simple Divide-and-Conquer Algorithms

Each of the divide-and-conquer algorithms examined in the last section led to a recurrence relation for its run time. There was a strong similarity between the two arguments, which we can generalize to derive a recurrence for the run time of a whole class of divide-and-conquer algorithms. The recurrences (often called divide-and-conquer recurrences) have three types of solutions, and we give examples of algorithms that illustrate each type.

In each of our two previous examples, the problem of polynomial multiplication was split into two subproblems of the same type that were half the original size. Here we generalize this slightly and assume that a problem of size n is split into several (say a) subproblems, each of size n/c for some constant c. For many divide-and-conquer algorithms $c = 2$ holds, but the same analysis works for general c.

In some cases the splitting process is easy and has negligible computational cost, but we want to allow for the possibility that splitting takes some time. Usually the time-consuming part of a divide-and-conquer algorithm is the combining step, when the answers to subproblems are used to compute the answer to the whole problem. Here is the general form of a divide-and-conquer algorithm:

The general form of a divide-and-conquer algorithm

```
PROCEDURE   DandC(DATA,n,SOLUTION)
   IF n is small
      THEN Solve by some special algorithm
      ELSE SPLIT(DATA,n) into (D₁,n/c)
```
$$\text{DandC}(D_1, n/c, S_1)$$
$$D_2 := f_2(\text{DATA}, D_1, S_1)$$
$$\text{DandC}(D_2, n/c, S_2)$$
$$D_3 := f_3(\text{DATA}, D_1, S_1, D_2, S_2)$$
$$\vdots$$
$$D_a := f_a(\text{DATA}, D_1, S_1, D_2, S_2, \ldots, D_{a-1}, S_{a-1})$$
$$\text{DandC}(D_a, n/c, S_a)$$
$$\text{SOLUTION} := f_{a+1}(\text{DATA}, D_1, S_1, D_2, S_2, \ldots, D_a, S_a)$$
```
      RETURN(SOLUTION)
```

Here we've allowed for the results of previous subproblems to be processed and then fed into the next subproblem. The function f_{a+1} performs the final combination of results to produce the solution.

An efficient algorithm should have a run time that is bounded by a function that is *polynomial* in the size of the input or perhaps the size of the output. For this reason we assume that the time for splitting and combining is given by a polynomial in n. Since we're interested in only the order of the run time, we can concentrate on the largest term of the polynomial and assume that the run time for the split and combine steps is bn^m for some positive constant b and some constant exponent m. (For this to be a polynomial, m must be an integer, but it costs nothing more to allow m to be any nonnegative real number.)

Under these assumptions the run time for a divide-and-conquer algorithm that splits a problem of size n into a subproblems of size n/c is given by the recurrence

$$(9.6) \qquad\qquad T(n) = aT(n/c) + bn^m$$

(compare this with (9.3) and (9.5).) The initial condition comes from the work done by the .algorithm that handles problems of small size. Since the small-size algorithm is run only on data of bounded size, we can bound its run time by some positive constant. This allows us to assume that the initial condition $T(1)$ is positive but otherwise unknown.

As in the last section, we essentially use logarithms to solve the recurrence in (9.6). Setting $n = c^k$ and $t_k = T(c^k)$, (9.6) becomes

$$t_k = at_{k-1} + bc^{km} \qquad \text{with } t_0 > 0,$$

a nonnegative difference equation with $\lambda_0 = a$. From Theorem 5.5.3 we know that the relative sizes of a and c^k yield three different orders of magnitude for the sequence $\langle t_k \rangle$,

$$
t_k = \begin{cases}
\Theta(c^{km}) & \text{for } a < c^m, \\
\Theta(kc^{km}) & \text{for } a = c^m, \\
\Theta(a^k) & \text{for } a > c^m.
\end{cases}
$$

In order to translate this back into the run time $T(n)$ we recall that $n = c^k$ and $t_k = T(c^k)$, and get

$$
T(n) = \begin{cases}
\Theta(n^m) & \text{for } a < c^m, \\
\Theta(\log_c(n)n^m) & \text{for } a = c^m, \\
\Theta(n^{\log_c(a)}) & \text{for } a > c^m.
\end{cases}
$$

(The last line follows from the observation that $a^k = c^{\log_c(a^k)} = c^{k\log_c(a)} = n^{\log_c(a)}$.) Therefore, there are three different types of behavior that can occur for the run time of a divide-and-conquer algorithm.

Run Time for Divide-and-Conquer Algorithms

Theorem 9.4.1. *Suppose that a divide-and-conquer algorithm splits a problem of size n into a subproblems, each of size n/c, and that the sum of the run times for* SPLIT *and* COMBINE *is a polynomial $p_m(n)$ of degree m. Assuming that $T(1)$ is positive, the run time for the divide-and-conquer algorithm satisfies*

$$
T(n) = aT(n/c) + p_m(n),
$$

and has order

$$
T(n) = \begin{cases}
\Theta(n^m) & \text{if } a < c^m, \\
\Theta(n^m \log(n)) & \text{if } a = c^m, \\
\Theta(n^{\log_c a}) & \text{if } a > c^m.
\end{cases}
$$

9.4.1   Example 1: A return to polynomial multiplication

In our two divide-and-conquer algorithms for polynomial multiplication, c was equal to 2 because the size of each subproblem was half the size of the original problem. Since the split and combine operations used only addition and subtraction of polynomials, their combined run time was a linear polynomial, giving $c^m = c^1 = 2$. The straightforward algorithm had $a = 4 > c^m$ while in the improved algorithm using (9.4), $a = 3 > c^m$, and the third case of the theorem applies for both. As we saw earlier, the algorithm with $a = 3$ is eventually faster than the one with $a = 4$, since the smaller value of $a = 3$ reduces the run time from $\Theta(n^2)$ to $\Theta(n^{\log_2(3)})$. If we could somehow further reduce the number of subproblems to $a = 2$, the second case would apply and we would have that $T(n) = \Theta(n \log n)$. No known divide-and-conquer algorithm for polynomial multiplication has $a = 2$, but in Section 9.5 we'll see that a completely different type of algorithm for polynomial multiplication does achieve the complexity of $\Theta(n \log n)$.

9.4.2 Example 2: Matrix multiplication

Another problem in which divide-and-conquer leads to a fast algorithm is matrix multiplication. If two $n \times n$ matrices are split into four $n/2 \times n/2$ matrices, we can compute the matrix product using

$$\begin{bmatrix} A_{11} & A_{12} \\ A_{21} & A_{22} \end{bmatrix} \begin{bmatrix} B_{11} & B_{12} \\ B_{21} & B_{22} \end{bmatrix} = \begin{bmatrix} C_{11} & C_{12} \\ C_{21} & C_{22} \end{bmatrix},$$

where

$$C_{11} = A_{11}B_{11} + A_{12}B_{21},$$
$$C_{12} = A_{11}B_{12} + A_{12}B_{22},$$
$$C_{21} = A_{21}B_{11} + A_{22}B_{21},$$
$$C_{22} = A_{21}B_{12} + A_{22}B_{22},$$

which uses eight half-size multiplications. Because addition of $n \times n$ matrices can be done in time proportional to n^2, the equation for the run time of this divide-and-conquer algorithm is

$$T(n) = 8T(n/2) + bn^2.$$

This equation has $a = 8$, $c = 2$, $m = 2$, giving $a > c^m$ and $T(n) = \Theta(n^{\log_2 8}) = \Theta(n^3)$, which is the same as the order of magnitude of the run time for the standard row-times-column algorithm for matrix multiplication.

A faster divide-and-conquer algorithm was designed by Strassen [156], using seven half-size multiplications. **Strassen's algorithm** computes the eight products

$$M_1 = (A_{12} - A_{22})(B_{21} + B_{22}), \quad M_2 = (A_{11} + A_{22})(B_{11} + B_{22}),$$

$$M_3 = (A_{11} - A_{21})(B_{11} + B_{12}), \quad M_4 = (A_{11} + A_{12})B_{22},$$

$$M_5 = A_{11}(B_{12} - B_{22}), \quad M_6 = A_{22}(B_{21} - B_{11}),$$

$$M_7 = (A_{21} + A_{22})B_{11},$$

and uses them in

$$C_{11} = M_1 + M_2 - M_4 + M_6, \quad C_{12} = M_4 + M_5,$$
$$C_{21} = M_6 + M_7, \quad C_{22} = M_2 - M_3 + M_5 - M_7,$$

to obtain the product matrix. It is a simple exercise in algebra to prove that this algorithm is correct; the real difficulty was discovering the algorithm in the first place! Because addition and subtraction of matrices can be carried out in time proportional to n^2, the run time of Strassen's algorithm obeys the equation

$$T(n) = 7T(n/2) + bn^2,$$

and $T(n) = \Theta(n^{\log_2 7})$.

9.4.3   Example 3: MERGESORT

A concrete example of the second type of divide-and-conquer behavior is the MERGESORT algorithm. In this algorithm the input is a one-dimensional array with n entries, and the output is the array in which the original entries are rearranged into increasing order. It is a divide-and-conquer algorithm, with the array divided into two subarrays that are individually sorted and then combined to give the sorted array. The following algorithm captures this approach.

```
PROCEDURE MERGESORT(A, n)
    IF n > 1
    THEN Split A into two arrays of size n/2, A₁ and A₂
         MERGESORT(A₁, n/2)
         MERGESORT(A₂, n/2)
         MERGE(A₁, A₂)
```

When $n = 1$ is reached there is only one item in the array, and no sorting needs to be done. All of the real work gets done in the MERGE algorithm, which combines two sorted arrays A_1 and A_2 into one sorted array A. Without going into the details, we'll say that MERGE can be accomplished with $n - 1$ comparisons of entries and n moves. This yields the recurrence

$$T(n) = 2T(n/2) + bn,$$

for some constant b. Since $a = 2$ and $c^m = 2$, we are in the second case, and $T(n) = \Theta(n \log n)$.

9.4.4   Example 4: Applications of Newton's method

For an example of the first type of behavior we use Newton's method to estimate values of the reciprocal function $g(x) = 1/x$ and the square root function $g(x) = \sqrt{x}$. These are special cases of estimating functions $g(x)$ at values $x = A$ when $g(x)$ is a one-to-one function in some neighborhood of A. The assumption that $g(x)$ is one-to-one ensures that it has a left inverse $f(x)$ with $f(g(x)) = x$ in a neighborhood of $x = A$. We further assume that the left inverse is also one-to-one and twice differentiable in this neighborhood. These conditions hold for each of $g(x) = \sqrt{x}$ and $g(x) = 1/x$ on $x > 0$. Since $A = f(g(A))$, any approximation to the solution of $f(x) - A = 0$ is close to $g(A)$.

We've already used Newton's method (refer to Section 5.4) to locate the roots of $f(x) = A$ by iterating

$$N(x) = x - \frac{f(x) - A}{f'(x)} \, .$$

For this to be computationally feasible, the right side should be in a form that's easy to compute, which for us means that it can be computed in time that is polynomial in the size of the input. Under reasonable hypotheses (refer to [20, 155]), when a good initial approximation is used, each iteration of Newton's method doubles the number of correct digits. (Also refer to Theorem 5.4.1.) For instance, if an approximation x agrees with the root on the first five bits, then $N(x)$ agrees on the first ten bits, and applying the iteration another time, $N^{(2)}(x)$ agrees on twenty bits, and so forth. Therefore, for input of size n the run time $T(n)$ satisfies the divide-and-conquer recurrence

$$T(n) = T(n/2) + p(n) \, ,$$

where the polynomial $p(n)$ records the difficulty of computing $N(x)$ for the particular function under consideration. Newton's method is thus a divide-and-conquer algorithm, provided $N(x)$ can be computed from x in polynomial time.

Consider the reciprocal function $g(x) = 1/x$ on $x > 0$, a one-to-one function that is its own inverse. Since $f'(x) = -1/x^2$, Newton's formula for the iteration is

$$N(x) = x - \frac{1/x - A}{-1/x^2} = x(2 - Ax) \, .$$

In this form, only two multiplications and one subtraction (and no divisions) of n-bit numbers are required, and the run time satisfies

(9.7) $$T(n) = T(n/2) + p_2(n) \, ,$$

for some quadratic polynomial $p_2(n)$. Here $a = 1$, $c^m = 2^2 = 4$, and we're in the first case, with run time $T(n) = \Theta(n^2)$. Since division of real numbers can be performed using the algorithm implicit in $\frac{a}{b} = a \cdot \frac{1}{b}$, this also yields an order $\Theta(n^2)$ algorithm for division.

Finding square roots can also be performed in $\Theta(n^2)$ time using Newton's method. The one-to-one function $g(x) = \sqrt{x}$ has the differentiable inverse function $f(x) = x^2$, which is one-to-one on $x > 0$. Since $f'(x) = 2x$, the Newton iteration is

$$N(x) = x - \frac{x^2 - A}{2x} = \frac{x}{2} + \frac{A/2}{x} \, ,$$

where $A/2$ is an easy one-time calculation.[2] The complicated part is the division by x, for which we use the quadratic-time algorithm above. Therefore, the run time again satisfies (9.7) for some quadratic polynomial $p_2(x)$, and $T(n) = \Theta(n^2)$.

It's worth noting that this discussion shows that both division and finding square roots can be accomplished in the same order of time as multiplication. Although we have used the standard $\Theta(n^2)$ multiplication algorithm, the same argument shows that any multiplication algorithm whose order is at least $\Theta(n)$ yields division and square root algorithms of the same order. The technique of the next section gives a multiplication whose order is only slightly worse than $\Theta(n)$, and this faster multiplication can be used to speed up both division and square root.

9.5   The Fast Fourier Transform

In our general discussion of divide-and-conquer algorithms we referred to an algorithm for polynomial multiplication that has order $\Theta(n \log n)$, and this section is devoted to a description and explanation of the method. It is based on the technique known as the **Fast Fourier Transform**, usually abbreviated as FFT. The FFT-based polynomial multiplication algorithm is designed for **dense polynomials**, polynomials in which almost all of the coefficients are non-zero. Many large-scale polynomial multiplication problems involve **sparse polynomials** and for such problems there are other algorithms that easily outperform the FFT. So there are practical issues that must be considered before the theoretically good FFT algorithm is used.

The key difference between the FFT method for polynomial multiplication and our earlier methods is that here a polynomial is represented by some of its functional values rather than by its coefficients. The basis for the technique is the fact that an $(n-1)^{\text{st}}$-degree polynomial is uniquely determined by its values at n distinct points. To see this, consider the evaluation of a polynomial $f(x) = a_0 + a_1 x + \cdots + a_{n-1} x^{n-1}$ at the n different

---

[2]Here it's helpful to represent numbers in binary notation, because $A/2$ is just a binary shift of A and so is computationally trivial.

complex numbers $x = \lambda_1, \lambda_2, \ldots, \lambda_n$. Writing this as a matrix equation, we have

$$(9.8) \quad (a_0, \ldots, a_{n-1}) \begin{bmatrix} 1 & \cdots & 1 \\ \lambda_1 & \cdots & \lambda_n \\ \vdots & \vdots & \vdots \\ \lambda_1^{n-1} & \cdots & \lambda_n^{n-1} \end{bmatrix} = (f(\lambda_1), f(\lambda_2), \ldots, f(\lambda_n)),$$

where the matrix is the **Vandermonde matrix** V associated with $\lambda_1, \ldots, \lambda_n$. (Refer to Chapter 2.) When the λ_i are distinct, we know that V is invertible, and accordingly, the coefficients of $f(x)$ satisfy

$$(9.9) \quad (a_0, a_1, \ldots, a_{n-1}) = (f(\lambda_1), f(\lambda_2), \ldots, f(\lambda_n))V^{-1},$$

and so the coefficients can be computed from n values of the polynomial. Such a process is called **interpolation**, and (9.9) is called an **interpolation formula**. Schematically, we have the bijections

$$\text{Coefficients} \underset{\text{Interpolation}}{\overset{\text{Evaluation}}{\rightleftarrows}} \text{Values}$$

defined by

$$(a_0, a_1, \ldots, a_{n-1}) \xrightarrow{\text{EVALUATION}} (f(\lambda_1), f(\lambda_2), \ldots, f(\lambda_n))$$
$$\xrightarrow{\text{INTERPOLATION}} (a_0, a_1, \ldots, a_{n-1}),$$

where EVALUATION and INTERPOLATION are the processes that take us from one representation to the other.

For the purpose of polynomial multiplication, encoding a polynomial using n of its functional values is superior to using its coefficients because to obtain the corresponding value for the product polynomial we need only multiply the respective values of the factor polynomials. But there are some problems. First, the product of two polynomials of degree $n-1$ is a polynomial of degree $2(n-1)$, and therefore the product polynomial is determined by $2n - 1$ values. More seriously, you might remember from our discussion of Horner's Method in Chapter 5 that in general it takes $\Theta(n)$ operations to perform one evaluation of a polynomial of degree $n - 1$ and hence $\Theta(n^2)$ operations to evaluate it n times. In fact, standard interpolation techniques take $\Theta(n^2)$ operations to interpolate a polynomial of degree $n - 1$ through n *generic* points. What saves us here is the freedom to choose the n points in a way that speeds up both evaluation and interpolation.

For example, evaluating $f(0)$ takes no work at all, and evaluating $f(1)$ takes n additions and no multiplications. Similarly, evaluation of $f(-1)$ also takes no multiplications, since it's the sum of the coefficients that have even indices minus the sum of the other coefficients. When $\lambda_1, \ldots, \lambda_n$ are chosen to equal to the nth roots of unity, we will give algorithms for both EVALUATION and INTERPOLATION that have run time $\Theta(n \log n)$.

9.5.1   The general form of the Fast Fourier Transform

Using terminology from Chapter 8, the set G_n of nth roots of unity is a *group* under multiplication and has n elements. The **principal** n^{th} root of unity,

$$\omega = \cos(\theta) + i \sin(\theta) \quad \text{for } \theta = \frac{2\pi}{n},$$

has order n in this group, which means that every element of G_n can be uniquely written as ω^j for some $j = 0, 1, \ldots, n-1$. Since $1, \omega, \omega^2, \ldots, \omega^{n-1}$ are all distinct, specializing to $\lambda_j = \omega^{j-1}$ in the Vandermonde matrix of (9.8) gives the $n \times n$ invertible (and symmetric) matrix

$$(9.10) \qquad V(\omega) = \begin{bmatrix} 1 & 1 & \cdots & 1 \\ 1 & \omega^1 & \cdots & \omega^{n-1} \\ 1 & \omega^{1 \cdot 2} & \cdots & \omega^{(n-1) \cdot 2} \\ 1 & \omega^{1 \cdot 3} & \cdots & \omega^{(n-1) \cdot 3} \\ \vdots & \vdots & \ddots & \vdots \\ 1 & \omega^{1 \cdot (n-1)} & \cdots & \omega^{(n-1) \cdot (n-1)} \end{bmatrix}.$$

Noting that for $0 \le i, j < n$ the dot product satisfies

$$(1, \omega^i, \ldots, \omega^{(n-1)i}) \cdot (1, \omega^{-j}, \ldots, \omega^{-j(n-1)})^T = 1 + \omega^{i-j} + \cdots + \omega^{(n-1)(i-j)},$$

from (4.36) on page 87 we have

$$(1, \omega^i, \ldots, \omega^{(n-1)i}) \cdot (1, \omega^{-j}, \ldots, \omega^{-j(n-1)})^T = \begin{cases} n & \text{if } i = j, \\ 0 & \text{if } i \ne j. \end{cases}$$

Therefore, the inverse of $V(\omega)$ is

$$V(\omega)^{-1} = \frac{1}{n} \begin{bmatrix} 1 & 1 & \cdots & 1 \\ 1 & \omega^{-1} & \cdots & \omega^{-(n-1)} \\ 1 & \omega^{-2} & \cdots & \omega^{-2(n-1)} \\ 1 & \omega^{-3} & \cdots & \omega^{-3(n-1)} \\ \vdots & \vdots & \ddots & \vdots \\ 1 & \omega^{-(n-1)} & \cdots & \omega^{-(n-1)^2} \end{bmatrix} = \frac{1}{n} V(\overline{\omega}),$$

and

$$(9.11) \qquad V(\omega)^{-1} = \frac{1}{n} V(\overline{\omega}) = \frac{1}{n} \overline{V(\omega)}.$$

Writing $f(x) = a_0 + a_1 x + \cdots + a_{n-1}x^{n-1}$, $A = (a_0, \ldots, a_{n-1})^T$, and $Y = (f(1), f(\omega), \ldots, f(\omega^{n-1}))^T$, equations (9.8) and (9.9) become

(9.12) $$V(\omega)A = Y$$

and

(9.13) $$A = \frac{1}{n} \cdot \overline{V(\omega)}Y ,$$

which respectively correspond to EVALUATION and INTERPOLATION. The very special structure of these matrices will allow us to construct quick divide-and-conquer algorithms when n is chosen to be a power of 2.

Historically, the term **Discrete Fourier Transform** (or DFT) was used for the general idea described above, and Fast Fourier Transform was reserved for the specific implementation with $n = 2^k$. Nowadays, this distinction is blurred, and both are referred to as the FFT.

9.5.2 The FFT when $n = 2^k$

We set $n = 2^k$ (where k is a natural number), and let V_k denote the Vandermonde matrix $V(\omega)$ constructed using the nth roots of unity as in (9.10). We assume that the powers of the principal n^{th} root of unity ω have been found and stored in a table. The FFT also uses the $(n/2)^{\text{th}}$ roots of unity, which are already in the table and are found by proceeding through the table in steps of size two. We now establish an iterative process for getting $V(\omega)$ from $V(\omega^2)$. This will allow us to construct fast divide-and-conquer algorithms for EVALUATION and INTERPOLATION.

The matrices for the first, second, and fourth principal roots of unity are respectively

$$V(1) = [1], \quad V(-1) = \begin{bmatrix} 1 & 1 \\ 1 & -1 \end{bmatrix}, \quad \text{and } V(i) = \begin{bmatrix} 1 & 1 & 1 & 1 \\ 1 & i & -1 & -i \\ 1 & -1 & 1 & -1 \\ 1 & -i & -1 & i \end{bmatrix}.$$

Note that the first column of $V(-1)$ is a stack of two $V(1)$'s, and its second column is $\begin{pmatrix} V(1) \\ -V(1) \end{pmatrix}$. Interchanging the second and third columns of $V(i)$ gives

(9.14) $$\begin{bmatrix} 1 & 1 & 1 & 1 \\ 1 & -1 & i & -i \\ 1 & 1 & -1 & -1 \\ 1 & -1 & -i & i \end{bmatrix},$$

whose first two columns is a matrix stack of two copies of $V(-1)$, and its last two columns is the stack $\begin{pmatrix} A \\ -A \end{pmatrix}$, where

$$A = \begin{bmatrix} 1 & 1 \\ i & -i \end{bmatrix} = \begin{bmatrix} 1 & 0 \\ 0 & i \end{bmatrix} V(-1).$$

Although it's somewhat of a stretch, from this it can be seen that $V(i)$ is built in two steps. First construct a block matrix of four 2×2 blocks, each of which is either $V(i^2)$ or a product of $V(i^2)$ and the diagonal matrix. Then $V(i)$ is obtained from this matrix by performing a suitable permutation of columns. The surprising news is that this is the general procedure for passing from $V(\omega^2)$ to $V(\omega)$!

In what follows, the indexing of rows and columns of $2^k \times 2^k$ matrices will begin with 0. We want to determine the permutation of the columns of $V(\omega)$ that is used. For $k = 2$, we use the binary representations of $0, 1, 2, 3$,

$$00, \ 01, \ 10, \ \text{and} \ 11,$$

to define the permutation Rev_2 obtained by reversing the bits,

$$00, \ 10, \ 01, \ \text{and} \ 11.$$

Applying Rev_2 to the column numbers of $V(i)$ interchanges the second and third columns and keeps the other two fixed, and *this* is the permutation we want. What happens for $k = 3$ is recorded in Table 9.3, where the first row contains the numbers 0 through 7, the second row their binary representations, the third row the bit reversals, and the fourth row the decimal equivalents of these reversed representations. Reading only the top and bottom rows, the permutation Rev_3 of $0, 1, \ldots, 7$ swaps 1 with 4 and 3 with 6.

TABLE 9.3. The bit reversal permutation on $0, \ldots, 7$

0	1	2	3	4	5	6	7
000	001	010	011	100	101	110	111
000	100	010	110	001	101	011	111
0	4	2	6	1	5	3	7

For any k the **bit reversal permutation** Rev_k of $\{0, 1, \ldots, 2^k - 1\}$ is obtained by reversing the bits in this way. The definition of Rev_k immediately gives that Rev_k^2 is the identity and that

$$(9.15) \qquad \text{Rev}_k(j) \text{ is an even integer} \iff 0 \leq j < 2^{k-1}.$$

Application of Rev_k to the column numbers of the $2^k \times 2^k$ identity matrix gives a **permutation matrix**, which we'll denote by P_k. You can check that the first three are

$$P_1 = \begin{bmatrix} 1 & 0 \\ 0 & 1 \end{bmatrix}, \; P_2 = \begin{bmatrix} 1 & 0 & 0 & 0 \\ 0 & 0 & 1 & 0 \\ 0 & 1 & 0 & 0 \\ 0 & 0 & 0 & 1 \end{bmatrix}, \; P_3 = \begin{bmatrix} 1 & 0 & 0 & 0 & 0 & 0 & 0 & 0 \\ 0 & 0 & 0 & 0 & 1 & 0 & 0 & 0 \\ 0 & 0 & 1 & 0 & 0 & 0 & 0 & 0 \\ 0 & 0 & 0 & 0 & 0 & 0 & 1 & 0 \\ 0 & 1 & 0 & 0 & 0 & 0 & 0 & 0 \\ 0 & 0 & 0 & 0 & 0 & 1 & 0 & 0 \\ 0 & 0 & 0 & 1 & 0 & 0 & 0 & 0 \\ 0 & 0 & 0 & 0 & 0 & 0 & 0 & 1 \end{bmatrix}.$$

Because Rev_k is the bit-reversal permutation, P_k has order two in the group of invertible $k \times k$ matrices and is its own inverse. Applying the permutation matrix P_k on the right of $V(\omega)$ gives $F(\omega) = V(\omega)P_k$, the matrix obtained by permuting the columns of $V(\omega)$ by Rev_k. For instance, $F(i)$ is the matrix in (9.14).

We refer to $F(\omega)$ as the **Fourier matrix of size** 2^k. Since the $(i,j)^{\mathrm{th}}$ entry of $V(\omega)$ is

$$v_{ij} = \omega^{ij},$$

then the $(i,j)^{\mathrm{th}}$ entry of $F(\omega)$ is

$$f_{ij} = \omega^{i\,\mathrm{Rev}_k(j)}.$$

We use this to show that $F(\omega)$ always has the form we've already noticed for $F(i)$.

Theorem 9.5.1. *Let ω be any primitive 2^k root of unity. Then $F(\omega)$ satisfies the recurrence*

$$F(\omega) = \begin{bmatrix} F(\omega^2) & DF(\omega^2) \\ F(\omega^2) & -DF(\omega^2) \end{bmatrix},$$

where D is the diagonal matrix with diagonal entries $1, \omega, \ldots, \omega^{2^{k-1}-1}$.

Proof. The indices in the upper left quadrant of $F(\omega)$ satisfy $0 \le i,j < 2^{k-1}$, and (9.15) implies that $\mathrm{Rev}_k(j)$ is even for such j. Because dividing $\mathrm{Rev}_k(j)$ by 2 removes the least significant bit,

$$f_{ij} = \omega^{i\,\mathrm{Rev}_k(j)} = (\omega^2)^{i\frac{\mathrm{Rev}_k(j)}{2}} = (\omega^2)^{i\,\mathrm{Rev}_{k-1}(j)},$$

and this proves that the upper left quadrant of $F(\omega)$ is $F(\omega^2)$.

In either lower quadrant, the row index satisfies $2^{k-1} \le i < 2^k$, which we can write as $i = I + 2^{k-1}$ for some $0 \le I < 2^{k-1}$. From this,

$$f_{ij} = \omega^{i\,\mathrm{Rev}_k(j)} = \omega^{I\,\mathrm{Rev}_k(j)} \cdot \omega^{2^{k-1}\,\mathrm{Rev}_k(j)} = f_{Ij} \cdot (-1)^{\mathrm{Rev}_k(j)}.$$

Because $\operatorname{Rev}_k(j)$ is even in both left quadrants, this proves that the two left quadrants are the same and also that the matrices that form the two right quadrants are negatives of each other.

It remains to show that the upper right quadrant is $DF(\omega^2)$. In that quadrant the indices satisfy $0 \le i < 2^{k-1} \le j$, and $j = J + 2^{k-1}$ for some $0 \le J < 2^{k-1}$. Since $\operatorname{Rev}_k(j) = 2\operatorname{Rev}_{k-1}(J) + 1$,

$$\frac{f_{ij}}{\omega^i} = \omega^{i(\operatorname{Rev}_k(j))}\omega^{-i} = \omega^{i(\operatorname{Rev}_k(j)-1)} = (\omega^2)^{i\operatorname{Rev}_{k-1}(J)},$$

which is the $(i, J)^{\text{th}}$ entry of $F(\omega^2)$. This completes the proof.   □

We will next show that the recursive structure of this Fourier matrix yields a divide-and-conquer strategy for both EVALUATION and INTERPOLATION.

9.5.3 Fast evaluation and fast interpolation

Recording the polynomial $f(x) = a_0 + a_1 x + \cdots + a_{n-1}x^{n-1}$ as the coefficient vector $A = (a_0, \ldots, a_{n-1})^T$, we recall that (9.12) can be used to obtain the evaluation vector $Y = (f(1), f(\omega), \ldots, f(\omega^{n-1}))^T$. We have encoded this in the matrix equation

$$(9.16) \qquad\qquad Y = F(\omega)P_k A,$$

and the reverse interpolation process given in (9.13) can be written as

$$(9.17) \qquad\qquad A = \frac{1}{n} \cdot \overline{F(\omega)}P_k Y.$$

So, we see that *INTERPOLATION is essentially the same process as EVALUATION*! This means that the analyses of INTERPOLATION and EVALUATION are the same.

To analyze EVALUATION, we first consider the time required to construct $P_k A$. Computing $\operatorname{Rev}_k(j)$ for any $0 \le j < 2^k = n$ can be done in time proportional to the length k, and computing the permutation Rev_k takes time $O(n \log n)$. Once Rev_k is known, applying it to A amounts to swapping pairs of indices (or pointers), and this takes time $O(n)$. Therefore, the time for finding $X = P_k A$ is $O(n \log n)$.[3]

The decomposition of the Fourier matrix $F(\omega)$ given in Theorem 9.5.1 suggests a divide-and-conquer strategy for computing $Y = F(\omega)X$. To see this, we write any $X \in \mathbb{C}^{2^k}$ as $X = \begin{pmatrix} X_1 \\ X_2 \end{pmatrix}$ where X_1 is the vector

---

[3] Many texts ignore the time necessary to compute the permutation, since in practice it takes very little time compared to the rest of the algorithm. Even if we had to apply a permutation in every iteration, the asymptotic run time is not affected (for this, refer to Exercise 9.20).

consisting of the first 2^{k-1} components of X, and X_2 contains the last 2^{k-1} components. The special form of $F(\omega)$ gives

$$F(\omega)X = F(\omega)\begin{pmatrix} X_1 \\ X_2 \end{pmatrix} = \begin{pmatrix} F(\omega^2)X_1 + DF(\omega^2)X_2 \\ F(\omega^2)X_1 - DF(\omega^2)X_2 \end{pmatrix}.$$

Noting that the products $F(\omega^2)X_1$ and $F(\omega^2)X_2$ need only be computed once, we conclude that this decomposition gives a divide-and-conquer algorithm for computing $F(\omega)X$. A schematic description is

$$A \xrightarrow{\text{permute}} P_k A =: \begin{pmatrix} X_1 \\ X_2 \end{pmatrix} \xrightarrow{\text{half-size}} \begin{pmatrix} F(\omega^2)X_1 \\ F(\omega^2)X_2 \end{pmatrix}$$

$$\xrightarrow{\text{combine}} \begin{pmatrix} F(\omega^2)X_1 + DF(\omega^2)X_2 \\ F(\omega^2)X_1 - DF(\omega^2)X_2 \end{pmatrix} = Y,$$

and likewise, INTERPOLATION can be described by the schematic

$$Y \xrightarrow{\text{conjugate and permute}} P_k\overline{Y} =: \begin{pmatrix} X_1 \\ X_2 \end{pmatrix} \xrightarrow{\text{half-size}} \begin{pmatrix} F(\omega^2)X_1 \\ F(\omega^2)X_2 \end{pmatrix}$$

$$\xrightarrow{\text{combine}} \begin{pmatrix} F(\omega^2)X_1 + DF(\omega^2)X_2 \\ F(\omega^2)X_1 - DF(\omega^2)X_2 \end{pmatrix} = V(\omega)\overline{Y}$$

$$\xrightarrow{\text{conjugate and scale by } 1/n} \frac{1}{n}\overline{V(\omega)Y} = A.$$

Note that in this process the matrix $F(\omega)$ never needs to be calculated, because all Fourier matrices in the scheme are simply recursive calls to a procedure. Also, the diagonal entries of D that are powers of ω have been stored in an array and can be pulled from the array when necessary.

We've already shown that the permutation steps can be done in $O(n \log n)$ time, and we now compute the run time of the divide-and-conquer part. Since the combine step involves multiplying by a diagonal matrix and then performing an addition and a subtraction, this step takes $\Theta(n)$ operations. Because the problem has been divided into two half-size subproblems, the divide-and-conquer recurrence has $a = c = 2$, and $T(n) = 2T(n/2) + bn$, which is the second case of the divide-and-conquer formula in Theorem 9.4.1, and so $T(n) = \Theta(n \log n)$. Finally, when n is not a power of two, appending zero coefficients just increases the implied constant in $\Theta(n \log n)$. This completes the proof that the FFT multiplies two polynomials of degree at most $n - 1$ in $\Theta(n \log n)$ operations.

9.5.4   The fast polynomial multiplication algorithm

The entire algorithm for computing the coefficients of the product of two polynomials is summarized in the following table.

FFT Polynomial Multiplication

Let $f, g \in \mathbb{C}[x]$ be polynomials with $\deg(f), \deg(g) < n$.

1. Place the coefficients of the polynomial f in the first n components of a vector A of length 2^{k+1}, where 2^k is the smallest power of 2 that is at least n. Set the remaining components of X to zero.

2. Place the coefficients of g into a vector B of length 2^{k+1}.

3. Permute both A and B, obtaining X and Y respectively.

4. For the principal 2^{k+1}-th root of unity ω, use the divide-and-conquer algorithm to compute both $F(\omega)X$ and $F(\omega)Y$.

5. Let Z be the componentwise product of these two vectors.

6. Permute and conjugate Z.

7. Use the divide-and-conquer algorithm to compute $F(\omega)Z$.

8. Conjugate each component of the result in Step 7 and divide by 2^{k+1}. This is the vector of coefficients in the product fg.

The computation is only *approximate*, because of the round-off in floating point operations and because we can only use finite approximations to the powers of ω. The first $2n - 1$ components of the output vector are approximations to the coefficients of the product polynomial. Although all higher coefficients in the product $f\,g$ are zero, some of these components may be non-zero in the output. The size of these extra components gives some indication of the accuracy of the computation.

Schematically, the algorithm is

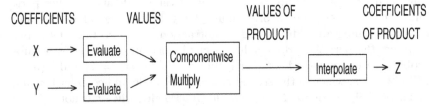

FIGURE 9.3. Schematic representation of FFT polynomial multiplication.

Any divide-and-conquer algorithm must specify what is meant by "small", the size of problem that is computed "by hand". In this algorithm, we could take as the "small case" the constant polynomials, compute their FFT using *no* operations, and proceed to the combine step. Other variations are possible. Since $f(\pm 1)$, $g(\pm 1)$, $f(\pm i)$, $g(\pm i)$, can all be computed using *no* multiplications, two component vectors or four component vectors can also be conveniently used as the base case. The choice of base case does not affect the Big Oh order of the run time, but this choice does affect the actual run time by changing the constant hidden by the Big Oh.

We illustrate the FFT algorithm by applying it to the simple example of squaring the polynomial $2 + x$. (In the following, we will write the vectors as rows, even though the description of our algorithm uses column vectors.) In Step 1, we form the vector $(2, 1, 0, 0)$, and Step 2 is the same for this example. In Step 3 we use Rev_2 to permute the vectors and obtain $(2, 0, 1, 0)$. In Step 3 the recursive routine is applied to the two vectors, which we break into the two vectors $(2, 0)$ and $(1, 0)$. Since

$$F(-1) = \begin{bmatrix} 1 & 1 \\ 1 & -1 \end{bmatrix},$$

premultiplication of $(2, 0)$ and $(1, 0)$ by $F(-1)$ gives $(2, 2)$ and $(1, 1)$. We next premultiply $(1, 1)$ by the diagonal matrix with 1 and i on the diagonal, yielding $(1, i)$. We now add and subtract this from $(2, 2)$ and place the results in a vector with four components and get

$$(2 + 1, 2 + i, 2 - 1, 2 - i) = (3, 2 + i, 1, 2 - i).$$

These numbers are supposed to be the values of the polynomial $f(x) = 2 + x$ at $x = 1$, i, -1, and $-i$, which indeed they are.

In Step 5 we multiply the two (in our case identical) "value vectors" componentwise. This amounts to squaring each component of the vector above, and this yields

$$Z = (9, 3 + 4i, 1, 3 - 4i),$$

whose coordinates are the values of the polynomial $4 + 4x + x^2$ at $x = 1$, i, -1, and $-i$. In Step 6 we permute using Rev_2 and conjugate to get the vector

$$(9, 1, 3 - 4i, 3 + 4i).$$

In Step 7 we apply the same recursive algorithm, starting at the above vector. It splits into $(9, 1)$ and $(3 - 4i, 3 + 4i)$, and these become $(10, 8)$ and $(6, -8i)$ in the base case of the recursion. We multiply $(6, -8i)$ by the diagonal matrix with 1 and i on its diagonal to get $(6, 8)$, and then $(10, 8)$ and $(6, 8)$ are added, subtracted, and placed in the vector

$$(10 + 6, 8 + 8, 10 - 6, 8 - 8) = (16, 16, 4, 0).$$

In Step 8 we conjugate (which has no effect, since our values at this point are real) and divide by $2^2 = 4$ to get the vector

$$(4, 4, 1, 0),$$

the coefficient vector for $f(x)g(x) = (2 + x)^2 = 4 + 4x + x^2$.

Notice that the FFT multiplies *complex* polynomials. Some improvement is possible if all coefficients in both polynomials are real. One minor change is that it is never necessary to conjugate in Step 6. More significantly, since the evaluation is done at $2^{k+1} \approx 2n$ complex points, one can pack a pair of real numbers into one complex number and design a real FFT, which evaluates at only 2^k instead of 2^{k+1} complex points. (For details on this, refer to Exercise 9.21 and [2].)

9.6   Average Case Analysis

Until now we've considered the run time of an algorithm to be a function of the input size. In practice, an algorithm might treat all inputs of the same size in the same way or it could handle some inputs more quickly than others. Because of this, we define the best case, worst case, and average case run times for an algorithm. The maximum run time over all inputs of the same size is called the **worst case run time**, and the minimum run time over all inputs of the same size is called the **best case run time**. Averaging the run time over all inputs of the same size is called the **average case run time**. When the probability associated with each of the various inputs of a particular size is unknown, it is difficult to calculate the average case time. For definiteness and simplicity of calculating the average case, it's often assumed that each input of a fixed size is equally likely to occur.

9.6.1   The LARGETWO algorithm

Consider the following algorithm, which finds the two largest entries in a one-dimensional array and assigns these values to the variables FIRST and SEC.

```
PROCEDURE   LARGETWO(C)
        FIRST:= C[1]
        SEC:= C[2]
        FOR I = 2 TO n DO
           IF C[I] > FIRST
              THEN SEC:=FIRST ; FIRST:= C[I]
              ELSE IF C[I] > SEC
                      THEN SEC:= C[I]
```

Note that this algorithm assumes that the array has at least two components, and so the results are unpredictable when it is used with a one-element array.

The ground rules for our run time analysis are that we'll count the number of comparisons of array entries and ignore any comparisons of numbers used to control the FOR loop. Each time through the FOR loop the algorithm makes either one or two comparisons, and so for an n-element array the algorithm uses at least $n - 1$ comparisons and at most $2(n - 1)$ comparisons. If the data happened to be arranged in order of increasing coordinates,

$$C[1] < C[2] < \cdots < C[n],$$

the algorithm performs exactly $n - 1$ comparisons, and this implies that the best case run time is $B(n) = n - 1$. The algorithm makes $2(n - 1)$ comparisons when $C[1]$ is the largest entry, and this is the worst case, with run time $W(n) = 2(n - 1)$.

What about the average case? Is it close to worst case, close to best case, or midway between the worst and best cases? Let $A(n)$ be the number of comparisons used on average by this algorithm. Since the algorithm begins with the first entry and proceeds in one direction across the array, we may reasonably assume that for the first $n - 1$ entries the algorithm on average uses $A(n-1)$ comparisons, the same number it uses if the last entry weren't there. For the last entry, it uses at least one comparison. If $C[n]$ is not the largest entry, then $C[n] > FIRST$ is false, and a second comparison must be made. If $C[n]$ *is* the largest entry, then $C[n] > FIRST$ is true, and only that one comparison is made. By our assumption of uniformity, the probability that $C[n]$ is the largest is $1/n$, and from these considerations, for all $n \geq 2$,

$$(9.18) \qquad A(n) = A(n - 1) + 2\left(1 - \frac{1}{n}\right) + 1 \cdot \frac{1}{n} = A(n - 1) + 2 - \frac{1}{n}.$$

When the array has two entries, uniformity implies that the maximum entry is equally likely to lie in either entry, and so two comparisons are required exactly half of the time, giving $A(2) = 3/2$. Since (9.18) is a first–order recurrence with eigenvalue $\lambda = 1$ and forcing function $\psi(i) = 2 - 1/i$, the solution is

$$A(n) = \sum_{i=2}^{n}\left(2 - \frac{1}{i}\right) = 2(n - 1) - \sum_{i=2}^{n}\frac{1}{i} \quad \text{for all } n \geq 2.$$

Using the result on the partial sums of the harmonic series given in Exercise 8.39, $A(n)$ is therefore asymptotically approximated by $2(n - 1) - \ln(n) + c$, where c is some constant. From this we see that the average case behavior of the algorithm is very close to the worst case, $W(n) = 2(n-1)$, especially as compared with the best case run time, $B(n) = n - 1$. Also, notice that the average of the worst and best case run times is $\frac{3}{2}(n - 1)$, a severe underestimate of the actual average case behavior.

Here's another analysis of the average case run time for LARGETWO. Let $C[j]$ be the maximal entry. From our assumption of uniformity, the probability that $C[j]$ is the largest entry is $1/n$. As noted above, $2(n-1)$ comparisons are made for $j = 1$. For $j \geq 2$ the algorithm performs $A(j-1)$ comparisons before the loop reaches $I = j$, then for $I = j$ the loop makes one comparison, and for each of $I = j+1, \ldots, n$ two comparisons are used. This gives

$$nA(n) = \sum_{j=2}^{n} [A(j-1) + 1 + 2(n-j)].$$

Expanding the sum gives

$$nA(n) = \sum_{j=2}^{n} A(j-1) + \sum_{j=2}^{n} 1 + 2 \sum_{j=1}^{n} (n-j)$$

$$= \sum_{j=1}^{n-1} A(j) + n - 1 + 2 \sum_{j=0}^{n-1} j$$

$$= \sum_{j=1}^{n-1} A(j) + n - 1 + (n-1)n$$

$$= \sum_{j=1}^{n-1} A(j) + n^2 - 1.$$

Therefore, for all $n \geq 2$,

(9.19)
$$nA(n) = \sum_{j=1}^{n-1} A(j) + n^2 - 1,$$

a recurrence that depends on all previous terms. Similarly to some recurrences in Chapter 4, its solution can be shown to satisfy a first–order recurrence. Since

$$(n-1)A(n-1) = \sum_{j=1}^{n-2} A(j) + (n-1)^2 - 1,$$

subtraction of equations gives

$$nA(n) - (n-1)A(n-1) = A(n-1) + n^2 - (n-1)^2 = A(n-1) + 2n - 1;$$
$$nA(n) = nA(n-1) + 2n - 1,$$

which is the same recurrence found earlier in (9.18).

9.6.2   The QUICKSORT algorithm

Next we consider QUICKSORT, an algorithm whose average case analysis is more complicated. The algorithm sorts an array S of numbers according to increasing size (and we allow repeated entries in S). The output of the algorithm is a sorted string of numbers, and the dots in the RETURN statement indicate concatenation of strings.

```
PROCEDURE   QUICKSORT(S)
    Pick an entry α of S at random.
    Divide S into three parts, S₁,S₂,S₃, where
        S₁ is the set of entries of S that are less than α.
        S₂ is the set of entries of S that equal α.
        S₃ is the set of entries of S that are greater than α.
    RETURN(QUICKSORT(S₁) · S₂· QUICKSORT(S₃))
```

This is a **random** algorithm, because it uses a random choice at some point, in this case for the choice of the entry α. The choice could be randomized by using a random number generator, but usually a rule is used in the hope that the distribution of the input data is not significantly correlated with the choice of rule. We'll assume that this correlation is low enough to allow us to treat the α's generated by the rule to be truly random. Three popular rules for choosing α are: always choose the first entry; always choose the last entry; always choose the middle entry. Another popular method is to choose the median of the first, last, and middle entries.

The run time of this algorithm on an array S of size n satisfies

$$T(S) = T(S_1) + bn + T(S_3),$$

where b accounts for the time used to place entries into the three sets and return the answer. The worst case occurs when S_1 or S_3 contains $n - 1$ entries and

$$W(n) = W(n - 1) + bn,$$

giving

$$W(n) = 1 + b \sum_{i=1}^{n} i = 1 + \frac{b}{2}n(n + 1) = \Theta(n^2).$$

The best case for QUICKSORT occurs when S_1 and S_3 each contain approximately $n/2$ entries. Then

$$B(n) = 2B(n/2) + cn,$$

for some constant c. This is a divide-and-conquer recurrence with $a = c = 2$ and $m = 1$, and $B(n) = \Theta(n \log n)$.

So, there is a large gap between the worst case, $W(n) = \Theta(n^2)$, and the best case, $B(n) = \Theta(n \log n)$. Does the algorithm on average behave closer to worst case or closer to best case? Notice that if the average case run time were the numerical average of worst case and best case, then the average would be $\Theta(n^2)$ and therefore close to the worst case. We show that the opposite occurs, and find that on average, the run time of QUICKSORT is $\Theta(n \log n)$ and hence behaves close to best case. This is intuitively plausible because nearly even splits should on average occur much more frequently than one-sided splits.

As usual, for $n \geq 1$ let $A(n)$ be the average case run time for an array with n entries and assume that α is equally likely to be any of the entries. Letting $A(0) = 0$, for $n \geq 1$,

$$A(n) = \frac{1}{n} \sum_{j=1}^{n} [A(j-1) + A(n-j) + b(n+1)],$$

where $b(n+1)$ is the time for split and combine (and is written in this form because it slightly simplifies the computation.) Therefore,

$$nA(n) = \sum_{j=1}^{n} A(j-1) + \sum_{j=1}^{n} A(n-j) + bn(n+1) = 2\sum_{j=1}^{n-1} A(j) + bn(n+1).$$

Using the technique of the previous example,

$$nA(n) - (n-1)A(n-1) = 2A(n-1) + 2bn,$$

giving

$$nA(n) = (n+1)A(n-1) + 2bn;$$
$$\frac{A(n)}{n+1} = \frac{A(n-1)}{n} + \frac{2b}{n+1}.$$

Setting $Z(n) = A(n)/(n+1)$, this becomes

$$Z(n) = Z(n-1) + \frac{2b}{n+1},$$

a first–order recurrence with forcing function $\psi(j) = 2b/(j+1)$, giving

$$Z(n) = c_1 + 2b \sum_{j=1}^{n} \frac{1}{j+1}$$

and

$$A(n) = c_1(n+1) + 2b(n+1) \sum_{j=1}^{n} \frac{1}{j+1},$$

for some constant c_1. Using Exercise 8.39 again gives $A(n) = \Theta(n \log(n))$.

This analysis shows that the average case behavior of QUICKSORT is close to the best case behavior, and suggests that QUICKSORT may be a good method to use to sort an array. However, remember that the analysis assumes that each permutation of the input behavior is equally likely. If that assumption were not true, then it isn't clear that QUICKSORT is a good algorithm to use. For instance, what if the data were in order except for a few items? If you always choose the first input for α, then the splits usually give one very large set, and it can be proved that the average run time of QUICKSORT for this type of data has order $\Theta(n^2)$.

A lesson to be learned from these two examples is that you know that the average case is *somewhere* between worst case and best case, but it can be anywhere between worst and best. Further, in many practical cases you'll have no idea about the distribution of the input data. The assumption of a uniform probability distribution often simplifies the average case calculation, but it is probably not applicable to a general input distribution.

9.7   Exercises

Ex 9.1. Use our perspective of moving the largest disk to derive a recurrence for the minimal number of moves needed to solve the n-disk Towers of Hanoi puzzle. Solve your recurrence with appropriate initial conditions and conclude that any algorithm that solves the n-disk Towers of Hanoi puzzle has run time $\Omega(2^n)$. (Refer to Chapter 1 for the meaning of Ω.)

Ex 9.2. Give an algorithm for the Towers of Hanoi puzzle that uses more than the minimal number of moves. Derive and solve a recurrence for your algorithm and show that it uses more moves than the minimum counted in the previous exercise. (If you don't see the usefulness of constructing algorithms that take longer than necessary, consider the problem of The Cab Driver and the Tourist.)

Ex 9.3. In the following algorithm, the towers are labeled 0, 1, and 2 rather than A, B, and C. The variable COUNT contains n bits numbered from 1 to n starting at the rightmost bit. The positions in COUNT alternate between odd and even (in that order).

```
PROCEDURE TOWERS(n)
    T:=0 (*Tower number computed modulo 3*)
    COUNT:=0
    P:={ +1   if n is even
       { -1   if n is odd
    WHILE true DO
        Move Disk 1 from T to T+P
        T := T+P
        COUNT := COUNT+1
        IF COUNT = all 1's THEN exit
        IF Rightmost 0 in COUNT is in an even position
            THEN move disk from T-P to T+P
            ELSE move disk from T+P to T-P
        COUNT := COUNT+1
    ENDWHILE
```

(a) Without using recurrences show that this algorithm uses the minimal number of moves.

(b) Even if incrementing COUNT and finding the rightmost 0 takes time cn for constant c, show that the run time of this algorithm is $\Theta(2^n)$.

Ex 9.4. (Refer to [19].) The following algorithm for the Towers of Hanoi problem has the towers arranged in a circle. By considering the situation in which the largest disk is moved, derive and solve a recurrence for the number of moves made and a recurrence for the run time of this algorithm.

```
PROCEDURE
    Move smallest disk one tower clockwise
    WHILE a disk other than the smallest disk can be moved
        DO move that disk
            move the smallest disk one tower clockwise
    ENDWHILE
```

(Notice that if the number n of disks is even, then the disks are moved one tower counterclockwise, while if n is odd, the disks are moved one tower clockwise.)

Ex 9.5. Which of the three algorithms for the Towers of Hanoi puzzle is fastest? The recursive algorithm and the algorithms from the last two exercises all have run time $\Theta(2^n)$. To determine which is really fastest, we suggest programming them and running them on a real computer. (You may want to make a small modification in the algorithm from the last exercise, always moving the disks to the same tower regardless of the parity of n.) Can you explain why one algorithm might be slower than the others in a real-world programming environment?

Ex 9.6. For a given one-dimensional array A, let $A[i]$ denote the content of the ith entry of A. Show that the following procedure has run time $\Theta(n)$ by deriving and solving the appropriate recurrence. Describe in English what the algorithm accomplishes.

```
    PROCEDURE  LARGEST(n)
        IF n > 1 THEN LARGEST(n − 1)
                    IF A[n − 1] > A[n]
                       THEN TEMP := A[n]
                            A[n] := A[n − 1]
                            A[n − 1] := TEMP
```

Ex 9.7. Using the result of the last exercise, give an inductive proof that the following algorithm sorts $(A[1], A[2], \ldots, A[n])$. Derive and solve a recurrence for the run time.

```
        PROCEDURE  SORT(n)
            IF n > 1 THEN LARGEST(n)
                          SORT(n − 1)
```

Ex 9.8. According to an old tale told by Édouard Lucas, the monks in a secret monastery in Hanoi are performing the moves for the Towers of Hanoi puzzle with 64 disks. When they finish, the world will end. If it takes them one minute to move a disk, should you worry about the end of the world? What if they can move one disk per second? After a wily computer salesperson convinces the monks to automate so that they can simulate moving a disk in one nanosecond, should you worry? (Also refer to the title story in *The Nine Billion Names of God: The Best Short Stories of Arthur C. Clarke*, Harcourt, Brace & World, New York, 1967.)

Ex 9.9. Let us (mis)construe the Towers of Hanoi to mean that an order of disks on a peg is acceptable as long as the disk on the bottom is the largest. Give an algorithm that moves a single stack of n disks from one peg to another in a minimum number of moves, and find a formula for that minimum number of moves. The disks are sorted from largest on bottom to smallest on top at the start, and are to be sorted in the same order, but on a different peg at the end (refer to [111]).

The following secure locking system is used in the next four exercises.

The locking system assumes that n locks are connected so that

1. Lock 1 may be changed from locked to unlocked or from unlocked to locked at any time.

2. For any $j > 1$, lock j may be changed from locked to unlocked (or vice versa) only if lock $j - 1$ is locked and locks 1 through $j - 2$ are unlocked.

Ex 9.10. One strategy for designing such a system involves thinking of unlocking all the locks in terms of unlocking a subset of locks, unlocking the last lock, re-locking the subset, and repeating this process until all locks have been unlocked. This strategy naturally leads to a design in which you have two mutually recursive procedures (or one recursive procedure with a switch that indicates whether the procedure is locking or unlocking.)
  (a) Use this strategy to design a recursive algorithm that gives the sequence of operations needed to unlock a series of n locks that are all initially locked.
  (b) Find a pair of recurrences that give the time and number of locking/unlocking operations used by your algorithm.
  (c) Solve the recurrences in part (b).

Ex 9.11. A second strategy rests on a fairly simple observation. The locks are in a specific configuration right after the last lock is unlocked, and to unlock the next-to-last lock, it is necessary to re-configure the locks. You can design a simple recursive procedure that takes the locks from the first configuration to the second configuration. This strategy leads to a design with two nested recursive procedures.
  (a) Use this strategy to design a recursive algorithm that gives the sequence of operations needed to unlock a series of n locks that are all initially locked.
  (b) Find the pair of recurrences that give the run time and number of locking/unlocking operations used by your algorithm.
  (c) To solve the recurrences in part (b), observe that the first depends on the second, but the second does not depend on the first. Thus, you can solve the second equation and plug its solution into the first equation and solve.

Ex 9.12. Note that the procedures in the previous two exercises use the same amount of time and number of moves because they do the same operations in reverse order. Show that this algorithm uses the minimal number of locking/unlocking operations.

Ex 9.13. To get a feel for the puzzle you've investigated in the last problems, go to your local toy store and get SPIN-OUT®. This is a physical embodiment of the puzzle with seven locks. Amaze your friends by solving SPIN-OUT® in the minimum number of moves.

Ex 9.14. Give a divide-and-conquer algorithm for finding the largest entry in a vector with n components and show that your algorithm uses $n - 1$ comparisons.

Ex 9.15. Design a divide-and-conquer algorithm to find the two largest entries in an array. Show that your algorithm is correct and calculate the number of comparisons it uses. Give an example that shows that your algorithm uses more comparisons than necessary.

Ex 9.16. Suppose you have a large number of coins and a two-pan balance whose pans are as large as needed. The balance tells you whether the coins in one pan weigh the same as the coins in the other pan or which set of coins is heavier. Among your coins is exactly one that has a different weight from the other coins.

(a) Assuming that the number of coins is a power of 3, design a divide-and-conquer algorithm to find the odd coin.
(b) Prove that your algorithm is correct.
(c) Find and solve a recurrence for the number of times your algorithm uses the balance.

Ex 9.17. Two finite strings C_1 and C_2 are said to **commute** if juxtaposing in either order gives the same string (that is, $C_1C_2 = C_2C_1$). Give a constructive proof that C_1 and C_2 commute iff there is a string w and two natural numbers k_1 and k_2 such that $C_1 = w^{k_1}$ and $C_2 = w^{k_2}$ (where w^k denotes the juxtaposition of k copies of w.) Give a constructive proof that such a w exists. Use your proof to construct an algorithm that finds w for two commuting strings.

Ex 9.18. Show how a divide-and-conquer strategy with the clever identity (9.4) can be used to construct an algorithm for multiplying large integers in run time $\Theta(n^{\log_2 3})$, where n is the number of bits in the larger factor.

Ex 9.19. Show that the power

$$\begin{bmatrix} 1 & 1 & 2 \\ 1 & 0 & 0 \\ 0 & 1 & 0 \end{bmatrix}^n$$

can be computed in $O(n)$ bit operations. Which special properties of this matrix allow this "fast" algorithm?

Ex 9.20. Consider a recurrence of the form $T(n) = 2T(n/2) + g(n)$, where $g(n)$ is a nonnegative function. Argue that the asymptotic order of growth of $T(n)$ is monotonic in $g(n)$ in the following sense: If $h(n) = O(g(n))$ and $S(n) = 2S(n/2) + h(n)$, then $S(n) = O(T(n))$. Then, using the technique of letting $t_k = T(2^k)$, show that $T(n) = O(n \log n)$ iff $g(n) = O(n \log n)$.

Ex 9.21. Set up and solve a recurrence for the number of *real* multiplications the FFT algorithm uses to multiply two real polynomials with n coefficients.

Ex 9.22.   (a) The number of multiplications used by the FFT depends on the size of the agreed upon "small-size" problem. Work out the difference in the number of multiplications among FFTs recursing to one component; two components; four components.

(b) Compare your answers in part (a) with the n^2 multiplications used by the standard algorithm for polynomial multiplication.

(c) Calculate the value of n_0 for which the FFT is faster than the standard method for polynomials with $n \geq n_0$ coefficients. Do you think that the FFT is useful in practice? If you had to multiply two polynomials with 2^{10} coefficients, which method would be faster?

(d) Assume that the time for all operations used is dominated by the time for multiplication. What is your prediction for the ratio of the run times of the FFT and the standard algorithm?

Ex 9.23. Do an average case analysis for the following procedure:

```
PROCEDURE  BIGTWO(C)
        FIRST:=C[1]
        SEC:=C[2]
        FOR I:=2 TO n DO
            IF C[I]≥ SEC
                THEN IF C[I] > FIRST
                        THEN SEC:=FIRST
                            FIRST:=C[I]
                        ELSE SEC:=C[I]
```

Is the average case run time for this algorithm nearer to its worst case or its best case? Which of the two procedures LARGETWO(C) or BIGTWO(C) would you use?

Ex 9.24. If the best case $B(n)$ of an algorithm satisfies

$$B\left(\frac{n}{2} - \gamma\right) + B\left(\frac{n}{2} + \gamma\right) + cn \leq B(n) \leq B\left(\frac{n}{2} - \alpha\right) + B\left(\frac{n}{2} + \alpha\right) + bn$$

for constants α, γ, b, c, show that $B(n) = \Theta(n \log n)$.

Ex 9.25. Show that if the matrix C is initialized to the zero matrix, then the following algorithm computes the $n \times n$ matrix product $C = A \times B$.

```
FOR i := 1 TO n DO
    FOR j := 1 TO n DO
        FOR k := 1 TO n DO
            c_ij = c_ij + a_ik * b_kj
```

Compute the run time of this algorithm. Explain why most computations use this algorithm rather than Strassen's algorithm from Section 9.4.2.

Ex 9.26. Show that if multiplication of two n-bit numbers can be performed in time $\Theta(n^{1+\epsilon})$ for some $\epsilon > 0$, then division of two n-bit numbers can also be carried out in time $\Theta(n^{1+\epsilon})$. (Refer to Section 9.4.4.)

Ex 9.27. Show that multiplication of two n-bit numbers can be performed in the same time order as division of two n-bit numbers.

Ex 9.28. Show that the square root of an n-bit number can be performed in the same time order as multiplication of two n-bit numbers.

Ex 9.29. Compute the product of the Fourier matrices $F(i)$ and $F(-i)$ to show that $F(\omega)^{-1} \neq F(\overline{\omega})$. Find a simple formula for $F(\omega)^{-1}$, the **inverse Fourier matrix**.

10

Some Nonlinear Recurrences

10.1 Some Examples

In previous chapters we have primarily discussed *linear* recurrences, or, said another way, only recurrences involving linear operators on the space of sequences. Recall that a function L on the space of sequences is a **linear operator** if it satisfies the following two conditions:

1. $L[x + y] = L[x] + L[y]$,

2. $L[cx] = cL[x]$,

for all sequences $x = \langle x_n \rangle$ and $y = \langle y_n \rangle$ and all constants c. (Notice that a square matrix is a linear operator on the space of vectors of appropriate size, since a matrix satisfies these conditions when x and y are any vectors and c is any scalar.) While the operation $+$ above is our usual addition of sequences, one could consider replacing $+$ by other operations. We do not follow this tack, but instead look at truly nonlinear equations. To keep things simple, we consider one-dimensional equations of the form

$$x_{t+1} = f(x_t).$$

For linear equations there is essentially only one linear one-dimensional equation, but to paraphrase Tolstoy, equations can be linear in only one way, but equations can be nonlinear in many different ways. So we should not expect to have a general theory for nonlinear equations. Rather, we hope to have different theories for different classes of nonlinear equations.

Let us consider the very simple nonlinear example

$$x_{t+1} = \frac{1}{x_t}.$$

If $x_0 = 2$, then $x_1 = 1/2$, $x_2 = 2$, $x_3 = 1/2$, and the sequence **oscillates** with period 2. For general non-zero x_0, the sequence is $x_0, 1/x_0, x_0, 1/x_0$, and the sequence generally has period 2. The only exceptions are the fixed points $x_0 = \pm 1$ and $x_0 = 0$. (Often we extend the reals to include ∞, and the undefined $1/0$ is taken to be $1/0 = \infty$ and $1/\infty = 0$. Then 0 is also a periodic point with period 2 oscillation.) Even though this equation is nonlinear, from this analysis we can guess that in some sense it is analogous to the linear equation

$$y_{n+1} = -y_n,$$

which generally has period 2, and 0 is the only fixed point. We continue this analogy with linear equations in Section 10.6.

As another simple example, consider

$$x_{n+1} = \sqrt{x_n}.$$

For this equation to make sense, we assume that $x_0 \geq 0$ and that $\sqrt{x}$ returns the nonnegative square root of x. We can calculate some iterates,

$$x_1 = x_0^{1/2},$$
$$x_2 = x_1^{1/2} = (x_0^{1/2})^{1/2} = x_0^{1/4},$$
$$x_3 = x_2^{1/2} = (x_0^{1/4})^{1/2} = x_0^{1/8},$$

and see that the solution is

$$x_n = x_0^{1/2^n}.$$

There are four cases:

1. $x_0 = 0$, and then $x_n = 0$ for all $n \geq 0$;

2. $0 < x_0 < 1$, and then $1 > x_{n+1} > x_n > 0$;

3. $x_0 > 1$, and then $1 < x_{n+1} < x_n$;

4. $x_0 = 1$, and then $x_n = 1$ for all $n \geq 0$.

We summarize these cases by saying that 0 and 1 are **fixed points** of the system, since $f(p) = p$ for $p = 0, 1$. The fixed point 0 is **unstable**, while 1 is a **stable fixed point** that attracts all solutions with $x_0 > 0$.

Our point, so far, is that *some* nonlinear equations are not difficult to analyze. But there are nonlinear equations that are more difficult. Consider the example

$$x_{n+1} = a \sin(x_n),$$

which can schematically be written as

$$x_n = a\sin(a\sin(\dots(x_0)\dots)),$$

a formula that gives no clear information about the behavior of the solutions. Some things are easy to see. For example, if x_0 is an integer multiple of π, then $x_1 = 0$ and $x_n = 0$ for all $n > 0$. On the other hand, it is not clear what sort of solutions arise for other values of x_0. In particular, what happens if x_0 is a small positive number? Is the solution attracted to 0? Does the solution converge to some positive value? Does the solution become periodic? In general, these questions are difficult to answer.

10.2   Nonlinear Systems

The preceding examples suggest that nonlinear difference equations may be difficult to analyze. In the remainder of this chapter we want to describe and give examples of some commonly used methods for analyzing these equations. Of course, in this relatively short chapter we cannot cover all techniques, and so we refer the interested reader to LaSalle [92], Devaney [51], and Parker and Chua [125], which include more information.

To keep things simple we concentrate on the one-dimensional equation

$$x_{t+1} = f(x_t),$$

where $f(x)$ is a nonlinear function and x is a real variable, which may be restricted to the positive reals or to the extended reals or to some interval on the real line. Such equations are sometimes called **discrete dynamical systems**, because they are the discrete analogs of dynamical systems occurring in physics. The function $f(x)$ is sometimes called a **map** to emphasize that this is a discrete system. Notice that these equations have zero input, and so the complexity comes from the nonlinear behavior of $f(x)$ rather than from an external source. Some complexity can come from the choice of initial condition.

The simplest way to deal with such equations is computation. If $f(x)$ is a reasonable function, then one should be able to write a computer program that, when given an initial value x_0, can compute as many values of the sequence $x_1, x_2, \dots$ as one desires. There is the usual caveat that computers do not compute with *real* real numbers, but instead they compute with *approximations* to real numbers. So a sequence generated by a computer may not be the sequence actually defined by the nonlinear recurrence. Of course, for reasonable functions one hopes that the real and the computed sequences are similar. In some cases this can be proved, but we do not tackle this problem of approximation here.

Since humans tend to understand pictures better than sequences of numbers, these computer calculations are often presented as graphs. (These are not the same kind of graphs we used in discussing nonnegative matrices.) The most straightforward graph is a plot of x_t as a function of the natural number t. This is often called a **time plot**. Figure 10.1 shows a time plot of the quadratic difference equation

$$x_{t+1} = x_t[1 + r(1 - x_t)]$$

with $r = 2.99$ and $x_0 = .35$. To make this plot look more like a continuous function, it is not plotted as a sequence of points, but rather each point (t, x_t) is connected by a straight line to the next point $(t + 1, x_{t+1})$. The time plot shows that the elements of the sequence are positive and don't get too big, but that they do jump around in an irregular manner.

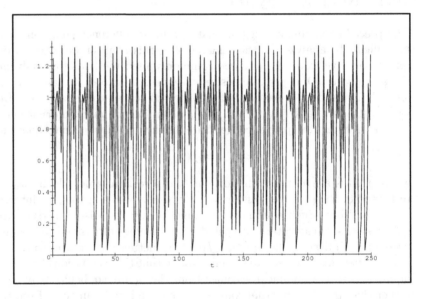

FIGURE 10.1. A time plot of a chaotic trajectory (quadratic model with $r = 2.99$).

Another plotting technique may be more revealing. This technique is called a **web plot**, and it is somewhat similar to the phase plane plot used in differential equations. The idea is to plot x_{t+1} as a function of x_t. While this does give some information, it does not show a solution sequence. To follow a solution it is useful to have both the *curve* $f(x)$ and the *line* $y = x$ plotted on the same axes. To follow a solution:

1. Start with $(x_0, f(x_0))$, which is a point on the *curve*.

2. Then draw a horizontal line from this point to the *line* $y = x$. The point of intersection on the line is $(f(x_0), f(x_0))$, i.e., (x_1, x_1).

3. Then draw a vertical line from this point to the *curve*. The point of intersection on the curve is $(x_1, f(x_1))$, i.e., (x_1, x_2).

4. Now simply repeat the steps going from *curve* to *line* and *line* to *curve* for as many steps as desired.

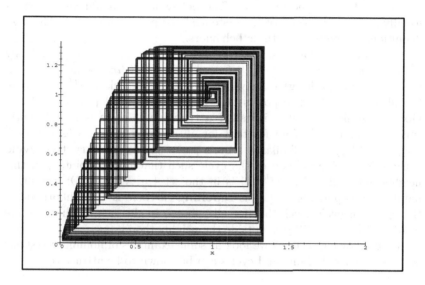

FIGURE 10.2. A web plot showing chaos (quadratic model with $r = 2.99$).

Figure 10.2 shows the web plot that corresponds to the time plot shown in Figure 10.1. This web plot may be more informative than the time plot. For example, the parabola $x[1 + r(1 - x)]$ is outlined in the web plot, which suggests that many or most values of x are visited on the trajectory, but this plot also suggests that some regions are more frequently visited than others. Further, the plot suggests that if the trajectory is periodic, then it has a very very long period. Also, the fixed point $x = 1$ is visible, and the plot suggests that this fixed point tends to repel nearby trajectories.

A full understanding of a nonlinear system might require computing solutions for essentially all initial conditions. Since this is infeasible, we would like some simple ways to summarize the possible behaviors. For arbitrary $f(x)$'s this is not possible, since there are functions that are complicated enough to simulate the behavior of Turing machines. For such functions there can be no algorithm that determines whether a given initial condition is eventually periodic [108], but the functions used in various applications tend to be simple enough or have enough reasonable properties to allow some sort of analysis. In particular, the assumption that $f(x)$ is continuous and sufficiently differentiable is usually enough to obtain some results.

We often focus on results that are invariant or asymptotic. A property is **invariant** if it remains the same along a trajectory. The most common example of this is a closed physical system in which the total energy is always the same. Even as the system's variables change, the total energy remains invariant. **Asymptotic properties** are those that hold in the limit as the time t increases. For example, a system might approach a "steady state" as time increases. The behavior of a nonlinear system is usually analyzed in terms of the system's fixed points and cycles. The next definitions define some of these behaviors.

As in previous chapters, the **iterates** of f are defined recursively by $f^{(0)}(x) = x$ and $f^{(i+1)}(x) = f(f^{(i)}(x))$. A string of *distinct* points $p_1, \ldots, p_K$ forms a **cycle of length** K, which we call a K-cycle if $f^{(K)}(p_i) = p_i$ for all $1 \leq i \leq K$. The points of a K-cycle are called **periodic** of **period** K. A fixed point p is called **attractive** if there is a sufficiently small neighborhood of p such that for all points q in this neighborhood, $\lim_{n \to \infty} f^{(n)}(q) = p$. Similarly, a cycle $p_1, \ldots, p_K$ is an **attractive cycle** if there is a neighborhood of the cycle such that for all points q in this neighborhood, $\lim_{n \to \infty} f^{(nK)}(q)$ is in the set $\{p_1, \ldots, p_K\}$. On the other hand, a fixed point (or cycle) is **repelling** if there is a neighborhood of the point (or cycle) such that for all points q in this neighborhood, there is an n such that $f^{(n)}(q)$ is outside the neighborhood.

In the following sections we look at some examples of nonlinear systems and see how fixed points and cycles can be shown to be attractive.

10.2.1 Sarkovskii's Theorem

Nonlinear systems may have many different co-existing cycle lengths, for example, $x_0, f(x_0), f^{(2)}(x_0)$ may all be distinct and $f^{(3)}(x_0) = x_0$, but $y_0 = f(y_0)$, and then there would be a cycle of length 3 and a cycle of length 1. Some common useful functions are the **interval maps**, which are functions defined on a bounded interval $[a, b]$ whose values are also in $[a, b]$. For the special case of continuous one-dimensional interval maps, the question of which size cycles can occur is answered by the following theorem. There is no analogous result for discontinuous maps or maps in higher dimensions. While the theorem dates from 1964, it was not known in the United States until many years later. In fact, for example, Li and Yorke [97] proved a special case in the 1970's, and Cull [32] and Rosenkranz [140] also proved special cases in the 1980's. The proof of the theorem is beyond the scope of this book, but it can be found in [51].

Theorem 10.2.1 (Sarkovskii's Theorem [144]). *If $f(x)$ is a continuous one-dimensional interval map that has a K-cycle, then $f(x)$ also has cycles of every length less than K, where "less" is defined by the following linear* *ordering:*

$$3 > 5 > 7 > \qquad \ldots\ldots \qquad > \ldots$$
$$> 2 \cdot 3 > 2 \cdot 5 > \qquad \ldots\ldots \qquad > \ldots$$
$$> 2^2 \cdot 3 > 2^2 \cdot 5 > \qquad \ldots\ldots \qquad > \ldots$$
$$\vdots \qquad\qquad \vdots$$
$$> 2^n \cdot 3 > 2^n \cdot 5 > \qquad \ldots\ldots \qquad > \ldots$$
$$\vdots \qquad\qquad \vdots$$
$$> 2^n > 2^{n-1} > \qquad \ldots\ldots \qquad > 2^2 > 2 > 1.$$

Note that there are two extreme cases: when $f(x)$ has a 3-cycle, then it has cycles of every length, and if $f(x)$ has any cycles other than fixed points, then it has a 2-cycle.

10.3   Chaos

Since the ground-breaking papers of Li and Yorke [97] and May [109, 110] in the mid-1970's, the importance of chaos in science has been evident. A variety of different meanings have been attached to the word *chaos*. For some, it simply means very complicated-looking behavior (refer to Figures 10.1 and 10.2), while for others, the **butterfly effect** is the signature of chaos. In poetic terms, the butterfly effect means that the flapping of a butterfly's wings in Borneo may cause a tropical storm in the Caribbean that devastates the sugar crop and leads to the downfall of Communism. In the original use of Li and Yorke [97], chaos meant the co-existence of cycles of every length. Other properties that some people like are that chaotic trajectories come near to every point in the space and that there is a measure that is invariant along trajectories. Detailed consideration of these issues are beyond the scope of this book, but we consider some nonlinear systems that have some of these properties

10.3.1   A simple chaotic system

Several properties of **chaos** are:
 (a) cycles of every length;
 (b) sensitive dependence on initial conditions;
 (c) the existence of bounded but aperiodic orbits;
 (d) for every open set A and every open set B there is an $x_0 \in A$ and a $K \in \mathbb{N}$ such that $f^{(K)}(x_0) \in B$.

In this section we consider a very simple example that displays all these chaotic properties. In order to define the system properly, we first need to discuss the term **equivalence mod 1**. Two real numbers x and y are said to be equivalent mod 1 if $x - y$ is an (integer) multiple of 1; that is, x and y differ by an integer. We symbolize this relationship by $x \equiv y$ mod 1, but in the following we will replace $\equiv$ by $=$ and write $x = y$ mod 1 . For example, $3.14 = .14$ mod 1. Equivalence mod 1 can be visualized as a clock face with circumference 1 that has 0 in the top position, and all the numbers between 0 and 1 are in their standard positions around the dial. Notice that 1 does *not* appear, because 1 mod $1 = 0$. That is, if you wrapped a string of length 1 around the clock face, both the beginning and the end of the string would be at 0. If your string had length $3/2$, wrapping it around the clock face puts the end at $1/2$, which agrees with $3/2$ mod $1 = 1/2$.

Let us consider some initial conditions for the system

(10.1) $$x_{t+1} = 2x_t \text{ mod } 1.$$

Clearly, 0 is a fixed point, and some points are attracted by 0. For example, if $x_0 = 1/2$, then $x_1 = 2(1/2)$ mod $1 = 0$. In fact, for any integer k, $1/2^k$ goes through $1/2^k, 1/2^{k-1}, \ldots, 1/2, 0$. This is rather tame behavior. We can also find periodic behavior. Let's try $x_0 = 2/3$. Then $x_1 = 2(2/3)$ mod $1 = 1/3$, and $x_2 = 2/3$, and we see that the trajectory of $x_0 = 2/3$ is a 2-cycle. More generally, for any k the initial condition $x_0 = 2^{k-1}/(2^k - 1)$ gives a k-cycle, since

$$x_1 = \frac{2^k}{2^k - 1} - 1 = \frac{1}{2^k - 1},$$
$$x_2 = \frac{2}{2^k - 1},$$
$$\vdots$$
$$x_k = \frac{2^{k-1}}{2^k - 1} = x_0.$$

This behavior can be seen more readily in binary notation. For example, multiplying $x_0 = \frac{2}{3} = .10101010\ldots$ by 2 and then reducing modulo 1 corresponds to shifting the binary point one place to the right and dropping any bit to the left of the binary point. So $2(\frac{2}{3})$ mod 1 is computed by taking $.10101010\ldots$, shifting the binary point to get $1.0101010\ldots$, and dropping the 1 to the left of the binary point to get $.0101010\ldots$, which is $\frac{1}{4} + \frac{1}{16} + \frac{1}{64} + \cdots = \frac{1}{4}(1 + \frac{1}{4} + \frac{1}{4^2} + \cdots)$. Taking the sum of the geometric series, we get $\frac{1}{4}(\frac{1}{1-\frac{1}{4}}) = \frac{1}{3}$. Similarly, starting with $.0101010\ldots$, shifting the binary point, and dropping the leading 0, we get $.10101010\ldots$, which is the number $\frac{2}{3}$ we started with two steps ago. Here we can *see* that the period 2 of the solution to the difference equation follows from the period 2 of the binary expansion of $\frac{2}{3}$.

Sensitive dependence on initial conditions is also easy to demonstrate for this equation. Since 0 is a fixed point of the equation, when 0 is chosen for the initial condition, the solution remains at 0 forever after. For a very small positive initial condition $x_0 = \epsilon > 0$, the trajectory begins at ϵ, goes to 2ϵ, to 4ϵ, and so forth. If ϵ is very small, then 2ϵ and 4ϵ are also very small, and so for some number of steps the two solutions—the one with initial condition 0 and the one with initial condition ϵ—are both nearly 0, and we don't see any divergence. But since ϵ is not zero, it has at least one 1 in its binary expansion; say the first 1 appears as the kth bit. Then after $k - 1$ steps the 1 will have moved to the first bit position, which means that $x_{k-1} \geq 1/2$, and the trajectory is now significantly far from the zero trajectory.

Another way to describe sensitive dependence is in terms of information. How many bits of the initial condition are needed to accurately predict x_k? For our example equation (10.1) we need about k bits to predict x_k. This means that if we could only measure the initial condition to 10 correct bits, we would be at a loss to predict the value of the solution after 11 steps. Sometimes this is explained as a loss of bits. If we know the initial condition to 10 correct bits and lose a bit of accuracy at each step, then we have no bits of accuracy left to predict the 11th value.

While our example is extreme for showing sensitive dependence, such dependence can arise in more realistic situations. Consider predicting the weather. One might measure a number of variables like pressure, tempera-ture, and wind direction to 10-bit accuracy (about one part in a thousand). Now assume that we are calculating the weather with a time step of $1/1000$ of a day. To compute tomorrow's weather we need to calculate about 1000 steps. If the weather conditions are *smooth*, we don't lose many bits per step. When we lose $1/1000$ bits per step in smooth conditions, we lose about 1 bit in tomorrow's forecast, and we expect our prediction to be accurate to about 9 bits. On the other hand, if in *turbulent* conditions we lose about $1/100$ bits per step, then in 1000 steps we will have lost about 10 bits, and so there would be no bits of accuracy left to make tomorrow's prediction. This is not just a story, since some of the original work on chaos was an attempt to describe why weather prediction is so difficult [100].

Returning to exhibiting sensitive dependence on initial conditions, let us note that there was nothing really special about looking at 0. For any two different but close initial conditions x_0 and $x_0 + \epsilon$, the kth term of the first sequence differs by about $1/2$ from the kth term of the second for $k \approx -\log \epsilon$.

Another characteristic of chaos is the existence of **bounded but ape-riodic orbits**. In our example, all orbits are bounded, since they are lim-ited to $[0, 1)$. Now we want to determine an initial condition x_0 such that $x_0, x_1, x_2, \ldots$ is an aperiodic sequence. The interpretation of the equation as a shift on binary sequences makes this easy, because all we need to do is find an aperiodic binary sequence. For example, $x_0 = .10100100010000\ldots$ (designed so the 1's are isolated and the number of separating 0's increases by one) is aperiodic because the number of 0's between consecutive 1's is increasing. Using x_0 as the initial condition, the solution starts above $1/2$, drops to a lower value, goes above $1/2$, drops to a lower value, builds up for 2 steps, goes above $1/2$, drops to a lower value, builds up for a few steps, goes above $1/2$, and so forth. Why do we write this initial condition as a binary sequence rather than as a ratio? Because no such ratio exists. That is, the number we have written is *not* a rational number, because it has an aperiodic binary expansion.

Next, for any two open sets A and B in $[0, 1)$ we want to show that there is an initial condition in A such that the trajectory visits B. Because A is open, A contains an open interval of the form $(a - 2^{-j}, a + 2^{-j})$. Let $b \in B$ and we will construct an element in A's open interval that is eventually mapped to b. (Notice that this is an even stronger condition than required.) Let $c = .a_1 a_2 \ldots a_j$ be the rational number formed from the first j bits in the binary expansion of a. Let $K = j + 1$ and $x_0 = c + 2^{-K} b$. We see that $x_0 \in A$ because $a - 2^{-j} < c \le a$, and by design, $f^{(K)}(x_0) = b$.

A similar construction yields a trajectory that visits *every* open set. (In what follows we refer to a open interval with two rational endpoints as a "rational interval.") Since every open set contains a rational interval, a trajectory can be shown to visit every open set by showing that it visits every rational interval. The rational numbers are **countable**, which means that they can be put into one-to-one correspondence with the natural numbers. There are many such correspondences, and any one of them gives an ordering of the rationals, and so we can speak of the first rational, the second rational, and so forth. But since we have just mapped the rationals to the natural numbers, we can map the upper and lower endpoints for a rational interval (and so the interval itself) to a pair of natural numbers. Hence, the set of rational intervals can be countably ordered. Since every rational interval contains a rational r with a terminating binary expansion, there is a natural number j such that $(r - 2^{-j}, r + 2^{-j})$ is inside the interval. If our trajectory visits each of these special intervals, it will visit all rational intervals and hence all open sets.

We are now ready to pick an initial condition x_0 whose trajectory visits every open set. For the sequence $\langle r_i \rangle$ of terminating rationals (ordered according to our ordering of rational intervals) define the initial condition x_0 by the binary expansion

$$x_0 = .r_1 \quad (j_1 + 1)\text{0's} \quad r_2 \quad (j_2 + 1)\text{0's} \quad r_3 \quad (j_3 + 1)\text{0's} \quad \ldots$$

(where $(j_i + 1)$0's is our abbreviation for $j_i + 1$ consecutive 0's.) By construction, x_0 lies in the first interval $(r_1, r_1 + 2^{-j_1})$; after some number of iterations, x_k lies in the second interval; and after an appropriate number of iterations, there is an x_n that lies in the interval $(r_i, r_i + 2^{-j_i})$. This means that the trajectory starting at this x_0 eventually visits every open set. With this we see that the system given in (10.1) satisfies all four of our given properties of chaos.

10.4   Local Stability

10.4.1   Local stability of a fixed point

For a fixed point x_0 we would like to know whether nearby values of x are attracted to it. In order to make this idea more precise, we say that a fixed point x_0 of $f(x)$ is **locally stable** if for every y in every small enough neighborhood of x_0, $f(y)$ is in the neighborhood; that is,

$$|f(y) - x_0| \le |y - x_0|,$$

and also $\lim_{K \to \infty} f^{(K)}(y) = x_0$.

Testing whether a point is locally stable may seem difficult, but when $f(x)$ is differentiable, there is the reasonably easy test given in the next theorem.

Theorem 10.4.1. *Let x_0 be a fixed point of $f(x)$ and assume that $f(x)$ is differentiable at x_0. Then x_0 is locally stable if $|f'(x_0)| < 1$, and if x_0 is locally stable, then $|f'(x_0)| \le 1$.*

Proof. If $f(x)$ is differentiable, there is a constant α such that for all y close enough to x_0,

$$\frac{|f(y) - f(x_0)|}{|y - x_0|} \le |f'(x_0)| + \alpha |y - x_0|.$$

If $|f'(x_0)| < 1$, then there is a $\delta > 0$ such that for all y very close to x_0, the right side of this inequality is less than $(1 - \delta)$. Since x_0 is a fixed point of $f(x)$, this gives

$$|f(y) - x_0| < (1 - \delta)|y - x_0| < |y - x_0|,$$

and $f(y)$ is in the same neighborhood as y. Further, letting

$$d_K = |f^{(K)}(y) - x_0| \quad \text{gives} \quad d_K \le (1 - \delta)d_{K-1} \le (1 - \delta)^K d_0,$$

and $\lim_{K \to \infty} d_K = 0$. Therefore, x_0 is locally stable. Conversely, if x_0 is locally stable, then for y close to x_0,

$$|f(y) - x_0| = |f(y) - f(x_0)| \le |y - x_0|,$$

and since $f(x)$ is differentiable,

$$|f'(x_0)| = \lim_{y \to x_0} \frac{|f(y) - f(x_0)|}{|y - x_0|} \le 1.$$

$\square$

Let's look a bit more closely at this proof. The basic idea is that when we are close enough to a fixed point we can approximate a nonlinear difference equation by a linear difference equation. Consider the difference equation $x_{t+1} = f(x_t)$ near a fixed point E. Define a new variable $d_K = |x_K - E|$ and consider the linear equation $d_K = (1 - \delta)d_{K-1}$. We have translated the fixed point E for x to the fixed point 0 for d. If $f(x)$ is differentiable with $|f'(x)| < 1$, Then the linear equation gives an upper bound on the behavior of $|x_K - E|$ in a small enough neighborhood of E, and the convergence of the linear equation to 0 implies the convergence of the nonlinear equation to E.

Linear approximations can also be made around non-fixed points, but we will not get as much useful information. Near a fixed point E, $f(E + \epsilon) \approx E + f'(E)\epsilon$, so nearby points tend to stay nearby. For non-fixed points, nearby points may not stay nearby. For fixed points we may be able to repeatedly use the same linear approximation, but for non-fixed points we will continually need new linear approximations. For numerical solutions to equations, linear approximations are often used, but we have to face the prospect that the linearly computed solution may be very far from the true solution.

Notice that the statement of this theorem has a small lacuna, since it does not say what happens when $|f'(x_0)| = 1$. Essentially this is because we are making a linear approximation to $f(x)$, and this linear approximation dominates the nonlinear behavior locally if $|f'(x_0)| < 1$, but the nonlinear terms are necessary to determine behavior when $|f'(x_0)| = 1$. For example, consider

$$x_{t+1} = f(x_t) = x_t[1 + r(1 - x_t)],$$

which has a fixed point at $E = 1$. The derivative is $f'(x) = 1 + r - 2rx$, and so $|f'(1)| \leq 1$ if $0 \leq r \leq 2$. By the theorem, $x = 1$ is locally stable when $0 < r < 2$. Let's see what happens in the special cases $r = 0$ and $r = 2$, that is, when $|f'(1)| = 1$. When $r = 0$, $f(x) = x$, and if the system starts near 1, it stays near 1 but the iterates do not converge to 1. On the other hand, when $r = 2$, $f(x) = 1 - (x - 1) - 2(x - 1)^2$, and starting at $1 - \epsilon$ gives $1 + \epsilon - 2\epsilon^2$, which is in the same ϵ neighborhood of 1. But starting at $1 + \epsilon$ gives $1 - \epsilon - 2\epsilon^2$, which is not in the same ϵ neighborhood of 1. In spite of this, $f(f(x))$ for x near 1 is always closer to 1, and $\lim_{K \to \infty} f^{(K)}(x) = 1$ for all x in $(0, 3/2)$. This suggests that our definition of "local stability" may not be ideal, and the interested reader should consult the literature for other definitions.

10.4.2   Local stability of a cycle

Stability for a cycle is similar to stability for a point. Recall that the system

$$x_{t+1} = f(x_t)$$

has a K-cycle $x_1, x_2, \ldots, x_K$ if $f^{(i)}(x_1) = x_{i+1}$ for all $i = 1, \ldots, K - 1$ and $f^{(K)}(x_1) = x_1$. It is convenient to note that each of the points $x_1, x_2, \ldots, x_K$

is a fixed point of the K-fold iterate $f^{(K)}(x)$, and we say that $x_1, x_2, \ldots, x_K$ is a **locally stable cycle** when each of the points $x_1, x_2, \ldots, x_K$ is a *locally stable* fixed point of $f^{(K)}(x)$.

We would like to be able to simplify things by checking for the local stability of only one of the points, but there is the worry that one of the points could be locally stable, while some of the other points are not. Luckily, when we assume that $f(x)$ is differentiable, these worries vanish because the stability condition at one point implies the stability condition at each of the other points. The reason for this is the Chain Rule, since the derivative is

$$D[f^{(K)}(x)] = f'(f^{(K-1)}(x))D[f^{(K-1)}(x)]$$
$$= f'(f^{(K-1)}(x))f'(f^{(K-2)}(x))D[f^{(K-2)}(x)]$$
$$\vdots$$
$$= f'(f^{(K-1)}(x))f'(f^{(K-2)}(x))\cdots f'(x),$$

and so

$$D[f^{(K)}(x)]|_{x=x_1} = f'(x_K)f'(x_{K-1})\cdots f'(x_1).$$

This is also the value of the derivative at *every* point in the cycle, because taking another point only results in the terms in the product appearing in a different order! Therefore, a sufficient condition for local stability of the K-cycle $x_1, x_2, \ldots, x_K$ is

$$|f'(x_K)|\,|f'(x_{K-1})|\cdots|f'(x_1)| < 1.$$

As an example, let us again consider

$$x_{t+1} = f(x_t) = x_t[1 + r(1 - x_t)],$$

this time with r slightly larger than 2. We will show that there is locally stable 2-cycle.

The condition for a 2-cycle for this function is

$$f(f(x)) = x = x[1 + r(1 - x)][1 + r(1 - f(x))].$$

We can eliminate the fixed point at $x = 0$ by dividing this equation by x. Simplifying the resulting equation by subtracting 1 and dividing by r gives

$$2(1 - x) - rx(1 - x) + r(1 - x)(1 - f(x)) = 0.$$

Now, the fixed point at $x = 1$ can be eliminated by dividing by $1 - x$, and after substituting for the remaining $f(x)$, this gives the quadratic equation

$$2 + r - r(2 + r)x + r^2 x^2 = 0,$$

which has two distinct real roots, since $r > 2$. (This can be seen by calculating the discriminant of the polynomial.) We could calculate these roots, x_1 and x_2, but this polynomial already tells us that

$$(10.2) \qquad x_1 x_2 = \frac{2+r}{r^2} \quad \text{and} \quad x_1 + x_2 = \frac{2+r}{r}.$$

Remember that the sufficient condition for local stability of the cycle is

$$|f'(x_1)f'(x_2)| < 1.$$

It is easy to compute

$$f'(x) = 1 + r - 2rx$$

and

$$f'(x_1)f'(x_2) = (1+r)^2 - 2r(1+r)(x_1+x_2) + 4r^2 x_1 x_2.$$

Using the formulas from (10.2) for $x_1 x_2$ and $x_1 + x_2$ gives

$$f'(x_1)f'(x_2) = (1+r)^2 - 2(1+r)(2+r) + 4(2+r) = 5 - r^2,$$

and the local stability condition becomes

$$2 < r < \sqrt{6}.$$

Hence, the system has a locally stable 2-cycle when r satisfies this inequality.

10.4.3 Local stability in two dimensions

We can use the techniques of the previous subsections to look at the local stability of a nonlinear system in more than one dimension. The only modifications are that we need to generalize the idea of neighborhood and the idea of derivative. A neighborhood in one dimension is an open interval $(x - \epsilon, x + \epsilon)$. For a **neighborhood** in more than one dimension we use the higher-dimensional ball of radius ϵ. So an ϵ-neighborhood of a point x is the set of all points y such that $|x - y| < \epsilon$, where we interpret the **absolute value** to mean Euclidean distance in the appropriate dimension.

A linear approximation to an n-dimensional function should consist of n linear functions, one for each dimension. Specifically, if

$$F(x, y) = \begin{pmatrix} f_1(x,y) \\ f_2(x,y) \end{pmatrix}$$

is a two-dimensional function, our analog of the derivative is

$$J(x, y) = \begin{bmatrix} \frac{\partial f_1}{\partial x} & \frac{\partial f_1}{\partial y} \\ \frac{\partial f_2}{\partial x} & \frac{\partial f_2}{\partial y} \end{bmatrix},$$

the **Jacobian matrix**, which is also important in vector calculus. The partial derivatives inside this matrix are functions that are evaluated at the point (x, y) at which we are making the linear approximation. For example, if $f_1(x, y) = xy$ and $f_2(x, y) = x + y^2$, For example, if

$$f_1(x, y) = xy$$
$$\text{and} \quad f_2(x, y) = x + y^2,$$

then the Jacobian matrix is

$$\begin{bmatrix} y & x \\ 1 & 2y \end{bmatrix}.$$

If we want a linear approximation near the point $(0, 0)$, then the Jacobian matrix evaluated at $(0, 0)$ becomes

$$\begin{bmatrix} 0 & 0 \\ 1 & 0 \end{bmatrix},$$

and the linear approximation at $(0, 0)$ is

$$\begin{pmatrix} f_1(x, y) \\ f_2(x, y) \end{pmatrix} \approx \begin{pmatrix} f_1(0, 0) \\ f_2(0, 0) \end{pmatrix} + \begin{bmatrix} 0 & 0 \\ 1 & 0 \end{bmatrix} \begin{pmatrix} x \\ y \end{pmatrix} = \begin{pmatrix} 0 \\ x \end{pmatrix}.$$

For the linear approximation at $(1, 0)$ we have

$$\begin{pmatrix} f_1(x, y) \\ f_2(x, y) \end{pmatrix} \approx \begin{pmatrix} f_1(1, 0) \\ f_2(1, 0) \end{pmatrix} + \begin{bmatrix} 0 & 1 \\ 1 & 0 \end{bmatrix} \begin{pmatrix} x - 1 \\ y - 0 \end{pmatrix} = \begin{pmatrix} y \\ x \end{pmatrix}.$$

As this example shows, we still have a problem, because the derivative in one dimension was a single number, but the Jacobian matrix is an array of numbers. The idea in one dimension was that around a fixed point the linear approximation can be viewed as a linear difference equation. And the solutions of this equation converge to the fixed point if the value of the derivative is less than 1 in absolute value, because the value of the derivative is also the eigenvalue of the associated difference equation. Similarly, the eigenvalues of the Jacobian matrix are the eigenvalues of a matrix linear difference equation around a fixed point, and the solutions converge to the fixed point if all the eigenvalues are less than one in absolute value. This result generalizes to all dimensions, as the following theorem indicates. (For a proof, refer to [92].)

Theorem 10.4.2. *If F is an n-dimensional differentiable function with fixed point X and J is the Jacobian matrix of F evaluated at X, then X is a locally stable fixed point if all eigenvalues of J have absolute value less than 1. If at least one of these absolute values is strictly greater than 1, the fixed point is unstable.*

As an example, consider the system

$$x_{t+1} = \sin(y_t),$$
$$y_{t+1} = \cos(x_t),$$

whose Jacobian matrix is

$$\begin{bmatrix} 0 & \cos(y) \\ -\sin(x) & 0 \end{bmatrix},$$

with characteristic polynomial $\mathrm{ch}(\lambda) = \lambda^2 + \sin x \cos y$ and eigenvalues $\pm\sqrt{-\sin x \cos y}$. Depending on the values of x and y, these roots may be real or complex. But we know that $|\sin x| \le 1$ and $|\cos y| \le 1$, and so the modulus of each of these roots is at most 1. More strongly, if (x, y) is a fixed point, then $|x| \le 1$ and $|y| \le 1$, but then $|\sin x| < 1$ and $|\cos y| \le 1$. So the eigenvalues are strictly less than 1 in absolute value, and all fixed points are locally stable.

The fixed points of this system can be approximated by plotting $\cos x$ along the x-axis and plotting $\sin y$ along the y-axis, and the intersections are the fixed points. From this one can see that there is only one fixed point, and its coordinates satisfy $0 < x < 1$ and $0 < y < 1$. This system has one fixed point, and it is locally stable.

Another simple example of a two-dimensional system is

$$x_{t+1} = x_t + x_t(1 - x_t - y_t)/6,$$
$$y_{t+1} = y_t(1 + x_t - y_t).$$

For fixed points $p = (x, y)$,

$$x = x + x(1 - x - y)/6,$$
$$y = y(1 + x - y),$$

and simplifying these gives

$$0 = x(1 - x - y),$$
$$0 = y(x - y),$$

which has the three solutions, $p = (0,0), (1,0), (1/2, 1/2)$. The matrix of partial derivatives for the system is

$$J(x, y) = \begin{bmatrix} 1 + (1 - x - y)/6 - x/6 & -x/6 \\ y & 1 + x - 2y \end{bmatrix}.$$

The Jacobian matrix for the fixed point $(0, 0)$ is

$$\begin{bmatrix} 7/6 & 0 \\ 0 & 1 \end{bmatrix},$$

which has an eigenvalue $7/6 > 1$, and so $(0,0)$ is unstable. The Jacobian matrix for the fixed point $(1,0)$ is

$$\begin{bmatrix} 5/6 & -1/6 \\ 0 & 2 \end{bmatrix},$$

and again there is an eigenvalue ($\lambda = 2$) larger than 1, and this fixed point is unstable. Finally, the fixed point $(1/2, 1/2)$ has the Jacobian matrix

$$\begin{bmatrix} 11/12 & -1/12 \\ 1/2 & 1/2 \end{bmatrix},$$

whose characteristic polynomial is $\text{ch}(x) = \lambda^2 - \frac{17}{12}\lambda + \frac{1}{2}$ and its eigenvalues are $2/3$ and $3/4$, and so this fixed point is locally stable. For this system, one could make the reasonable guess that if the system were started at any point that is not a fixed point, the trajectory eventually would approach the fixed point $(1/2, 1/2)$. While this guess is reasonable, one still needs to rule out other possible behaviors. For example, there could be cycles or aperiodic orbits that simply don't show up in a fixed point analysis.

10.5   Global Stability

A locally stable point attracts trajectories within a small neighborhood of the point, but one is usually interested in larger neighborhoods. At the extreme, the neighborhood of interest could be the whole space.

We encapsulate this by saying that a fixed point p of $x_{t+1} = f(x_t)$ is **globally stable** for $f(x)$ on the set B if $\lim_{K \to \infty} f^{(K)}(b) = p$ holds for each initial condition b in B.

Usually the set B is not specified, since by context one can tell that B is the reals or the positive reals or the interval $[0,1]$ or some other reasonable set. One would also like to say that a cycle or other invariant set is globally stable, but this is difficult, since fixed points and the special trajectories that lead to them might not converge to the cycle. People usually solve this problem by saying that the cycle is globally stable without mentioning the existence of the *relatively* few points that do not lead to the cycle.

Unlike local stability, there is no nice characterization of global stability even if the function is differentiable. The classical technique to show global stability is to find a "Liapunov" or "energy" function, and show that this function is nonnegative and equal to 0 only at the fixed point, and that the "energy" decreases along each trajectory. (See LaSalle [92] for details.) Unfortunately finding a Liapunov function and showing that it has the required properties is a formidable task. In Section 10.6 we will see that some simple functions can be used to prove global stability in one dimension. Of course, in higher dimensions global stability is harder to deal with.

10.5.1   *Staircase convergence*

In Figure 10.2 we saw a rather complicated-looking web plot. But we might hope that for some systems that display global stability, the web plot would be much simpler. In particular, some systems have a web plot that looks like a staircase that leads up and down to the globally stable fixed point. We'll first give a staircase theorem and then apply it to several examples.

Theorem 10.5.1 (Staircase Theorem). *Let $f(x)$ be a continuous function on the interval (a, b). If $x < f(x) \le p$ on (a, p) and $p \le f(x) < b$ on (p, b), then $\lim_{n \to \infty} f^{(n)}(x_0) = p$ for all x_0 in (a, b).*

A formal proof of this theorem would argue that for x_0 in (a, p), $f^{(n)}(x_0)$ forms an increasing but bounded sequence, so this sequence has a limit, and argue from the continuity of f that any limit must be a fixed point of f. A similar argument for x_0 in (p, b) would then show that p is the limit for every x_0. A more visual argument simply follows the web plot. For x_0 in (a, p), the "staircase" starts at (x_0, x_0). The first "riser" goes up to $(x_0, f(x_0))$. The "stair" goes across to $(f(x_0)), f(x_0))$. Of course, this last point is closer to (p, p) than (x_0, x_0) was. Continuing in this fashion, that the staircase builds up toward (or hits) (p, p) is evident. Notice that the convergence is **monotone**. If the sequence starts below the fixed point, it is always increasing until it hits the fixed point, and if the sequence starts above the fixed point, it is always decreasing until it hits the fixed point.

With this theorem's assumptions, the difference equation

$$x_{t+1} = f(\, x_t \,)$$

obeys $\lim_{t \to \infty} x_t = p$ for every choice of initial condition in (a, b). For example, the linear difference equation

$$x_{t+1} = \frac{1}{2}\, x_t$$

obeys $\lim_{t \to \infty} x_t = 0$ for every choice of initial condition in $(-\infty, \infty)$, because $x < \frac{1}{2} x < 0$ on $(-\infty, 0)$, and $0 < \frac{1}{2} x < x$ on $(0, \infty)$.

For the nonlinear difference equation $x_{t+1} = f(\, x_t \,)$, where

$$f(x) = \begin{cases} x(2 - x) & \text{on } (0, 1)\,, \\ 2 - 1/x & \text{on } (1, 2)\,, \end{cases}$$

the theorem says that $\lim_{t \to \infty} x_t = 1$ for all $x_0 \in (0, 2)$. Notice that while $f(x)$ is continuous, $f'(x)$ is not continuous at $x = 1$, but neither differentiability nor monotonicity of $f(x)$ is required by the theorem. For example, the nonlinear difference equation $x_{t+1} = f(x_t)$ with

$$f(x) = \begin{cases} 4x & \text{on } (0, 1/4), \\ 1 - (x - 1/4) & \text{on } (1/4, 1/2), \\ 3/4 + (x - 1/2) & \text{on } (1/2, 3/4), \\ 1 & \text{on } (3/4, 5), \end{cases}$$

satisfies the theorem's assumptions with $p = 1$, and $\lim_{t \to \infty} x_t = 1$ for all $x_0 \in (0, 5)$.

As another example, let's consider $x_{t+1} = e^{2(x_t - 1)}$. we claim that there exists a fixed point p, where $0 < p < 1$ and

(10.3) $\begin{aligned} x &< f(x) < p & \text{for } x \in (0, p), \\ p &< f(x) < x & \text{for } x \in (p, 1). \end{aligned}$

Even though we don't know the exact value of p, we can still argue that p exists. Let $g(x) = e^{2(x-1)} - x$. Clearly, $g(0) > 0$ and $g(1/2) < 0$, so $e^{2(x-1)}$ has a fixed point in $(0, 1/2)$. Computing $g'(x) = 2e^{2(x-1)} - 1$ and $g''(x) = 4e^{2(x-1)}$, we find that $g'(x)$ is an increasing function that is negative at $x = 0$, and positive at $x = 1$. Hence $g(x)$ has two roots, one at $x = 1$ and one at $x = p$ with $0 < p < 1$. Further, this argument shows that the bounds (10.3) needed for the staircase theorem hold, and starting at any $x_0 \in (0, 1)$, $\lim_{t \to \infty} x_t = p$. The pleasant conclusion is that even though we don't know the value of p, we can use the iterates of the difference equation $x_{t+1} = e^{2(x_t - 1)}$ to calculate an approximate value of p.

10.5.2 Nonmonotonic convergence

Surprisingly enough, the staircase theorem can be used to show global stability for functions that do *not* satisfy its hypotheses. The "trick" is that even if $f(x)$ does not satisfy the hypotheses, the iterate $f(f(x))$ might.

Take a look at Figure 10.3, where $f(x) = x e^{2(1-x)}$. Clearly, $f(x)$ does not satisfy the hypotheses, because there is a point p_0 with $f(p_0) = 1$ and $p_0 < 1$, so that $f(x) > 1$ for all $x \in (p_0, 1)$. But for all $x \in (p_0, 1)$, $f(f(x)) < 1$. Now if we can show that $f(f(x)) > x$ for these x's, we will have half the staircase hypothesis for $f^{(2)}$. We want $f(f(x)) = x e^{2(1-x)+2(1-f)} > x$, but dividing by x and taking logarithms this is equivalent to

$$2 - x > f,$$

and you can see from Figure 10.3 that this inequality holds. For the interval $(1, 2 - p_0)$, from the figure, $f(x) > 2 - x > p_0$ and $f(f(x)) > f(p_0) = 1$. Also, for this interval, $f(f(x)) = x e^{2(1-x)+2(1-f)} < x$, because $2 - x < f(x)$. So the second iterate $f(f(x))$ satisfies the staircase hypotheses for the interval $(p_0, 2 - p_0)$, and hence $\lim_{n \to \infty} f^{(2n)}(x_0) = 1$ holds for every x_0 in this

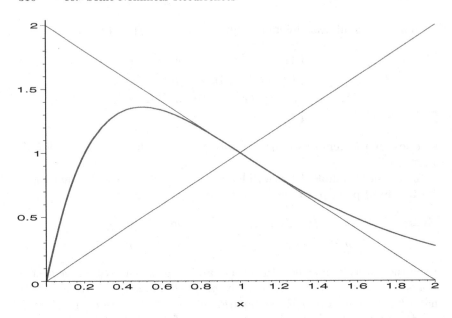

FIGURE 10.3. Showing global stability for a simple model. The curve $f(x)$ is bounded by the line $y = 2 - x$.

interval. Further, since $f(x_0)$ is also in this interval, $\lim_{n\to\infty} f^{(2n+1)}(x_0) = 1$, and $\lim_{m\to\infty} f^{(m)}(x_0) = 1$ holds for all x_0 in $(p_0, 2 - p_0)$.

We still have a little cleaning up to do. If $x_0 \in (0, p_0)$, then for some k, $f^{(k)}(x_0) \in (p_0, 1)$, and if $x_0 > 2 - p_0$, then $f(x_0) \in (0, 1)$. So for any $x_0 \in (0, \infty)$, some iterate falls into the interval where the staircase theorem applies, and $\lim_{n\to\infty} f^{(n)}(x_0) = 1$ for all $x_0 \in (0, \infty)$.

In a sense, we have just shown global stability by extending local stability from small enough neighborhoods to larger intervals. For local stability, we approximated a difference equation by a *linear* difference equation. Said another way, we try to bound a curve locally above and below by a straight line. If this straight line has slope at most 1 in absolute value, we may be able to use the line to show local stability. In our global stability example, we bounded the function $f(x)$ above and below by the straight line $2 - x$. Since this line has slope -1, following the difference equation $x_{t+1} = 2 - x_t$ for two steps brings us back to to the same point. But following $x_{t+1} = f(x_t)$ for two steps will bring us nearer to the fixed point. Since the bounding by $2 - x$ holds over a large interval, we can argue that the solutions to the nonlinear difference equation will converge to the fixed point. In our specific example, the form of $f(x)$ made checking the bounding easy, but the bounding was all that was needed for global stability. We summarize this discussion with the following theorem.

Theorem 10.5.2. *If $x < f(x) < 2 - x$ for x in $(0, 1)$ and if $x > f(x) > 2 - x$ for $x > 1$, then $x = 1$ is globally stable for $(0, \infty)$.*

In the next section, we consider a class of functions that generalize the linear functions and show that global stability for the difference equation $x_{t+1} = f(x_t)$ can be shown by bounding $f(x)$ by one of the functions from this special class.

10.6   Linear Fractional Recurrences

Among the simplest nonlinear recurrences are the linear fractional recurrences, which are simply the ratios of two linear recurrences. These recurrences are worth studying here because:
(a) they are simple;
(b) they give examples of how nonlinear recurrences can work;
(c) techniques from linear recurrences can be used;
(d) they can be used to study other nonlinear recurrences.

We write a **linear fractional recurrence** in the form

$$x_{t+1} = \frac{ax_t + b}{cx_t + d},$$

where we assume that a, b, c, d are real constants and that x_t is a real variable. Later we will consider the special case in which the constants and initial values are rational numbers. The usual questions asked about nonlinear systems include the existence of fixed points, the existence of cycles of various lengths, the asymptotic behavior of the system (e.g., are the fixed points or cycles attractive in some sense?), local and global stability of fixed points and cycles, chaos or chaotic-like behavior, and average behavior for a distribution of initial conditions. Pleasantly enough, all these questions can be answered in a relatively easy fashion for linear fractional systems.

First we should note that a linear fractional system degenerates into a linear one when $c = 0$, since

(10.4)
$$x_{t+1} = \frac{\hat{a}x_t + \hat{b}}{0x_t + \hat{d}} = ax_t + b.$$

As we've already seen, the solution to such a linear recurrence is

$$x_t = a^t x_0 + b \sum_{i=0}^{t-1} a^i = \begin{cases} x_0 + bt & \text{if } a = 1, \\ a^t x_0 + b\frac{a^t - 1}{a - 1} & \text{if } a \neq 1. \end{cases}$$

It is usual to analyze such a system in terms of the size of a. Generally, when $|a| > 1$, the solutions increase in absolute value, while when $|a| < 1$, the solutions decrease in absolute value. In fact, using the change of variable $y = x + \frac{b}{a-1}$, the linear system (10.4) can be rewritten as $y_{t+1} = ay_t$, and the analysis in terms of growth rate is perfectly appropriate. Of course, this transformation does not make sense when $a = 1$, since in the original (10.4) all solutions diverge, while in the transformed system every point is a fixed point.

If we analyze (10.4) as a nonlinear system, we get a different picture. First, there is a fixed point at $-b/(a-1)$, which is attractive when $|a| < 1$ for all real initial conditions. If we extend the real numbers by including ∞, the linear system always has ∞ as a fixed point, which is attractive when $|a| > 1$. In summary, a linear system can be seen to have two fixed points, one at $-b/(a-1)$ and the other at ∞. Which one is attractive depends on whether $|a| < 1$ or $|a| > 1$. For the case $a = 1$, the two fixed points degenerate into a single fixed point at ∞, which is an attractive fixed point. When one of the two fixed points is attractive, the system exponentially converges to that fixed point, but the convergence is only linear when there is one fixed point.

Periodic behavior occurs when $a = -1$, since all non-fixed points have period 2, and the 2-cycles are not attractive. They are sometimes called **neutrally stable** because if the system starts at a point near a cycle, the system always stays near the cycle but the system does not approach any closer to the cycle. For the very special case $a = 1$ and $b = 0$, every point becomes a neutrally stable fixed point. Finally, in the extremely singular case $a = 0$, all trajectories go to b in one step, and b is called a **superstable point**.

Behavior identical to the linear system (10.4) can also be found in the nonlinear system

$$z_{t+1} = \frac{z_t}{bz_t + a}.$$

The only difference is that x has been replaced by $1/x$, and so the fixed points have been shifted from $-b/(a-1)$ and ∞ to $(a-1)/-b$ and 0. This observation suggests that linear fractional systems may have many features in common with linear ones, but that their analysis depends on functions of several parameters and that these functions should be invariant under such transformations as taking reciprocals of the system's variable.

10.6.1 Asymptotic behavior

As for other nonlinear systems, the first step in analyzing linear fractional systems is to find their fixed points and determine their stability. Because linear fractional systems are closely related to linear systems, stability will depend on the eigenvalues of an associated matrix. Since the fixed points and eigenvalues may be irrational even when the linear fractional system's parameters are rational, we will make use of two simpler functions of the parameters. These are the **determinant**, $Det = ad - bc$, and the **discriminant,**, $Disc = (a-d)^2 + 4bc$.

The fixed-point equation is

$$p = \frac{ap + b}{cp + d}.$$

This is equivalent to the polynomial equation $cp^2 + (d - a)p - b = 0$. By the quadratic formula, the roots of this equation are

$$p = \frac{1}{2}\left[a - d \pm \sqrt{(a-d)^2 + 4bc}\right]$$

$$= \frac{1}{2}\left[a - d \pm \sqrt{(a+d)^2 - 4(ad - bc)}\right].$$

So, the number of fixed points will depend on the discriminant, $Disc$.

Let us now assume that the linear fractional system is nonlinear,

$$x_{t+1} = \frac{ax_t + b}{cx_t + d}, \qquad \text{where } c \neq 0.$$

When the determinant $Det = ad - bc$ of the recurrence is zero, we can use $ad = bc$ to get

$$f(x) = \frac{ax + b}{cx + d} = \frac{adcx + bcd}{cd(cx + d)} = \frac{bc(cx + d)}{cd(cx + d)} = \frac{b}{d}.$$

In one iteration all trajectories go to the superstable point b/d. (The superstable point is a/c if $d = 0$.) So, the determinant tells us something about stability.

We now know what happens when Det is zero. We still have to investigate the behavior when Det is positive, and when Det is negative.

From the fixed point equation $p = \frac{1}{2}\left[a - d \pm \sqrt{(a+d)^2 - 4(ad - bc)}\right]$, we see that if Det is negative, then $Disc$ is positive, and this equation will have two *real* roots. On the other hand, if Det is positive, the sign of $Disc$ is not determined. If $Disc$ is positive, there will be two *real* fixed points. If $Disc$ is zero, there is only one fixed point. (In essence, the two fixed points have coalesced into one point.) If $Disc$ is negative, there are no *real* fixed points. (In this case, there are two *complex* fixed points, but we will not see them because all our operations are in the reals.)

A linear fractional system can be represented in a linear fashion by

$$\begin{bmatrix} a & b \\ c & d \end{bmatrix}\begin{pmatrix} x_0 \\ 1 \end{pmatrix},$$

where x_0 is the initial point and the ratio of the components of the produced vector gives the value of x_1. So, x_n can be computed by performing the linear iteration

$$\begin{bmatrix} a & b \\ c & d \end{bmatrix}^n \begin{pmatrix} x_0 \\ 1 \end{pmatrix}$$

and then taking a ratio of components to obtain x_n. The eigenvalues of the matrix are called the **eigenvalues** of the linear fractional system, and they are the roots of the equation $\lambda^2 - (a+d)\lambda + ad - bc = 0$. Notice that the discriminant of this polynomial is $(a+d)^2 - 4(ad - bc)$, which equals our previously defined $Disc$. So, if $Disc$ is positive, the matrix will have two distinct *real* eigenvalues. As above, $Det < 0$ implies $Disc$ is positive, and also implies that there is one positive and one negative eigenvalue. The extra assumption that $a+d \neq 0$ implies that the eigenvalues have different absolute values. On the other hand, when $a + d = 0$, the two eigenvalues will have the same absolute value: one eigenvalue will be the negative of the other eigenvalue.

Let us assume that $Disc$ is positive. Since the eigenvalues are distinct, powers of the matrix can be computed by diagonalizing the matrix, and the nth linear iteration is calculated as

$$(10.5) \quad \frac{b}{\lambda_2 - \lambda_1} \begin{bmatrix} 1 & 1 \\ \frac{\lambda_1 - a}{b} & \frac{\lambda_2 - a}{b} \end{bmatrix} \begin{bmatrix} \lambda_1^n & 0 \\ 0 & \lambda_2^n \end{bmatrix} \begin{bmatrix} \frac{\lambda_2 - a}{b} & -1 \\ -\frac{\lambda_1 - a}{b} & 1 \end{bmatrix} \begin{pmatrix} x_0 \\ 1 \end{pmatrix}.$$

If we assume that $|\lambda_1| > |\lambda_2|$, then taking a ratio and a limit gives

$$\lim_{n \to \infty} x_n = \frac{b}{\lambda_1 - a},$$

unless $x_0 = b/(\lambda_2 - a)$. It is easy to check that these two points are in fact the fixed points, and so the linear fractional system has one unstable fixed point and one globally attractive fixed point when $Disc > 0$ and $|\lambda_1| > |\lambda_2|$. (The given formulas are indeterminate when $b = 0$, but then they can be written in the equivalent forms $(\lambda_1 - d)/c$ and $(\lambda_2 - d)/c$.) A special case arises when $Disc > 0$ and $|\lambda_1| = |\lambda_2|$, but for this to occur, both $Det < 0$ and $a + d = 0$ are required. For this special case, a simple calculation shows that $f(f(x)) = x$, which means that every point except for the two fixed points will have period 2, and of course, neither of the fixed points will be attractive.

A more interesting case occurs when $Disc$ is zero, which means that there is only one fixed point. Geometrically, this says that the line $y = x$ is tangent to the curve $y = f(x)$ and the point of tangency is the fixed point. It can be seen (see Figure 10.5) that every point above the fixed point iterates to a point still above but nearer to the fixed point. While a point below the fixed point first iterates away from the fixed point, eventually one of its iterates jumps across the discontinuity and then will jump to a point above the fixed point. As an example, consider $f(x) = x/(x+1)$. (Refer to Figure 10.5.) Here, $x_n = x_0/(nx_0 + 1)$, and every trajectory converges to the fixed point 0. The jumping between branches is hidden by this formula, but can be displayed by following an example. The trajectory starting at $-3/4$ gives $-3/4, -3, 3/2$ and then converges through positive values to 0. Notice that the convergence is different in the one and two fixed point cases.

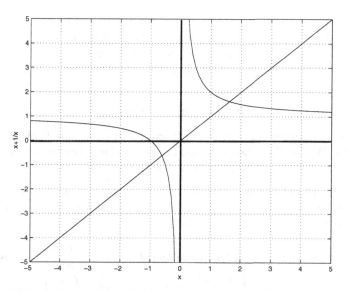

FIGURE 10.4. A plot showing the linear fractional system $x_{t+1} = (x_t + 1)/x_t$ with two fixed points. All iterates converge to the upper fixed point.

For two fixed points, $x_n = p + O(\gamma^n)$, where p is the stable fixed point and $|\gamma| < 1$, while in the one fixed-point case, $x_n = p + O(1/n)$. Figure 10.4 shows the geometry of a two fixed-point case and Figure 10.5 shows the geometry of a one fixed-point case.

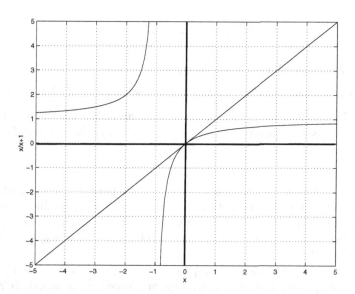

FIGURE 10.5. A plot showing the linear fractional system $x_{t+1} = x_t/(x_t + 1)$ with one fixed point. Convergence to the fixed point is from above.

So far, linear fractional systems have not behaved very differently from linear systems. The remaining cases, in which $Disc$ is negative (and so $4Det = (a+d)^2 - Disc$ is positive), will display more nonlinear behavior. We will discuss these cases in the following subsections.

10.6.2 Rational coefficients and periodicity

Like other nonlinear systems, linear fractional systems can display periodic behavior. But unlike other systems, linear fractional systems do not allow the co-existence of cycles of different lengths.

Theorem 10.6.1. *If for a linear fractional map f there exists a point x such that $f^{(K)}(x) = x$, then either $K = 1$ (and x is a fixed point) or for all y, $f^{(K)}(y) = y$ (all points are periodic).*

Proof. As above, the linear fractional map $f(x) = \dfrac{ax+b}{cx+d}$ can be considered as a 2×2 matrix $A = \begin{bmatrix} a & b \\ c & d \end{bmatrix}$ acting on a vector $\begin{pmatrix} x \\ 1 \end{pmatrix}$. Since A is a 2×2 matrix, its characteristic polynomial is quadratic, and for every natural number K, there exist scalars α_K and β_K such that $A^K = \alpha_K A + \beta_K I$, and

$$A^K \begin{pmatrix} x \\ 1 \end{pmatrix} = \alpha_K A \begin{pmatrix} x \\ 1 \end{pmatrix} + \beta_K \begin{pmatrix} x \\ 1 \end{pmatrix}.$$

So if $f^{(K)}(x) = x$, then

$$\frac{\alpha_K(ax+b) + \beta_K x}{\alpha_K(cx+d) + \beta_K} = x.$$

There are two possibilities for this equation:

(a) $\alpha_K = 0$, and so $A^K = \beta_K I$, and for every y, $A^K \begin{pmatrix} y \\ 1 \end{pmatrix} = \beta_K \begin{pmatrix} y \\ 1 \end{pmatrix}$,
   which is equivalent to $\beta_K y / \beta_K = y$, so all points have period K ;
(b) $\alpha_K \neq 0$, and then $ax + b = (cx+d)x$, which is the equation for fixed points.

$\square$

This theorem allows for the co-existence of periodic points and fixed points. When the linear fractional system has real eigenvalues, two real fixed points will occur, and periodic behavior will occur when the two eigenvalues have the same magnitude. In this case, all non-fixed points will have period 2.

When the eigenvalues are complex, there will be no real fixed points, and the magnitudes of the eigenvalues are forced to be equal. Here, let $e^{i\theta} = \lambda_1 / \lambda_2$. If θ is a rational multiple of π, then $e^{i\theta}$ is a root of unity, and there will be a least positive integer K such that $A^K = \beta_K I$. In this case, every point will be periodic and have period K. If θ is not a rational multiple of π, the system will not be periodic, and all points will fail to be periodic. We will consider this situation in more detail in the next section.

In general, all values of θ are possible, since a direct calculation of λ_1/λ_2 shows that any desired complex number of norm 1 can be produced by appropriate choice of the parameters a, b, c, d for the linear fractional map. Of course, these parameters can be chosen as real numbers, but in realistic uses of linear fractional systems one would like to assume that the parameters have finite representations. In particular, one might like to assume that these parameters are rational, and then of course the same linear fractional map can be represented using integer parameters. We would like to know which periods are possible for linear fractional systems with rational parameters. The following theorem gives the answer.

Theorem 10.6.2. *A linear fractional system with rational (or integer) parameters can only have periods 1, 2, 3, 4, and 6, and there are examples of rational linear fractional systems with each of these periods.*

Proof. Let $\gamma = \lambda_1/\lambda_2$, the ratio of the eigenvalues, and γ is a root of a polynomial $\gamma^2 - e\gamma + 1$, where e is a rational function of a, b, c, d. If the linear fractional system has period K, then γ is a primitive K^{th} root of unity; that is, $\gamma^K = 1$ but $\gamma^J \neq 1$ for all $0 < J < K$. For each natural number K, it can be shown [141, Section 1.2] that the **cyclotomic polynomial**

$$\Phi_K(x) = \Pi(x - \zeta), \quad \text{where } \zeta \text{ ranges over all primitive } K^{\text{th}} \text{ roots of unity},$$

has integer coefficients and its degree is $\phi(K)$, where ϕ is Euler's Phi Function used in Chapter 8. Further, this polynomial is minimal in the sense that it does not have any smaller-degree rational polynomial factors.

Recalling that $\phi(K)$ counts the number of positive integers $i \leq K$ with $\gcd(K, i) = 1$, it can be shown that every integer $K > 6$ has $\phi(K) \geq 3$, because 1 and $K - 1$ and a third number are relatively prime to K. We find this third number in each of three cases. One, if K is odd and $K > 5$, then 4 is also relatively prime to K. Two, if $K > 6$ and $K = 2m$ with m odd, then $m - 2$ is relatively prime to K because

$$\gcd(2m, m - 2) = \gcd(m - 2, 4) = 1,$$

since $m - 2$ is odd. (We need $K > 6$ because for $K = 6$, $m - 2 = 1$.) Three, if $K > 6$ and K is a multiple of 4, then $K/2 - 1$ is relatively prime to K because

$$\gcd(K, K/2 - 1) = \gcd(K/2 - 1, 2) = 1,$$

since $K/2 - 1$ is odd.

Returning to the problem at hand, γ is a root of the cyclotomic polynomial $\Phi_K(x)$, and γ also satisfies a rational *quadratic* polynomial. The minimality of Φ_K implies that $\phi(K)$ must equal 1 or 2. Hence, the only possible values for K are $1, 2, 3, 4, 6$, since we have ruled out all $K \geq 7$, and it's easy to check that $\phi(5) = 4$.

We now give an example of a rational linear fractional system for each of these five periods.

Period 1: $f(x) = x$. This is a degenerate linear fractional system in which all points have period one.

Period 2: For $f(x) = (x - 2)/(2x - 1)$ there are complex eigenvalues and no fixed points. An example period is $2 \longleftrightarrow 0$.

For $f(x) = (5x - 2)/(4x - 5)$, there are real eigenvalues and two fixed points at $(5 \pm \sqrt{17})/4$. All other points have period 2, for example $1 \longleftrightarrow -3$.

Period 3: $f(x) = (x - 1)/x$.
Example period is $3 \longrightarrow 2/3 \longrightarrow -1/2$.

Period 4: $f(x) = (x - 1)/(x + 1)$.
Example period is $2 \longrightarrow 1/3 \longrightarrow -1/2 \longrightarrow -3$.

Period 6: $f(x) = (2x - 1)/(x + 1)$.
Example period is $3 \longrightarrow 5/4 \longrightarrow 2/3 \longrightarrow 1/5 \longrightarrow -1/2 \longrightarrow -4$.

$\square$

A pleasant outcome of this analysis is that it is easy to test for periodicity in *rational* linear fractional systems. One simply tries an initial condition and checks to see whether it gives periodic behavior of length at most 6. The only minor problem is that in the case of real eigenvalues one could chance on a fixed point and other points would need to be tested for periodicity.

10.6.3   Chaotic-like behavior

Three characteristics of chaos are:
(a) cycles of every period;
(b) sensitive dependence on initial conditions;
(c) for every open set A and every open set B there is an $x_0 \in A$ and a $K \in N$ such that $f^{(K)}(x_0) \in B$.

For linear fractional systems, (a) is simply false, but we will see that (b) and (c) do hold when the ratio of the eigenvalues is not a root of unity. Recall that sensitive dependence means that regardless of how close two different trajectories are when they begin, eventually they are far apart. For linear fractional systems, the pole at $p = -d/c$ forces trajectories to diverge from one another. For example, if one chose to consider trajectories starting at $p - \epsilon$ and $p + \epsilon$, then $|f(p+\epsilon) - f(p-\epsilon)| = 2|Det/c^2\epsilon|$, a quantity that can be made as large as one likes by taking ϵ small. For some linear fractional systems, this is not a problem because trajectories are attracted to stable fixed points and any initial divergence disappears in the long term. For *periodic* linear fractional systems, an initial difference is in essence maintained and will not increase. But we will see that there are linear fractional systems with complex eigenvalues for which the dependence on initial conditions does not die out, since every trajectory eventually comes close to the pole and then will be thrown far away. So two trajectories that start close together are eventually far apart, and there is no tendency for them to come close again. To see what happens in this case, we first show that (c) holds. For this, a sequence $\langle S_n \rangle$ is called a **source** for A if for every $\alpha \in A$ and for every $\epsilon > 0$ there is a K such that $|\alpha - S_K| < \epsilon$.

Lemma 10.6.3. *If $\langle S_n \rangle$ is a source for A and $g(x)$ is a function from A onto B such that every preimage has a neighborhood of continuity, then $g(\langle S_n \rangle)$ is a source for B.*

Proof. If w is the desired point in B with desired closeness δ, then since g is onto, there is a preimage $v \in A$ such that $g(v) = w$. Since there is a neighborhood of continuity around v, there exists an $\epsilon > 0$ such that if $|x - v| < \epsilon$, then $|g(x) - g(v)| = |g(x) - w| < \delta$. But $\langle S_n \rangle$ is a source for A, which means that there is a K such that $|S_K - v| < \epsilon$ and so $|g(S_K) - w| < \delta$. This proves that $g(\langle S_n \rangle)$ is a source for B.   □

Theorem 10.6.4. *If $\langle x_n \rangle$ is a trajectory of a linear fractional system in which the ratio of its eigenvalues is $\lambda_1/\lambda_2 = e^{i\theta}$, where θ is an irrational multiple of π, then $\langle x_n \rangle$ is a source for $(-\infty, \infty)$.*

Proof. The nth term of $\langle x_n \rangle$ can be written as

$$x_n = \frac{-ax_0 - b + x_0[\cos\theta - \sin\theta \cot n\theta]}{a - cx_0 - [\cos\theta + \sin\theta \cot n\theta]}.$$

Since θ is an irrational multiple of π, the sequence $\langle n\theta/2\pi \mod 1 \rangle$ is a source for $[0, 1]$ (refer to [120, Chapter 3]). But the transformation from $n\theta$ to x_n satisfies the hypotheses of the lemma, and so $\langle x_n \rangle$ is a source for $(-\infty, \infty)$.   □

Hence, these linear fractional systems obey (c), and they also obey (b), because any two nearby trajectories eventually hit a small neighborhood of the pole and then are thrown far apart.

Figure 10.6 shows the web diagram for the linear fractional map $f(x) = (x - 5)/(x + 1)$. One can see that with the 100 iterates used in the diagram calculation, almost all points are filled in. Figure 10.7 shows sensitive dependence on initial conditions in that the two displayed trajectories are often very close but are occasionally far apart.

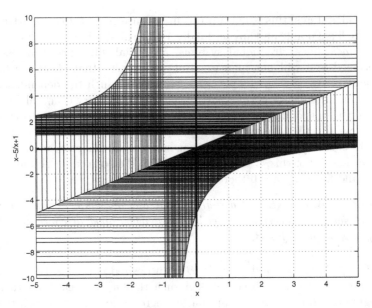

FIGURE 10.6. A web plot of the model $\frac{x-5}{x+1}$, showing that most points are visited.

10.6.4 Invariant distributions

When a system has wandering trajectories that do not converge to cycles or fixed points, one might ask whether perhaps other properties are conserved. Natural questions to ask are how long a trajectory stays in a region and how often a trajectory visits a region. It may be difficult to answer these questions and perhaps unenlightening, because we want to know how the *system* behaves rather than how a specific trajectory behaves. So, we consider putting a probability distribution on the space and then asking how the distribution is changed by the system. In particular, we would like to know whether there is a very special distribution that remains the same after the system acts on it. If we could prepare enough copies of a system and start these copies in accord with this invariant distribution, then we would later (even a long time later) still see the same distribution of states. Although the copies that started in particular states are no longer in those states, other copies are in those states in proportion to the invariant density. We could also hope for some sort of convergence. It might be possible that the invariant distribution would be attractive, if not for all initial distributions, then at least for a class of initial distributions. By attractive here we mean that under some reasonable definition of distance between distributions, the distance between the system's distribution at time t and the invariant distribution converges to 0 as t increases. We will show that linear fractional systems do have invariant distributions, but these distributions are not attractive.

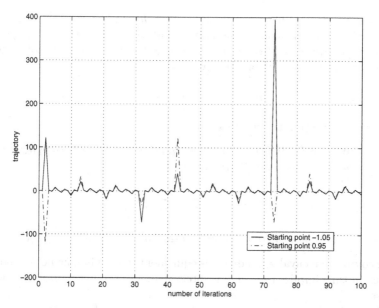

FIGURE 10.7. Two trajectories for $f(x) = \frac{x-5}{x+1}$. One trajectory starts at -1.05 and the other at -0.95. These two trajectories are often very close, but occasionally they are far apart.

Although we spoke of probability *distributions*, it may be easier to work with probability *densities*. A **density function** $g(x)$ is a nonnegative function defined on the real numbers with the property that every interval (α, β) has probability $\int_\alpha^\beta g(x)dx$. We normalize $g(x)$ to have $\int_{-\infty}^\infty g(x)dx = 1$. Now let us see how the mass on the interval (α, β) is transformed by the mapping f. First, the interval (α, β) is transformed to $(f(\alpha), f(\beta))$, and then the mass that was on (α, β) is spread over this new interval. But if $g(x)$ is an invariant density, then the density on the new interval must also be $g(x)$. If we let $\hat{g}(x)$ be the distribution on the new interval, we have $\int_\alpha^\beta g(x)dx = \int_{f(\alpha)}^{f(\beta)} \hat{g}(x)dx$, and using $x = f(y)$ and $dx = f'(y)dy$ this becomes $\int_\alpha^\beta g(x)dx = \int_\alpha^\beta \hat{g}(x)f'(y)dy$. Since α and β are arbitrary, differentiating with respect to β gives $g(x) = \hat{g}(f(x))f'(x)$ at least for those x's at which these functions are continuous. If $g_I(x)$ is an invariant density, then $g_I(x) = g_I(f(x))f'(x)$. Since we want $g_I(x)$ to have a finite integral over $(-\infty, \infty)$, $g_I(x)$ should be close to 0 when x is large in absolute value, and so it may be easier to look at $h(x) = 1/g_I(x)$. Then $h(x) = h(f(x))/f'(x)$. Using a linear fractional map for f, which has been normalized to have $Det = ad - bc = 1$, then

$$h(x) = (cx + d)^2 h\left(\frac{ax + b}{cx + d}\right).$$

Assuming that $h(x)$ has a power series, then $h(x)$ must be a polynomial of degree 2. The coefficients of the polynomial can be found by solving a set of linear equations. Up to an unknown scale factor, the unique solution to the set of three linear equations gives

$$h(x) = cx^2 + (d - a)x - b,$$

and we obtain the following theorem.

Theorem 10.6.5. *If $f(x)$ is a linear fractional map with complex eigenvalues, then there is a unique invariant density*

$$g(x) = \frac{\gamma}{cx^2 + (d - a)x - b},$$

where γ is determined from the normalization condition $\int_{-\infty}^{\infty} g(x)dx = 1$.

The complex eigenvalue condition is equivalent to $g(x)$ having no real poles, which in turn implies that $g(x)$ is integrable, and the normalization makes sense.

Conveniently enough, the integral of the density function can also be written in closed form as

$$\int_{-\infty}^{z} g(x)dx = \frac{1}{\pi} \arctan \frac{2cz + d - a}{\sqrt{-(d-a)^2 - 4bc}} + \frac{1}{2}.$$

If one could choose the states of an ensemble of systems to satisfy the density $g_I(x)$, then as the states evolve under the application of f, the same density must be maintained. One might hope that every density evolves to $g_I(x)$ under f. The clue that this is not the case is in the theorem's condition that all one needs for an invariant distribution is that the eigenvalues are complex. But for instance, any linear fractional system with eigenvalues i and $-i$ has period 2. For such a system, *any* initial distribution repeats after two steps, and the density does not converge to the invariant density. More specifically, one can prove the following theorem.

Theorem 10.6.6. *For a linear fractional system with complex eigenvalues, the invariant density is not attractive even within the class of densities whose reciprocals are quadratic polynomials. For such densities, the discriminant is a conserved (invariant) quantity.*

The proof of this theorem consists in showing that the coefficients of the new density can be computed from the coefficients of the old density by applying a matrix to the vector of coefficients. The invariant density corresponds to the 1-eigenvector (the eigenvector associated with the eigenvalue 1). One can then show that the other two eigenvalues of the matrix also have absolute value equal to 1. Hence, even though a density converts to a new density, the difference from the 1-eigenvector does not decrease, and the 1-eigenvector corresponding to the invariant density is not attractive. In fact, one can also show that the discriminant is preserved. Starting with a density that is the reciprocal of quadratic polynomial and iterating using f results in a density of the same kind, and the discriminant of this density is identical to the discriminant of the starting density.

What does a typical aperiodic trajectory look like? One way to describe such a trajectory is to use a histogram, that is, to break up the range into small bins and then count the number of times the trajectory visits each bin. One can then normalize by the number of iterates and hope that a limiting histogram exists.

Let $\langle f^{(n)}(x_0) \rangle$ be the sequence of iterates of f starting at x_0. If this trajectory is well behaved, then there is an associated histogram H such that $H((a,b))$ is the frequency with which the trajectory visits the interval (a,b). Then

$$H((a,b)) = \lim_{N \to \infty} \frac{1}{N} \sum_{i=1}^{N} I_{(a,b)}[f^{(i)}(x_0)],$$

where $I_{(a,b)}[z]$ is the indicator function that gives 1 when its argument is in (a,b) and 0 when its argument is not in (a,b). Assuming that these limits exist and that H is smooth, then

$$H(x) = \lim_{\epsilon \to 0} H((x - \epsilon, x + \epsilon))$$

should exist and behave like a probability density.

Assuming that all of this is true, what should $H(x)$ look like? Since we are looking at limiting behavior, it should not matter whether the trajectory starts at x_0 or at $f(x_0)$. Hence, we expect the limiting histogram to be an invariant density. But by Theorem 10.6.5 there is a unique invariant density, so $H(x)$ should look like $g_I(x)$. Figure 10.8 shows a histogram for 1000 iterates of $(x - 5)/(x + 1)$. This histogram looks quite smooth and agrees reasonably with the invariant density $1/(x^2 + 5)$.

These results tell us that in spite of a seemingly irregular trajectory, a fairly simple property is maintained. If one looks at any individual trajectory, then the long-run histogram should look like a fairly simple function. On the other hand, if one took multiple copies of the same linear fractional system, assigned initial conditions with the probabilities given by the histogram, and then looked at the distribution of states after one or several time intervals, the probability density would still be the same as the initial density. This is a sort of **ergodic theorem**, which says that the average over one trajectory is the same as the appropriate average over an ensemble of systems.

Notice that if one picked a density $g(x)$ and picked the initial conditions for an ensemble according to $g(x)$, one would expect a different density, say $g_1(x)$, to occur after one time step, and then densities $g_2(x), g_3(x), g_4(x), \ldots$ for subsequent time steps. There is no reason to expect this sequence to

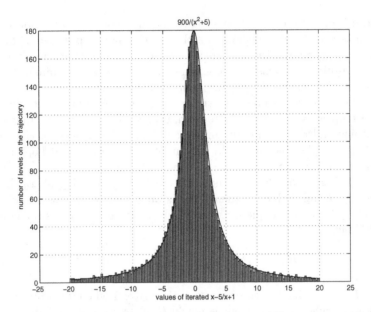

FIGURE 10.8. A histogram for a trajectory of $(x - 5)/(x + 1)$ showing good agreement with the density $1/(x^2 + 5)$.

converge to a single density. In fact, Theorem 10.6.6 says that this sequence does *not* converge, but the sequence of averages does converge. That is, for $G_0(x) = g(x)$ and $G_N(x) = \frac{1}{N} \sum_{i=0}^{N-1} g_i(x)$, then $\lim_{N \to \infty} G_N(x) = g_I(x)$.

10.6.5 *Proving global stability*

Our initial interest in linear fractional systems came from an application to population models. For some years, it has been known that the usual one-dimensional population models are globally stable exactly when they are locally stable [63, 32, 148]. We were pleased to find that the usual population models were "enveloped" by linear fractional maps of the special form

$$\phi(x) = \frac{1 - \alpha x}{\alpha - (2\alpha - 1)x}, \quad \text{where } \alpha \in [0, 1) \ .$$

These special linear fractional systems all have period 2, and so by a variation on Sarkovskii's Theorem (refer to Theorem 10.2.1) we were able to prove the following.

Theorem 10.6.7. *Let $\phi(x)$ be a monotone decreasing function that is positive on $(0, x_-)$ and such that its second iterate is the identity; that is, $\phi(\phi(x)) = x$. If $f(x)$ is a continuous function such that*

- $\phi(x) > f(x)$ *on* $(0, 1)$,
- $\phi(x) < f(x)$ *on* $(1, x_-)$,

- $f(x) > x$ on $(0, 1)$,

- $f(x) < x$ on $(1, \infty)$,

- $f(x) > 0$ on $(1, x_\infty)$,

then for all $x \in (0, x_\infty)$, $\lim_{k \to \infty} f^{(k)}(x) = 1$.

This theorem enabled us to show that local stability implies global stability for the following seven population models:

- $x_{t+1} = x_t e^{r(1-x_t)}$,

- $x_{t+1} = x_t[1 + r(1 - x_t)]$,

- $x_{t+1} = x_t[1 - r \ln x_t]$,

- $x_{t+1} = x_t\left(\frac{1}{b + cx_t} - d\right)$,

- $x_{t+1} = \frac{(1+ae^b)x}{1+ae^{bx}}$,

- $x_{t+1} = \frac{(1+a)^b x}{(1+ax)^b}$,   with $a > 0$, $b > 0$,

- $x_{t+1} = \frac{rx}{1+(r-1)x^c}$.

Details of these results appear in [38, 39, 45].

10.6.6 Summary

Linear fractional systems form a fairly simple class of nonlinear systems, yet they display many of the possible behaviors for nonlinear systems. In particular, they have both stable and unstable fixed points and exhibit both periodic and chaotic-like behavior. Table 10.1 gives a summary of the possible behaviors and the corresponding conditions on the parameters.

In contrast to most nonlinear systems, linear fractional systems can be analyzed using techniques from elementary mathematics and linear systems. Further, we have shown that linear fractional systems can be used to analyze more complicated systems, and we suggest that linear fractional systems should be a standard part of the toolbox for studying nonlinear systems.

TABLE 10.1. The possible asymptotic behavior of a linear fractional system $f(x) = \frac{ax+b}{cx+d}$ (we assume $c \neq 0$ to get a nonlinear system) and set $Det = ad - bc$, $Disc = (a - d)^2 + 4bc$.

$Det = 0$		$f(x) = b/d$	Superstable fixed point
$Det < 0$ Two real fixed points	$d + a \neq 0$	One stable fixed point One unstable fixed point Convergence $x_n = F.P. + O(\lambda^n)$, $|\lambda| < 1$	
	$d + a = 0$	All points except the fixed points have period 2 neutral stability	
$Det > 0$	$Disc = 0$	One globally stable fixed point Convergence $x_n = F.P. + O(\frac{1}{n})$	
	$Disc > 0$	One stable fixed point One unstable fixed point Convergence $x_n = F.P. + O(\lambda^n)$, $|\lambda| < 1$	
	$Disc < 0$	Periodic, all points have the same period	Rational coefficients - possible periods 1,2,3,4,6
			Irrational coeffs - any period possible
		Chaotic-like	No periodic points All open sets visited Sensitive dependence on initial conditions Invariant density - Non-attractive

10.7   Conclusion

As we said at the beginning of this chapter, nonlinear systems can be non-linear in many different ways. We have considered some very simple types, and our examples have usually been one-dimensional. Realistic systems often have high dimension, and the simple analyses possible in one dimension do not apply. On the other hand, knowing that chaos is possible in one dimension both warns us that higher-dimensional systems may behave in very complicated ways, and suggests that we may be able to explain this complicated behavior by finding a chaotic one-dimensional subsystem. We have limited our examples to the reals or rationals, but some systems may be easier to understand if one extends the analysis to the complex numbers. For example, a linear fractional map may or may not have fixed points when restricted to the real numbers, but such systems always have fixed points in the complex numbers. Other nonlinear systems may be composed of simple components with finite state sets like neural nets, or have an infinite but locally finite state set like cellular automata, or have state sets like the natural numbers or strings over a finite alphabet like Turing machines and other models of computation. Analyses of such systems can be shown to be impossible [77]. Even when limited to finite state sets, the analysis of such systems can be shown to be possible, but practically unreasonable [2].

We close with the warning that realistic systems may be highly nonlinear and highly complicated, but also with the hope that as in the past, future generations of scientists and engineers will find simple enough models to solve societal problems.

10.8   Exercises

Ex 10.1. Show that if $\lim_{k \to \infty} f^{(k)}(x_0) = p$ and f is a continuous function, then p is a fixed point of f.

Ex 10.2. Let $x_{t+1} = f(x_t)$, where

$$f(x) = \begin{cases} \frac{x}{2} + \frac{1}{4} & \text{if } x < \frac{1}{2}, \\ \frac{x}{2} + \frac{1}{2} & \text{if } \frac{1}{2} < x \leq 1, \\ 1 & \text{if } x > 1. \end{cases}$$

Show that for $x_0 < 1/2$, $\lim_{K \to \infty} f^{(K)}(x_0) = 1/2$, but for $x_0 > 1/2$, $\lim_{K \to \infty} f^{(K)}(x_0) = 1$. Does this contradict the results of the previous exercise? Can this system be enveloped by a linear fractional system? Does this violate Theorem 10.6.7?

Ex 10.3. Show that $x_{t+1} = \sin x_t$ has one fixed point, which is both locally and globally stable.
Hint: A local linear approximation is insufficient. Use a local nonlinear approximation to show local stability.

Ex 10.4. Analyze the local and global stability of $x = 1$ in the system

$$x_{t+1} = x_t[1 + 2(1 - x_t)].$$

Ex 10.5. Analyze the local and global stability of $x = 1$ in the system

$$x_{t+1} = x_t[1 + 2.00001(1 - x_t)].$$

Ex 10.6. Show that

$$x_{t+1} = x_t[1 + r \ln x_t]$$

has a locally stable period 2-cycle when r is slightly larger than 2.

Ex 10.7. Let $g(x) = \sqrt{|x|}$. Show that Newton's method oscillates with period 2 for non-zero initial values.

Ex 10.8. Let $x_{t+1} = f(x_t)$, where

$$f(x) = \begin{cases} -\sqrt{x} & \text{if } x \geq 0, \\ \sqrt{|x|} & \text{if } x < 0. \end{cases}$$

Show that this system has a repelling fixed point and an attracting 2-cycle. Is this 2-cycle globally stable?

Ex 10.9. Let $x_{t+1} = f(x_t)$, where

$$f(x) = \frac{x^2 + 2}{2x - 1}.$$

Show that $x_0 = 1/2$ separates the behavior of this system in the sense that if $x_0 < 1/2$, there is one type of behavior, and if $x_0 > 1/2$, there is a different behavior.

Ex 10.10. Consider the system

$$x_{t+1} = x_t e^{r(1-x_t-y_t)},$$
$$y_{t+1} = y_t e^{r(1-x_t-y_t)}.$$

Find a quantity that is *invariant* along trajectories. Discuss the convergence properties of this system in light of this invariant. Are there fixed points? Are any of these fixed points locally or globally stable?

Ex 10.11. For the difference equation $x_{t+1} = 2x_t \bmod 1$, plot x_{t+1} as a function of x_t. Find the fixed points in your diagram. Does the plot show a discontinuity? Plot x_{t+2} as a function of x_t. Can you find the oscillations of period 2?

Ex 10.12. For $x_{t+1} = 2x_t \bmod 1$, plot x_{t+3} as a function of x_t. How does this show that there are 2 distinct period 3 oscillations? Find these oscillations.

Ex 10.13. A sequence is **eventually periodic** if there is an r such that $x_r, x_{r+1}, \ldots$ is periodic of period p. Show that $x_{t+1} = 2x_t \bmod 1$ has many eventually periodic solutions by giving a procedure that takes r and p as input, and outputs an initial condition that gives a solution of eventual period p after a run-in of r steps.

Ex 10.14. Following our construction for period K, find an oscillation of period 5 for $x_{t+1} = 2x_t \bmod 1$. Show that this is *not* the only possible construction by finding another solution of period 5 for this equation.

Ex 10.15. Show that $x_{t+1} = 2x_t \bmod 1$ has at most $\lfloor (2^K - 1)/K \rfloor$ oscillations of period K. Can you find a better formula for the number of distinct oscillations of period K?

Ex 10.16. Is

$$x_{t+1} = 2x_t \bmod 1$$

a linear equation? Refer to the definition and notice that something is missing. Show that by filling in the missing part of the definition in different ways, you can declare the equation to be linear or you can show that it is nonlinear.

Ex 10.17. Consider the one-dimensional system

$$x_{t+1} = f(x_t) = x[1 + r(a - x)]$$

for some fixed positive a, r with $ar > 2$. Show that the system has two fixed points and at least one locally stable 2-cycle when $ar < \sqrt{6}$.

Ex 10.18. Consider the two-dimensional system

$$x_{t+1} = f_1(x_t, y_t) = x(5 - x - y)/3,$$
$$y_{t+1} = f_2(x_t, y_t) = y(3y - x)/3.$$

Show that $(2, 0)$ is a locally stable fixed point of the system and $(3/4, 5/4)$ is a fixed point that is not locally stable.

Ex 10.19. Show that the two-dimensional system

$$x_{t+1} = f_1(x_t, y_t) = x(1 + x + y)/3,$$
$$y_{t+1} = f_2(x_t, y_t) = y(1 - x + y)/2,$$

has four fixed points, but only one is locally stable.

Ex 10.20. Show that

$$x_{t+1} = \frac{3x_t + 2}{2x_t + 1}$$

has two fixed points, $\lambda_0 > 0$ and $\lambda_1 < 0$, and that λ_0 is locally stable but λ_1 is repelling. Further, show that the initial value $x_0 = 0$ gives $x_n = f_{3n}/f_{3n-1}$, where f_i is the ith Fibonacci number. Give a good estimate for how close x_n is to λ_0. How does the recurrence behave for other initial values?

Ex 10.21. Following the staircase method from Section 10.5.1, set up an iteration to calculate the fixed $1/(1 - 2 \ln x)$ inside the interval $(0, 1)$. Also, show that your iteration converges to this fixed point.

Ex 10.22. Find the invariant density for the linear fractional $f(x) = \frac{x - 1}{x + 1}$. Show by example that this invariant density is *not* attractive. Does the histogram for the trajectory starting at 0 look like the invariant density?

Ex 10.23. Pick an aperiodic linear fractional system with complex eigenvalues. Find the invariant density for your system. Pick an initial point and calculate the histogram for the trajectory starting at this initial point. Does the histogram look like the invariant density?

Ex 10.24. Show that the fixed point $x = 1$ is globally stable for $x_{t+1} = f(x_t)$ with

$$f(x) = \begin{cases} 6x & 0 \le x < 1/2, \\ 5 - 4x & 1/2 \le x < 1, \\ 1 & 1 \le x. \end{cases}$$

Further show that there is *no* linear fractional system that bounds $f(x)$ as required by Theorem 10.6.7.

Ex 10.25. Find a value for α such that

$$f(x) = x[1 + 2(1 - x)]$$

is bounded by a linear fractional system as in the hypotheses of Theorem 10.6.7. Use this to show that $x = 1$ is a globally stable fixed point for

$$x_{t+1} = x[1 + r(1 - x_t)]$$

for all $r \in (0, 2]$.

Appendix A
Worked Examples

In the first chapters of this book we consider k^{th}–**order linear recurrences**, that is, equations of the form

(L) $\qquad s_n - c_1 s_{n-1} - c_2 s_{n-2} - \cdots - c_k s_{n-k} = \psi(n) \qquad$ for $n \geq k$,

where c_i are complex scalars with $c_k \neq 0$, and ψ is a complex-valued function. There we found a nice solution for the special case in which the forcing function has the form

$$\psi(n) = \lambda^n p(n),$$

where λ is a fixed scalar and p is a polynomial with complex coefficients. The initial value problems we consider in this appendix have this form. Our principal tool is Theorem 3.3.1, which we use to find a particular solution $\langle v_n \rangle$. Throughout, $\lambda_1, \ldots, \lambda_t$ are the *different* eigenvalues of the recurrence.

A.1 All Simple Roots

All examples in this section are second–order linear equations whose characteristic polynomial is

$$ch(x) = x^2 - x - 6 = (x - 3)(x + 2),$$

which has the simple roots $\lambda_1 = 3$, $\lambda_2 = -2$, and the general solution of the *homogeneous* system has the form $s_n = a_1 3^n + a_2(-2)^n$ for some $a_1, a_2 \in \mathbb{C}$, and for any ψ the equation

$$s_n = s_{n-1} + 6s_{n-2} + \psi(n)$$

has the general solution

$$s_n = a_1 3^n + a_2(-2)^n + v_n,$$

for some particular solution v_n. The constants a_1, a_2 are determined by the initial conditions.

Theorem 3.3.1 can be used to find a particular solution to the *nonhomogeneous* equation

$$s_n = s_{n-1} + 6s_{n-2} + \lambda^n p(n),$$

where p is a non-zero polynomial and λ is a scalar. For these equations the following special case of the theorem applies.

Solving (L) when there are no multiple roots

When $ch(x)$ has no multiple roots, then $\langle s_n \rangle$ is a solution to the equation

(A.1) $$s_n - c_1 s_{n-1} - c_2 s_{n-2} - \cdots - c_k s_{n-k} = \lambda^n p(n),$$

if and only if

(A.2) $$s_n = \sum_{i=1}^{k} a_i \lambda_i^n + \lambda^n n^\delta q(n),$$

where $a_1, \ldots, a_k \in \mathbb{C}$, $q(x)$ is a polynomial with $\deg(q) = \deg(p)$ and

(A.3) $$\delta = \begin{cases} 0 & \text{if } \lambda \notin \{\lambda_1, \ldots, \lambda_t\} \\ 1 & \text{if } \lambda \in \{\lambda_1, \ldots, \lambda_t\} \end{cases}.$$

A particular solution is $\lambda^n n^\delta q(n)$.

Let's first analyze the homogeneous initial value problem.

Example A.1.1. For any initial value problem with equation

$$s_n = s_{n-1} + 6s_{n-2}$$

we have $s_n = a_1 3^n + a_2(-2)^n$, and the initial conditions s_0, s_1 give

(A.4) $$s_0 = a_1 + a_2 \quad \text{and} \quad s_1 = 3a_1 - 2a_2.$$

Multiplying the first equation by 3 and subtracting that from the second equation, we obtain

$$s_1 - 3s_0 = -5a_2 \quad \text{and} \quad a_2 = \frac{3s_0 - s_1}{5}.$$

Inserting this value for a_2 into the first equation of (A.4), we have

$$a_1 = s_0 - \frac{3s_0 - s_1}{5} = \frac{2s_0 + s_1}{5},$$

and the solution to the homogeneous initial value problem is

$$s_n = \frac{2s_0 + s_1}{5}3^n + \frac{3s_0 - s_1}{5}(-2)^n.$$

Example A.1.2. Consider the second–order equation

$$s_n = s_{n-1} + 6s_{n-2} + 2^n,$$

which has $\deg(p) = 0$ and $\lambda = 2 \notin \{3, -2\}$, and so $\deg(q) = 0$ and $\delta = 0$. The above formula gives the particular solution $v_n = c\,2^n$, and the recurrence can be used to solve for c:

$$v_{n-1}+6v_{n-2}+2^n - v_n = c2^{n-1}+6c2^{n-2}+2^n - c2^n = 2^{n-1}(c+3c+2-2c).$$

Therefore, $2c + 2 = 0$, which gives the particular solution $v_n = -2^n$, and

(A.5) $s_n = a_1 3^n + a_2(-2)^n - 2^n$ for some $a_1, a_2 \in \mathbb{C}$.

We can write this in terms of the initial values s_0, s_1, namely,

(A.6) $s_0 = a_1 + a_2 - 1$ and $s_1 = 3a_1 - 2a_2 - 2$,

which can be solved for a_1, a_2 as in the previous example. Multiplying the first equation by 2 and adding it to the second, we have

$$2s_0 + s_1 = 5a_1 - 4 \text{ and } a_1 = \frac{2s_0 + s_1 + 4}{5}.$$

Substituting this value of a_1 into the first equation of (A.6), we obtain

$$a_2 = s_0 - a_1 + 1 = s_0 - \frac{2s_0 + s_1 + 4}{5} + 1 = \frac{3s_0 - s_1 + 1}{5},$$

and (A.5) becomes

$$s_n = \frac{2s_0 + s_1 + 4}{5}3^n + \frac{3s_0 - s_1 + 1}{5}(-2)^n - 2^n.$$

To satisfy ourselves that this has been done correctly, let us use the last expression to verify that

$$s_2 = \frac{2s_0 + s_1 + 4}{5}9 + \frac{3s_0 - s_1 + 1}{5}4 - 4 = s_1 + 6s_0 + 4,$$

which is consistent with the recurrence.

Example A.1.3. For the second–order equation

(A.7) $$s_n = s_{n-1} + 6s_{n-2} + n2^n,$$

let us first check that there is no constant c for which $w_n = c\,2^n$ is a particular solution. We have

$$w_{n-1} + 6w_{n-2} + n2^n - w_n$$
$$= c2^{n-1} + 6c2^{n-2} + n2^n - c2^n$$
$$= 2^{n-1}(c + 3c + 2n - 2c) = 2^n(c + n),$$

which cannot be zero for all n when c is a constant. Rather, since $\deg(p) = 1$ and $\lambda = 2 \notin \{3, -2\}$, the formula says that there exists a particular solution of the form $v_n = (an + b)2^n$ for some constants a, b. The recurrence gives

$$v_{n-1} + 6v_{n-2} + n2^n - v_n$$
$$= (an - a + b)2^{n-1} + 6(an - 2a + b)2^{n-2} + n2^n - (an + b)2^n$$
$$= 2^{n-1}((an - a + b) + 3(an - 2a + b) + 2n - 2(an + b))$$
$$= 2^{n-1}((2a + 2)n + (-7a + 2b)),$$

which must equal zero for all n; that is, $2a + 2 = 0$ and $-7a + 2b = 0$. This gives $a = -1$, $2b = 7a = -7$, and $v_n = -(2n + 7)2^{n-1}$ is a particular solution of (A.7), and the general solution has the form

$$s_n = a_1 3^n + a_2(-2)^n - (2n + 7)2^{n-1}, \text{ where } a_1, a_2 \text{ are constants in } \mathbb{C}.$$

For initial values s_0, s_1, we obtain

$$s_0 = a_1 + a_2 - 7/2 \text{ and } s_1 = 3a_1 - 2a_2 - 9.$$

Simultaneously solving this system of equations yields

$$a_1 = \frac{2s_0 + s_1 + 16}{5} \text{ and } a_2 = \frac{6s_0 - 2s_1 + 3}{10},$$

which gives

$$s_n = \frac{2s_0 + s_1 + 16}{5}3^n - \frac{6s_0 - 2s_1 + 3}{5}(-2)^{n-1} - (2n + 7)2^{n-1}.$$

Verifying this calculation for $n = 2$ shows that

$$s_2 = \frac{18s_0 + 9s_1 + 144}{5} + \frac{12s_0 - 4s_1 + 6}{5} - 22 = s_1 + 6s_0 + 8,$$

as required.

Example A.1.4. For
$$s_n = s_{n-1} + 6s_{n-2} + 3^n,$$
$\deg(p) = 0$ and $\lambda = \lambda_1$, which means that $\delta = 1$, and the formula gives $v_n = 3^n b n$ for some $b \in \mathbb{C}$. (Note that $w_n = c3^n$ is a solution to the homogeneous equation and so cannot be a particular solution here. Also, $v_n = 3^n(bn + a)$ could be used here, but the constant term can be absorbed into the earlier coefficient of 3^n.) To find b, use
$$v_{n-1} + 6v_{n-2} + 3^n - v_n = 3^{n-1}b(n - 1 + 2n - 4 - 3n) + 3^n = 3^{n-1}(-5b + 3),$$
and from this, $b = 3/5$ and $v_n = n3^{n+1}/5$. Therefore,
$$s_n = a_1 3^n + a_2(-2)^n + \frac{n3^{n+1}}{5},$$
and $s_0 = a_1 + a_2$, $s_1 = 3a_1 - 2a_2 + 9/5$ gives
$$a_1 = \frac{2s_0 + s_1 - 9/5}{5} \quad \text{and} \quad a_2 = \frac{3s_0 - s_1 + 9/5}{5}.$$
Hence,
$$s_n = \frac{1}{5}\Big((2s_0 + s_1 - 9/5 + 3n)3^n + (3s_0 - s_1 + 9/5)(-2)^n\Big).$$

Example A.1.5. For
$$s_n = s_{n-1} + 6s_{n-2} + n(-2)^n,$$
$\deg(p) = 1$ and $\lambda = \lambda_2$, from which we obtain $\delta = 1$ and $v_n = (-2)^n q(n)$, where $q(n) = cn^2 + bn$ for some $b, c \in \mathbb{C}$. Then
$$v_{n-1} + 6v_{n-2} + n(-2)^n - v_n$$
$$= (-2)^{n-1}\big(q(n-1) - 3q(n-2) - 2n + 2q(n)\big)$$
equals zero for all integers $n \geq 2$. In particular, from $n = 2$ and $n = 3$ we obtain
$$q(1) - 3q(0) + 2q(2) = 4,$$
$$q(2) - 3q(1) + 2q(3) = 6,$$
which give $5b + 9c = 4$ and $5b + 19c = 6$, and $c = 1/5$, $b = 11/25$. Therefore, $q(n) = \frac{5n^2 + 11n}{25}$ and
$$s_n = a_1 3^n + \left(a_2 + \frac{5n^2 + 11n}{25}\right)(-2)^n.$$
Substituting for $n = 0$ and $n = 1$, we have
$$s_0 = a_1 + a_2 \quad \text{and} \quad s_1 = 3a_1 - 2a_2 - 32/25,$$
and from this we obtain
$$a_1 = \frac{2s_0 + s_1 + 32/25}{5} \quad \text{and} \quad a_2 = \frac{3s_0 - s_1 - 32/25}{5}.$$

A.2   One Multiple Root

Solving (L) when there is a single eigenvalue

When there is a single eigenvalue, λ_1, it has multiplicity k, and the rule for particular solutions simplifies to
$v_n = \lambda^n n^\delta q(n)$ *is a particular solution*
where q is a polynomial with $\deg(q) = \deg(p)$ and

(A.8)
$$\delta = \begin{cases} 0 & \text{if } \lambda \neq \lambda_1 \\ k & \text{if } \lambda = \lambda_1 \end{cases}.$$

If $\lambda \neq \lambda_1$, the solution is

$$s_n = \lambda_1^n q_1(n) + \lambda^n q(n)$$

where $\deg(q) = \deg(p)$, $\deg(q_1) \leq k$ and the coefficients of q_1 are determined from the initial conditions.
If $\lambda = \lambda_1$, the solution is

$$s_n = \lambda_1^n q_1(n) + \lambda^n n^k q(n) = \lambda^n Q(n)$$

and $\deg(Q) = k + \deg(p)$.

The five examples in this section are second–order equations, and each has the characteristic polynomial

$$ch(x) = x^2 - 4x + 4 = (x - 2)^2.$$

This has the double root $\lambda_1 = 2$, which means that the general solution is

$$s_n = (a_1 n + a_2)2^n + v_n,$$

where v_n is a particular solution. As in the last section, v_n depends on ψ, and a_1, a_2 can be calculated from the initial conditions.

Example A.2.1. For any initial value problem with recurrence

$$s_n = 4s_{n-1} - 4s_{n-2},$$

$s_n = (a_1 n + a_2)2^n$, where the constants $a_1 = (s_1 - 2s_0)/2$ and $a_2 = s_0$ can be computed from the initial conditions s_0, s_1, and the solution is

$$s_n = ((s_1 - 2s_0)n + 2s_0)2^{n-1}.$$

Example A.2.2. Consider the second–order equation

$$s_n = 4s_{n-1} - 4s_{n-2} + 3^n,$$

where $\deg(p) = 0$ and $\lambda = 3$ is not an eigenvalue of the recurrence. From the formula, a particular solution is $v_n = c3^n$ for some constant c, and

$$4v_{n-1} - 4v_{n-2} + 3^n - v_n = 3^{n-2}(12c - 4c + 9 - 9c) = 3^{n-2}(9 - c),$$

which equals zero for $c = 9$. This gives $v_n = 3^{n+2}$ and general solution

$$s_n = (a_1 n + a_2)2^n + 3^{n+2}, \text{ for some } a_1, a_2 \in \mathbb{C}.$$

For the initial conditions s_0, s_1, computation gives

$$s_n = \left(\frac{s_1 - 2s_0 - 9}{2} n + (s_0 - 9) \right) 2^n + 3^{n+2}.$$

Example A.2.3. The second–order equation

$$s_n = 4s_{n-1} - 4s_{n-2} + 3^n n$$

has $\deg(p) = 1$, and $\lambda = 3$ is again not an eigenvalue. Therefore, there are constants a, b such that $v_n = (an + b)3^n$ is a particular solution. Also,

$$4v_{n-1} - 4v_{n-2} + n3^n - v_n$$
$$= 3^{n-2}(12(an - a + b) - 4(an - 2a + b) + 9n - 9(an + b))$$
$$= 3^{n-2}((9 - a)n - (4a + b)),$$

which equals zero for all integers $n \geq 2$ when $a = 9$ and $b = -4a = -36$. Therefore, $v_n = (9n - 36)3^n = (n - 4)3^{n+2}$ is a particular solution, and the general solution has the form

$$s_n = (a_1 n + a_2)2^n + (n - 4)3^{n+2}.$$

For initial values s_0, s_1 we obtain

$$s_n = \left(\frac{s_1 - 2s_0 + 9}{2} n + (s_0 + 36) \right) 2^n + (n - 4)3^{n+2}.$$

Example A.2.4. The second–order equation

$$s_n = 4s_{n-1} - 4s_{n-2} + 2^n$$

has $\deg(p) = 0$, and $\lambda = 2$ is an eigenvalue. Since its multiplicity is 2, $\delta = 2$ holds in the formula, and $v_n = an^2 2^n$ is a particular solution for some $a \in \mathbb{C}$. Then

$$4v_{n-1} - 4v_{n-2} + 2^n - v_n$$
$$= 2^n(2a(n - 1)^2 - a(n - 2)^2 + 1 - an^2)$$
$$= 2^n(-2a + 1),$$

implying $a = \frac{1}{2}$, which means that $v_n = n^2 2^{n-1}$ is a particular solution. The general solution therefore has the form

$$s_n = (n^2 + a_1 n + a_2) 2^{n-1},$$

and from the initial values s_0, s_1 we obtain

$$s_n = (n^2 + (s_1 - 1 - 2s_0)n + 2s_0) 2^{n-1}.$$

Example A.2.5. The second–order equation

$$s_n = 4s_{n-1} - 4s_{n-2} + 2^n n$$

has $\deg(p) = 1$, and $\lambda = 2$ is again the double eigenvalue of the recurrence, which means that $v_n = 2^n q(n)$ is a particular solution for some $q(n) = an^3 + bn^2$. Then

$$4v_{n-1} - 4v_{n-2} + n2^n - v_n$$
$$= 2^n [2q(n-1) - q(n-2) + n - q(n)],$$

which must equal zero for all values of $n \geq 2$. From

$$2q(n-1) - q(n-2) + n - q(n) = 0 \text{ for } n = 2, 3$$

we obtain $a = 1/6$ and $b = 3a = 1/2$. Therefore, $v_n = (\frac{1}{3}n^3 + n^2)2^{n-1}$ is a particular solution, and the general solution has the form

$$s_n = \left(\frac{n^3}{3} + n^2 + a_1 n + a_2 \right) 2^{n-1}.$$

Initial values s_0, s_1 give

$$s_n = \left(\frac{n^3}{3} + n^2 + (s_1 - 2s_0 - \frac{4}{3})n + 2s_0 \right) 2^{n-1}.$$

Since from (A.8) the solution is $2^n Q(n)$ with $Q(n)$ a polynomial of degree 3, the same solution could have been obtained by solving a system of four linear equations to find the coefficients of $Q(n)$. To obtain these coefficients as functions of only s_0 and s_1, the recurrence (A.2.5) could be used twice to give s_2 and s_3 in terms of s_0 and s_1.

A.3   One Multiple Root, Several Simple Roots

Solving (L) when there is a multiple eigenvalue

When $m_1 \geq 2$ and $m_2 = \cdots = m_t = 1$ are the respective multiplicities of the distinct eigenvalues $\lambda_1, \lambda_2, \ldots, \lambda_t$ of the recurrence (L), then $v_n = \lambda^n n^\delta q(n)$ is a particular solution where q is a polynomial with $\deg(q) = \deg(p)$ and

(A.9)
$$\delta = \begin{cases} 0 & \text{if } \lambda \notin \{\lambda_1, \ldots, \lambda_t\} \\ m_1 & \text{if } \lambda = \lambda_1 \\ 1 & \text{if } \lambda \in \{\lambda_2, \ldots, \lambda_t\} \end{cases}.$$

Since each of the last five examples had only one (double) root, The last alternative of (A.9) did not occur. The next example illustrates this case.

Example A.3.1. The second–order equation

$$s_n = 4s_{n-1} - 5s_{n-2} + 2s_{n-3} + 2^n$$

has $\deg(p) = 0$ and $\lambda = 2$. Since the characteristic polynomial factors as

$$ch(x) = x^3 - 4x^2 + 5x - 2 = (x-1)^2(x-2),$$

$\lambda = 2$ is a simple root of $ch(x)$ and $\delta = 1$, which gives the particular solution $v_n = an2^n$. Then

$$4v_{n-1} - 5v_{n-2} + 2v_{n-3} + 2^n - v_n$$
$$= 2^{n-2}(8a(n-1) - 5a(n-2) + a(n-3) + 4 - 4an) = 2^{n-2}(4-a),$$

and $v_n = 4n2^n = n2^{n+2}$ is a particular solution, and the general solution is

$$s_n = (a_1 n + a_2) + a_3 2^n + n2^{n+2} = (a_1 n + a_2) + (a_3 + 4n)2^n.$$

For initial values s_0, s_1, s_2 we obtain

$$s_0 = a_2 + a_3 \text{ and } s_1 = a_1 + a_2 + 2a_3 + 8; \; s_2 = 2a_1 + a_2 + 4a_3 + 32,$$

which implies

$$s_n = (-2s_0 + 3s_1 - s_2 + 8)n + (2s_1 - s_2 + 16) + (s_0 - 2s_1 + s_2 - 16)2^n + n2^{n+2}.$$

A.4   The Input is $\gamma_1^n p_1(n) + \gamma_2^n p_2(n)$

For this form of input, one can find a particular solution by taking the sum of particular solutions to two related equations. Of course, there is nothing special about 2, so if the input is a sum of j terms, one can find a particular solution by taking the sum of particular solutions to j related equations.

Example A.4.1. For the second–order equation

$$(A.10) \qquad\qquad s_n = 4s_{n-1} - 4s_{n-2} + 3^n\, n + 2^n,$$

we break this equation into two equations, one for the input $3^n\, n$ and one the input 2^n

$$(A.11) \qquad\qquad v_n = 4v_{n-1} - 4v_{n-2} + 3^n\, n,$$

$$(A.12) \qquad\qquad w_n = ws_{n-1} - 4w_{n-2} + 2^n.$$

From Example A.2.2, we have that

$$v_n = (n-4)\,3^{n+2}$$

is a particular solution to (A.11), and from Example A.3.1,

$$w_n = n^2\,2^{n-1}$$

is a particular solution to (A.12). Combining these we have that

$$(n-4)\,3^{n+2} + n^2\,2^{n-1}$$

is a particular solution to (A.10). The general solution to (A.10) is then

$$s_n = (\alpha_1 n + \alpha_2)2^n + (n-4)\,3^{n+2} + n^2\,2^{n-1},$$

where α_1 and α_2 depend on the initial conditions s_0 and s_1. Solving for α_1 and α_2, we find that

$$s_n = \left((\tfrac{s_1}{2} - s_0 + 4)\,n + s_0 + 36 \right) 2^n + (n-4)\,3^{n+2} + n^2\,2^{n-1}.$$

Appendix B
Complex Numbers

Because the square of a real number cannot be negative, there is no real number whose square is -1, and this means that the simple quadratic equation $z^2 + 1 = 0$ has no real roots. Using notation introduced by Leonhard Euler in 1777, we reserve i to mean a symbol for which $i^2 = -1$ holds, and then define the set of complex numbers to be

$$\mathbb{C} = \{a + bi \ : \ a, b \in \mathbb{R} \ \}.$$

There is a one-to-one correspondence between each element $a + bi$ in $\mathbb{C}$ and the point (a, b) in the set $\mathbb{R}^2$. In this bijection, real numbers a correspond to the points of the form $(a, 0)$, and **imaginary numbers** $0 + bi$ correspond to the points on the y-axis. If the op-

In *La Géometrie* (1637) René Descartes introduced the terms real and imaginary numbers. In 1936, a US mathematician named Arnold Dresden suggested that the term "imaginary" be changed to "normal," since the imaginary axis is normal (that is, perpendicular) to the real axis. That term never caught on, and instead has been used for something different: a real number is called **normal** if the digits of its base b–expansion behave in a suitably random manner for every (positive integer) base b.

erations of addition and scalar multiplication (using real scalars) on $\mathbb{C}$ are defined "coordinatewise," namely, by

$$(a_1 + b_1 i) + (a_2 + b_2 i) = (a_1 + a_2) + (b_1 + b_2)i$$

and

$$c(a + bi) = (a + bi)c = (ca) + (cb)i,$$

the correspondence is actually a vector space isomorphism, and $\mathbb{C}$ is a two dimensional real vector space under these operations. In particular, $\mathbb{C}$ is an **abelian group** under addition. This means that the operation of addition on the set $\mathbb{C}$ is a commutative, associative operation for which the real number 0 is the additive identity and every element of $\mathbb{C}$ has an additive identity.

How is multiplication of two complex numbers defined? If we want multiplication to be associative and commutative and also to distribute over addition, then it turns out that there is only one way to define multiplication! This is because for any $a, b, c, d \in \mathbb{R}$,

$$
\begin{aligned}
(a + bi)(c + di) &= (a + bi)c + (a + bi)(di) & \text{distributive law in } \mathbb{C} \\
&= (a + bi)c + ((a + bi)d)i & \text{associative law in } \mathbb{C} \\
&= c(a + bi) + (d(a + bi))i & \text{commutative law in } \mathbb{C} \\
&= (ca + cbi) + (da + dbi)i & \text{scalar multiplication in } \mathbb{R}^2 \\
&= (ac + bci) + (ad + bdi)i & \text{commutative law in } \mathbb{R} \\
&= (ac + bci) + (adi + bdi^2) & \text{distributive law in } \mathbb{C} \\
&= (ac + bci) + (-bd + adi) & i \text{ satisfies the identity } i^2 = -1 \\
&= (ac - bd) + (ad + bc)i & \text{definition of addition in } \mathbb{C}.
\end{aligned}
$$

This shows that because $\mathbb{C}$ is a real vector space, the only associative, commutative, and distributive multiplication on $\mathbb{C}$ that extends the scalar multiplication by elements of $\mathbb{R}$ is

$$
(a + bi)(c + di) = (ac - bd) + (ad + bc)i.
$$

(Also note that $(a + 0i)(c + 0i) = ac$, which says that this multiplication is consistent with multiplication on the subset $\mathbb{R}$.) This definition of multiplication satisfies the commutative law for multiplication, and the real number 1 is the multiplicative identity. As you might already know, the set $\mathbb{C}$ with the operations of addition and multiplication defined above is the **field of complex numbers**. Also, a is called the **real part** and b the **imaginary part** of the complex number $a + bi$.

> On the surface, the multiplication of two complex numbers $a + bi, c + di$ seems to require the four real multiplications ac, bd, ad, bc. Check that the auxiliary multiplication $(a + b)(c + d)$ can be used to reduce the number of multiplications to three.

Until now, our geometric representation of complex numbers has been in *rectangular* coordinates. Thinking in terms of polar coordinates, we can rewrite the complex number $z = a + ib$ as $z = |z|(\cos(\theta) + i\sin(\theta))$, where $|z| = \sqrt{a^2 + b^2}$ is called its **modulus**, and its **argument** θ is the angle between the positive x-axis and the vector $a + bi$. For the moment, we write $e(\theta) = \cos(\theta) + i\sin(\theta)$, and then

$$z = |z|e(\theta) \text{ for some } 0 \le \theta < 2\pi.$$

For instance, two values are $e(\pi) = \cos(\pi) = -1$ and $e(2\pi) = \cos(2\pi) = 1$.

Let's consider the function $e(\theta)$ from another point of view. From calculus you know that the functions $\cos(\theta)$ and $\sin(\theta)$ are functions of the real variable θ whose power series are

$$\cos(\theta) = \sum_{k \ge 0} \frac{(-1)^k}{(2k)!}\theta^{2k} \quad \text{and} \quad \sin(\theta) = \sum_{k \ge 0} \frac{(-1)^k}{(2k+1)!}\theta^{2k+1},$$

which converge at every real number θ. Therefore, for any real number y the complex number $e(y) = \cos(y) + i\sin(y)$ equals

$$e(y) = \sum_{k \ge 0} \frac{1}{(2k)!}(iy)^{2k} + \sum_{k \ge 0} \frac{1}{(2k+1)!}(iy)^{2k+1},$$

since $(-1)^k = i^{2k}$. If we can add these two series, then

(B.1)
$$e(y) = \sum_{k \ge 0} \frac{(iy)^k}{k!},$$

which is reminiscent of the power series expansion of $e^x = \sum_{k \ge 0} x^k/k!$. In fact we will show that these series *can* be added and it then makes sense to define the **complex exponential function** e^z of the complex variable $z = x + iy$ as

(B.2)
$$e^z = e^x(\cos(y) + i\sin(y)).$$

Once this is done, we can drop the notation $e(\theta)$ and write $z = |z|e^{i\theta}$, where θ is the argument of z.

We've gotten ahead of ourselves here, because we haven't yet defined what we mean by convergence of power series with complex coefficients. As with the reals, for any sequence $\langle a_i \rangle$ of complex numbers and any complex number α the power series $\sum_{k \ge 0} a_k(z - \alpha)^k$ is said to **converge** at the complex number $z = z_0$ if the sequence of partial sums $\sum_{k=0}^{N} a_k(z_0 - \alpha)^k$ converges, where complex modulus is used for the absolute value. We will show that every power series has a **radius of convergence** R (it might be infinite), which has the property that the series $\sum_{k \ge 0} |a_k(z - \alpha)^k|$ converges for all $|z - \alpha| < R$ and diverges for all $|z - \alpha| > R$. Because $e^x = \sum_{k \ge 0} x^k/k!$ converges for all real numbers, this complex power series cannot have a finite radius of convergence. This means that e^z is *absolutely convergent* for all complex numbers z and so can be rearranged, and e^z as defined in (B.2) is a well-defined function on $\mathbb{C}$.

Remember that we want to prove that each power series $\sum_{k\geq 0} a_k(z-\alpha)^k$ has a radius of convergence. We give a proof of this fact by showing that the **supremum** of the set

$$\mathcal{S} = \{r \geq 0 : \text{there exists } M > 0 \text{ such that} |a_k| r^k \leq M \text{ for every } k\}$$

is the radius of convergence. To do this we show that for $R = \sup(\mathcal{S})$ the series $\sum_{k\geq 0} |a_k(z-\alpha)^k|$ converges for all $|z-\alpha| < R$ and the series $\sum_{k\geq 0} a_k(z-\alpha)^k$ diverges for all $|z-\alpha| > R$. If the complex number z satisfies $|z-\alpha| = r > R$, then r is not in $\mathcal{S}$, which means that $|a_k||z-\alpha|^k$ is unbounded and the series must diverge. On the other hand, for $r = |z-\alpha| < R$ we can choose ρ such that $r < \rho < R$. Then ρ is an element of $\mathcal{S}$, and there exists $M > 0$ such that $|a_k\rho^k| < M$ for every k. This gives

$$|a_k(z-\alpha)^k| = |a_k| r^k = |a_k| \rho^k \left(\frac{r}{\rho}\right)^k < M\left(\frac{r}{\rho}\right)^k,$$

where the ratio of the bounding geometric series $\sum_{k\geq 0} (r/\rho)^k$ satisfies $0 \leq r/\rho < 1$. Therefore, by comparison, the series $\sum_{k\geq 0} a_k(z-\alpha)^k$ is absolutely convergent on $|z-\alpha| < R$, and $R = \sup(\mathcal{S})$ is the radius of convergence of the power series.

Any power series $\sum_{k\geq 0} a_k(z-\alpha)^k$ whose radius of convergence R is positive can be shown to be a *differentiable* function on its disk of convergence. Also, its derivative is the power series $\gamma = \sum_{k\geq 0} ka_k(z-\alpha)^{k-1}$, and γ has the *same* radius of convergence as the original series. To see this, let R_1 be the radius of convergence of the series γ, and $R_1 \leq R$ follows from $|ka_k| \geq |a_k|$. To show that $R_1 \geq R$, we'll prove that $R_1 \geq r$ for all $0 < r < R$. The construction of R implies that for any ρ with $r < \rho < R$ there exists $M > 0$ such that $|a_k|\rho^k \leq M$ for all k, and

$$|ka_k| r^{k-1} = \frac{|a_k\rho^k|}{r} \cdot k\left(\frac{r}{\rho}\right)^k \leq \frac{M}{r} \cdot k\left(\frac{r}{\rho}\right)^k.$$

Since $0 \leq r < \rho$, then $\lim_{k\mapsto\infty} k(r/\rho)^k = 0$ and, $|ka_k| r^{k-1}$ is bounded, giving $R_1 \geq r$ for all $0 < r < R$. From this we see that $R_1 \geq R$, as required.

To summarize, if we let $f(z) = \sum_{k\geq 0} a_k(z-\alpha)^k$ for any power series $\sum_{k\geq 0} a_k(z-\alpha)^k$ then, $f(z)$ is a differentiable complex-valued function on its disk of convergence. Further, its derivative is $f'(z) = \sum_{k\geq 0} D(a_k(z-\alpha)^k)$, where D is the differentiation operator on the space of polynomials. Repeating this argument, $f'(z)$ must also be differentiable on the same disk, and $f''(z) = \sum_{k\geq 0} D^2(a_k(z-\alpha)^k)$. From this we see that $f(z)$ is *infinitely* differentiable on its disk of convergence.

Before leaving power series, we define the **Taylor series** of f about $z = \alpha$ as

$$\sum_{k\geq 0} \frac{D^k(f)(\alpha)}{k!}(z-\alpha)^k,$$

where $D^k(f)(\alpha)$ is the kth derivative of $f(z)$ evaluated at $z = \alpha$. It can be proved that the Taylor series of f converges to $f(z)$ on some disk about α, provided $\sum_{n \geq 0} a_n(z - \alpha)^n$ has a non-zero radius of convergence.

Returning to the complex exponential function, from the formulas for the sum of two angles we have

$$e^{i(\theta_1 + \theta_2)} = [\cos(\theta_1)\cos(\theta_2) - \sin(\theta_1)\sin(\theta_2)] + i[\sin(\theta_1)\cos(\theta_2) + \sin(\theta_2)\cos(\theta_1)].$$

The definition of multiplication allows us to unwrap this identity to get

$$e^{i(\theta_1 + \theta_2)} = e^{i\theta_1} e^{i\theta_2},$$

and iteration of this process for any positive integer m gives

$$e^{im\theta} = e^{i\theta + i(m-1)\theta} = e^{i\theta} e^{i(m-1)\theta} = \cdots = (e^{i\theta})^m.$$

This proves the laws of exponents

$$e^{z_1 + z_2} = e^{z_1} e^{z_2} \quad \text{and} \quad e^{mz} = (e^z)^m.$$

When $\theta_0 = 2\pi/m$, this formula becomes $e^{i\theta_0 m} = e^{i2\pi} = 1$, which means that $z = e^{i\theta_0}$ is a root of the equation $z^m - 1 = 0$, which is called the **principal m^{th} root of unity.** (The complex numbers satisfying $z^m - 1 = 0$ are called the mth **roots of unity.**) Writing the equation $z^m - 1 = 0$ in the form $z^m = 1$, we see that every power of the principal m^{th} root of unity is also an mth root of unity, and since $e^{i\theta} = e^{i(\theta + 2\pi)}$ always holds, this gives m different roots $e^{i\theta}$, where

$$\theta = 0, \frac{2\pi}{m}, \frac{4\pi}{m}, \ldots, \frac{(m-1)\pi}{m}.$$

Since these roots are complex numbers that are equally separated on the unit circle $|z| = 1$, they are often called **cyclotomic** (or "circle dividing"). Because a polynomial equation of degree m has at most m roots in any *field* (refer to Exercise 8.23), these are all the roots of $z^m - 1 = 0$ in $\mathbb{C}$. Geometrically, they form the vertices of a regular m-gon inscribed in the complex unit circle with one vertex anchored at 1. Algebraically, they form a group $\mathcal{G}_m$ under multiplication, and this group is isomorphic to the integers modulo m under the map

$$e^{ik2\pi/m} \mapsto k \bmod m.$$

Among these roots are the **primitive** m^{th} roots of unity, the ones that are not roots of $z^k - 1 = 0$ for any $0 < k < m$. (The principal root of unity is an example of a primitive root of unity.) Every primitive root of unity generates the group $\mathcal{G}_m$, in the sense that the powers

$$\zeta, \zeta^2, \ldots, \zeta^m$$

of any primitive root ζ are all different, and so every m^{th} root of unity can be written as a power of ζ. For instance, every fifth root of unity except 1 is primitive, whereas there are only two primitive sixth roots of unity, the principal root $e^{i2\pi/6}$ and its multiplicative inverse $e^{-i2\pi/6}$. The roots of unity occur naturally in many contexts throughout this book.

> The complex exponential function gives one of the most remarkable formulas in all of mathematics:
>
> $$e^{\pi i} = -1.$$
>
> This formula, sometimes called **Euler's formula**, relates four of the fundamental mathematical constants, namely: e, the base of the natural logarithms; π the ratio of the circumference of a circle to its diameter; i the complex unit; and -1 the basic negative number.

The fact that every polynomial of the form $f(z) = z^m - 1$ has m complex roots is not an anomaly, since *every* polynomial of degree m has m complex roots (where here roots are counted according to their multiplicity). This statement is usually referred to as the **Fundamental Theorem of Algebra** and is an example of a statement that is very easy to state but quite difficult to prove. Although a version of this result was conjectured sometime in the sixteenth century, Newton was probably the first to state the Fundamental Theorem in fairly modern terms when he wrote that every non-constant polynomial with real coefficients can be factored into a product of linear and quadratic polynomials with real coefficients. The theorem was finally proved by Gauss in his 1799 dissertation, and over his lifetime Gauss published at least three other proofs of the Fundamental Theorem. Each proof is an *existence* proof, since it proves only that such roots exist and doesn't give a method for explicitly constructing any of the roots. The roots of most polynomials are difficult or impossible to determine explicitly, but we've ignored this important practical problem by choosing polynomials with relatively obvious factorizations.

Appendix C
Highlights of Linear Algebra

This appendix contains a review of the linear algebra we use in this book. It is essentially a synopsis of an introductory linear algebra course with a number of topics omitted and some new ones added.

C.1 Vector Spaces and Subspaces

Every pair of elements in a vector space V can be added together, and also each element can be multiplied by scalars. For us, the scalars are either real or complex numbers, and V is called a real vector space or a complex vector space according to whether $\mathbb{R}$ or $\mathbb{C}$ is used. (In general, the set of scalars must be a *field* as defined in Section 8.2.) In order for V to be a **vector space**, the operations of addition and scalar multiplication must satisfy a list of axioms requiring that both operations are associative, addition is commutative, and scalar multiplication distributes over addition. Also, there must be an additive identity and every element of V must have an additive inverse in V. The basic examples of vector spaces are $\mathbb{R}^n$ and $\mathbb{C}^n$ (with scalars from $\mathbb{R}$ and $\mathbb{C}$ respectively).

A nonempty subset W of V is called a **subspace** if W is also a vector space under the operations of V. Given a set S of vectors, we can form **linear combinations** of elements $v_1, \ldots, v_n$ from S, sums of the form

$$\alpha_1 v_1 + \cdots + \alpha_n v_n, \text{ where the } \alpha_i \text{ are scalars.}$$

Every subspace is closed under formation of linear combinations, and it can be proved that any nonempty subset of a vector space that is closed under addition and scaling is a subspace. The subspace of all linear combinations of elements from S is called the **span** of S, which we will denote by $\mathrm{Span}(S)$; it is the smallest subspace containing S. A set S is called a **spanning set** for V when $\mathrm{Span}(S) = V$, which means that every element of V can be written as a linear combination of elements from S.

Although we will principally concentrate on subspaces of $\mathbb{R}^n$ and $\mathbb{C}^n$, there is a more general class of vector spaces that we encounter, the vector spaces of real-valued or complex-valued functions defined on some set X. Because a sequence is a function defined on the natural numbers, the complex vector space of all sequences of complex numbers is an example of such a vector space. Both addition and scalar multiplication are defined "componentwise". In other words, to add two sequences we add the nth term of the first sequence to the nth term of the second, and multiplying every element of a sequence by the scalar α gives the scaled sequence. Considering an infinite sequence $\langle s_n \rangle$ as the function $s(n) = s_n$ defined on $\mathbb{N}$, these componentwise vector space operations are the usual operations for functions,

$$(s + t)(n) = s(n) + t(n) \quad \text{and} \quad (\alpha s)(n) = \alpha s(n).$$

As we said above, in general, the set of all functions from any set X into $\mathbb{R}$ or $\mathbb{C}$ forms a vector space under the usual operations of function addition and scaling. When $X = \mathbb{Z}$ we obtain the set of all doubly infinite sequences of real or complex numbers.

C.2   Linear Independence and Basis

How many different linear combinations yield the same vector? A nonempty set S of vectors is called **linearly independent** if every element in $\mathrm{Span}(S)$ can be expressed as a linear combination of elements of S in only one way. Since $\mathrm{Span}(S)$ is a subspace of V, it contains the zero vector, and so linear independence means that whenever

$$\alpha_1 v_1 + \cdots + \alpha_n v_n = 0 = \beta_1 v_1 + \cdots + \beta_m v_m$$

holds we have $\alpha_i = \beta_i$ for all i. Since the choice of all $\beta_i = 0$ gives the zero vector, linear independence means that

$$\alpha_1 v_1 + \cdots + \alpha_n v_n = 0 \iff \alpha_i = 0 \text{ for all } i.$$

The last condition is often taken as the definition of linear independence.

A nonempty subset $\mathcal{B} \subseteq V$ is called a **basis** for V if it is a linearly independent spanning set for V. In other words, $\mathcal{B}$ is a basis exactly when every vector can be expressed as a linear combination of elements from $\mathcal{B}$ in a unique way. When V has a basis with n elements, we can encode every element of V as an n-tuple. For instance, it can be checked that the set $\mathcal{B} = \{(1,2), (-1,1)\}$ is a basis for $V = \mathbb{R}^2$. Because the vector $v = (-2,3)$ satisfies

$$v = \frac{1}{3}(1,2) + \frac{7}{3}(-1,1),$$

its encoding in terms of the basis $\mathcal{B}$ is

$$v_{\mathcal{B}} = \begin{pmatrix} 1/3 \\ 7/3 \end{pmatrix}.$$

In general, when $\mathcal{B} = \{v_1, \ldots, v_n\}$ is a basis for V and $v = \alpha_1 v_1 + \cdots + \alpha_n v_n$, we write the **coordinate vector of v relative to $\mathcal{B}$** as $v_{\mathcal{B}} = (\alpha_1, \ldots, \alpha_n)^T$.

The **dimension** of a vector space is defined to be the number of elements in a basis. This definition makes sense because any two bases for the same vector space must have the same number of elements. The proof of this fact can be broken into two cases, whether or not the vector space has a *finite* spanning set. In the first case the vector space is called **finite-dimensional**. The proof for this case proceeds by proving that if V has a spanning set with n elements, then any subset that has more than n elements must be linearly dependent. From this we know that any basis can have at most n elements. If $\mathcal{B}$ and $\mathcal{B}'$ are two bases for V, then $\mathcal{B}$ is a spanning set, and the linear independence of $\mathcal{B}'$ implies that the number of elements in $\mathcal{B}'$ is bounded above by the number of elements in $\mathcal{B}$. Reversing the roles of $\mathcal{B}$ and $\mathcal{B}'$ yields the desired result, namely, that $\mathcal{B}$ and $\mathcal{B}'$ have the same size. If there is no *finite* spanning set for V, the vector space V is called **infinite-dimensional**. The above argument is too simplistic for this case, but there is a more sophisticated argument that works. We don't give that argument here.

When V is a finite-dimensional vector space, any subspace W is also finite-dimensional, and the dimension of any *proper* subspace $W \neq V$ is strictly less than the dimension of V.

C.3 Linear Transformations

A linear transformation is a map between two vector spaces (they must have the same field of scalars) that preserves addition and scalar multiplication. In other words, a function $T : V \to W$ is a **linear transformation** if

$$T(\alpha v + v') = \alpha T(v) + T(v') \quad \text{for all } v, v' \in V \text{ and all scalars } \alpha.$$

Because of linearity, a linear transformation is defined by its action on a basis.

Linear transformations that are functions from a vector space to itself are called **operators**. We can represent a linear operator on an n-dimensional vector space V with respect to a basis $\mathcal{B} = \{b_1, \ldots, b_n\}$ by the $n \times n$

matrix whose i^{th} column is $T(b_i)$ as represented in the basis $\mathcal{B}$. We denote this matrix by $[T]_\mathcal{B}$. For example, if T is the operator on $\mathbb{R}^3$ defined by $T(x, y, z) = (3x + 2y, z, -z)$ and $\mathcal{B} = \{(1, 0, 0), (1, -1, 0), (0, 0, 1)\}$, then $T(b_1) = (3, 0, 0) = 3b_1$; $T(b_2) = (1, 0, 0) = b_1$; $T(b_3) = (0, 1, -1) = b_1 - b_2 - b_3$, and

$$[T]_\mathcal{B} = \begin{bmatrix} 3 & 1 & 1 \\ 0 & 0 & -1 \\ 0 & 0 & -1 \end{bmatrix}.$$

Notice that we are assuming an inherent order to the elements in the basis, and the term **ordered basis** is normally used to emphasize this. For $A = [T]_\mathcal{B}$ we have the helpful identity

$$T(v)_\mathcal{B} = Av_\mathcal{B} \text{ for all } v \in V,$$

and T is an invertible operator iff A is an invertible matrix.[1] From this it follows that if we are given two pieces of data, an ordered basis $\mathcal{B}$ for an n-dimensional vector space and a linear operator T, then from this information we can obtain an $n \times n$ matrix A with the property that a coordinate vector for $T(v)$ relative to $\mathcal{B}$ is the result of multiplying A by the coordinate vector for v relative to $\mathcal{B}$. If $\mathcal{C} = \{c_1, \ldots, c_n\}$ is another ordered basis for V, consider the matrix P whose i^{th} column is c_i represented in the basis $\mathcal{B}$. This matrix P is often called the **change of basis matrix** from $\mathcal{C}$ to $\mathcal{B}$. Since the associated linear operator (the identity operator) is invertible, the matrix P is invertible and

$$T(v)_\mathcal{C} = P^{-1}APv_\mathcal{C},$$

and the representation of T in the basis $\mathcal{C}$ is the matrix $P^{-1}AP$. Two matrices A and B, that are related by a matrix equation of the form

$$B = P^{-1}AP$$

are called **similar matrices**, and they represent the same linear operator with respect to two different bases.

C.4   Eigenvectors

A non-zero vector $v \in V$ is called an **eigenvector** of the linear operator T if $T(v) = \lambda v$, and the scalar λ is called the associated **eigenvalue**. Sets of eigenvectors corresponding to different eigenvalues are linearly independent.

When V is finite-dimensional and $A = [T]_\mathcal{B}$, the matrix of T relative to any convenient basis $\mathcal{B}$, then any eigenvector v with associated eigenvalue

---

[1] In this book, "iff" is shorthand for "if and only if".

λ satisfies $(A - \lambda I)v_{\mathcal{B}} = 0$. Since v is non-zero (and so $v_{\mathcal{B}}$ is non-zero), this implies that $(A - \lambda I)$ is a singular matrix and $\det(A - \lambda I)$ is zero. From this we see that the eigenvalues of T (which are also called the eigenvalues of the matrix A) are the roots of the polynomial $ch_A(x) = \det(A - xI)$, the **characteristic polynomial** of A. Finding the roots of any polynomial is a difficult problem, but our examples have been constructed so that roots of the characteristic polynomials are relatively easy to find. Then finding the eigenvectors associated with an eigenvalue λ amounts to using Gaussian elimination to solve the homogeneous system of equations $(A - \lambda I)v = 0$.

When the vector space has a (finite) basis $\mathcal{B}_1$ of eigenvectors, the matrix representing T in the basis $\mathcal{B}_1$ is a diagonal matrix (and the diagonal elements are the respective eigenvalues). If A is again the matrix of T relative to any basis $\mathcal{B}$ and P is the change of basis matrix from $\mathcal{B}_1$ to $\mathcal{B}$ then $P^{-1}AP$ is a diagonal matrix and A is called **diagonalizable**. We summarize this very nice situation in the following theorem whose proof can be found in [78, Chapter 6].

Theorem C.4.1. *Let A be an $n \times n$ matrix with complex entries and* **characteristic polynomial** $ch_A(x) = \det(A - Ix)$. *Let $\lambda_1, \lambda_2, \ldots, \lambda_n$ be the eigenvalues of A, the complex roots of $ch_A(x) = 0$ repeated according to multiplicity. Then:*

(a) The matrix A is diagonalizable iff V has a basis of eigenvectors.

(b) If all the λ_i are distinct then A is diagonalizable.

(c) Suppose A is diagonalizable and $\{v_1, \ldots, v_n\}$ is a basis of eigenvectors. If P is the $n \times n$ matrix whose i^{th} column is v_i then $P^{-1}AP$ is the diagonal matrix D with diagonal entries $\lambda_1, \ldots, \lambda_n$.

(d) Real symmetric matrices are always diagonalizable.

For us the most useful application of diagonalizability is that it allows the powers of A to be computed quickly, since $P^{-1}AP = D$ yields $A^n = PD^nP^{-1}$ for all $n \geq 0$.

Not all square matrices are diagonalizable, but every square matrix A with complex entries has a **Jordan form**. This means that there exists an invertible matrix P with the property that $P^{-1}AP$ is composed of **Jordan blocks** on the diagonal, where a Jordan block is a bidiagonal matrix of the form

$$\begin{bmatrix} \lambda_i & 1 & 0 & & \ldots & 0 \\ 0 & \lambda_i & 1 & 0 & \ldots & 0 \\ & & \ddots & \ddots & & \\ 0 & 0 & & & & 1 \\ 0 & 0 & 0 & 0 & \ldots & \lambda_i \end{bmatrix} \text{ for some } \lambda_i.$$

For example, the matrix on the right is in Jordan form. It has three Jordan blocks: one of size 2 with $\lambda_1 = 3$, one of size 1 with $\lambda_2 = 3$, and one of size 1 with $\lambda_3 = 2$. The Jordan form of A is unique up to the order of the individual Jordan blocks, the λ's appearing in the blocks of the Jordan form are

$$\begin{bmatrix} 3 & 1 & 0 & 0 \\ 0 & 3 & 0 & 0 \\ 0 & 0 & 3 & 0 \\ 0 & 0 & 0 & 2 \end{bmatrix}$$

the eigenvalues of A, and the minimal and characteristic polynomials of A can be found from J. The proofs of all these facts and more information on Jordan form can be found in [78, Section 7.3]. Since $A^n = PJ^nP^{-1}$, powers of A are relatively easy to compute because the nth power of a Jordan block is

(C.1)

$$\begin{bmatrix} \lambda & 1 & 0 & & \cdots & 0 \\ 0 & \lambda & 1 & 0 & \cdots & 0 \\ & & \ddots & \ddots & & \\ 0 & 0 & & & & 1 \\ 0 & 0 & 0 & 0 & \cdots & \lambda \end{bmatrix}^n = \begin{bmatrix} \lambda^n & n\lambda^{n-1} & \cdots & \binom{n}{l}\lambda^{n-l} \\ 0 & \lambda^n & \cdots & \binom{n}{l-1}\lambda^{n-l+1} \\ & \ddots & \ddots & \vdots \\ 0 & 0 & \ddots & \vdots \\ 0 & 0 & \cdots & \lambda^n \end{bmatrix}$$

C.5 Characteristic and Minimal Polynomials

Recall that the characteristic polynomial of an $n \times n$ matrix A is defined as $ch_A(x) = \det(A - Ix)$. For instance, the characteristic polynomial of $A = \begin{bmatrix} 0 & -1 \\ 1 & 0 \end{bmatrix}$ is $ch_A(x) = x^2 + 1$, and we note that

$$ch_A(A) = A^2 + I = \begin{bmatrix} -1 & 0 \\ 0 & -1 \end{bmatrix} + \begin{bmatrix} 1 & 0 \\ 0 & 1 \end{bmatrix} = \begin{bmatrix} 0 & 0 \\ 0 & 0 \end{bmatrix}.$$

This property holds in general.

Before stating the theorem, let us make sure that the computation makes sense. Addition of matrices, like addition of vectors is componentwise, that is, if A, B, and C are $n \times n$ matrices and $C = A + B$, the the $(i, j)^{\text{th}}$ entry in C is the sum of the $(i, j)^{\text{th}}$ entry from A and he $(i, j)^{\text{th}}$ entry from B. Scalar multiplication is also componentwise. So if c is a complex number and A is an $n \times n$ matrix, then cA is an $n \times n$ matrix whose $(i, j)^{\text{th}}$ entry is c times the $(i, j)^{\text{th}}$ entry of A. The product of two $n \times n$ matrices is the $n \times n$ matrix which represents the linear transformation obtained by applying one matrix and then the other. Specifically, if $C = A * B$, then $c_{i,j} = \sum_{k=1}^{n} a_{i,k} b_{k,j}$. In the special case when we are computing powers of A, order doesn't matter. Because taking powers, multiplying by a constant (scalar), and addition, all make sense for $n \times n$ matrices, evaluating a polynomial at a matrix makes sense.

Theorem C.5.1 (The Cayley–Hamilton Theorem). *For any* $n \times n$
complex matrix A, *let* $ch_A(x)$ *be* A's *characteristic polynomial, then*

$$ch_A(A) = 0,$$

that is, evaluating A's *characteristic polynomial at* A *results in the* $n \times n$
matrix which has every entry equal to zero.

There is another important polynomial associated with a square matrix
A. From the Cayley–Hamilton Theorem we know that $ch_A(x)$ is a non-
zero polynomial $p(x)$ for which $p(A)$ is the zero matrix. The **minimal
polynomial** $\min_A(x)$ of A is defined to be the non-zero polynomial of *least*
degree (with leading coefficient equal to 1) such that $p(A) = 0$. Dividing
$ch_A(x)$ by $\min_A(x)$ gives polynomials $q(x), r(x)$ such that

$$ch_A(x) = q(x) \min_A(x) + r(x) \text{ where } \deg(r(x)) < \deg(\min_A(x)) .$$

Since $ch_A(A) = 0 = \min_A(A)$, then also $r(A) = 0$ and the minimality of
$\deg(\min_A)$ forces the remainder $r(x)$ to be the zero polynomial. This proves
the useful fact that the minimal polynomial always divides the characteristic polynomial.

C.6   Exercises

Ex C.1. Use induction to verify formula (C.1) for the powers of Jordan
blocks.

Ex C.2. Show that a companion matrix is diagonalizable if and only if it
has distinct eigenvalues.

Ex C.3. Prove that the Jordan form of a companion matrix A contains
exactly one Jordan block for each eigenvalue of A.

Ex C.4. If A is the companion matrix of a polynomial $f(x) \in \mathbb{C}[x]$ show
that $f(x)$ is the characteristic polynomial of A.

Ex C.5. Suppose A is an $k \times k$ real matrix with k distinct real eigenvalues.
Then there is a complex matrix P such that $P^{-1}AP$ is diagonal. Can you
always find a real matrix P with this property?

Ex C.6. Consider computing the characteristic polynomial of a $k \times k$
matrix A. Then $A - Ix$ is a matrix whose off-diagonal entries are complex
constants, and each diagonal matrix is a monic linear polynomial.
(a) Let M be a $k \times k$ matrix which has $k - m$ rows in which every entry
is a constant and in the remaining m rows all but one entry is a constant
and that entry is a monic linear polynomial. If no column contains two
polynomial entries, show that $det(M)$ is a polynomial of degree m.
(b) Construct an inductive argument to show the degree of the characteristic polynomial of a $k \times k$ matrix A is always k.

Appendix D
Roots in the Unit Circle

Our analysis of difference equations shows that the general solution of a recurrence converges to zero when all roots of the characteristic polynomial are less than 1 in absolute value. Since some roots of the characteristic polynomial might be nonreal, this condition means that all roots lie within the unit circle in the complex plane. In what follows we describe a method of Morris Marden [105, Chapter X] for calculating the number of roots of a polynomial within the unit circle. Here we specialize his technique to polynomials with real coefficients and recast it as an algorithm.

Before we discuss the general method, we would like to consider the special example of nonnegative polynomials, which is the leading case for applications. Recall that a **nonnegative polynomial** is a polynomial p of the form

$$p(x) = x^k - c_1 x^{k-1} - c_2 x^{k-2} - \cdots - c_{k-1} x - c_k ,$$

where all $c_i \geq 0$ and $c_k > 0$. In Section 5.1 we prove some special properties of nonnegative polynomials. Among these, Theorem 5.1.3 says that any nonnegative polynomial has exactly one positive real root λ_0, which is **dominant** in the sense that any other root λ satisfies $|\lambda| \leq |\lambda_0|$. Further, from Corollary 5.1.5, when p is **primitive** (that is, when $\gcd\{i \mid c_i > 0\} = 1$), λ_0 is **strictly dominant**, the *only* root whose modulus has the maximum value. This result can be applied to construct polynomials whose roots lie inside the unit circle. For instance, if $q(x)$ is a real polynomial for which

$$p(x) = (x - 1)q(x) = x^k - c_1 x^{k-1} - c_2 x^{k-2} - \cdots - c_{k-1} x - c_k$$

is nonnegative, then $\lambda_0 = 1$ must be the dominant root of p. When p is a primitive polynomial, then $\lambda_0 = 1$ is a strictly dominant root of p, and all roots of q therefore lie inside the unit circle. As an example, for $q(x) = x^2 + \frac{1}{2}x + \frac{1}{2}$, $p(x) = (x-1)q(x) = x^3 - \frac{1}{2}x^2 - \frac{1}{2}$ is a primitive nonnegative polynomial. Therefore, all roots of q lie inside the unit circle, and the general solution of the associated recurrence $s_n = \frac{1}{2}s_{n-1} + \frac{1}{2}s_{n-3}$ converges to zero. Of course, in this simple example it's easy to check that q has two complex roots and that the absolute value of each is $1/\sqrt{2}$.

A more complicated example is the **generalized Fibonacci polynomial** , $p(x) = x^k - x^{k-1} - \cdots - x - 1$ for integer $k \geq 2$. The dominant root λ_0 lies in the open interval $(1,2)$ because $p(1) \leq -1 < 0 < p(2)$. Dividing $p(x)$ by $(x - \lambda_0)$ gives the polynomial

$$q(x) = x^{k-1} + (\lambda_0 - 1)x^{k-2} + \cdots + (\lambda_0^{k-1} - \cdots - 1).$$

Setting

$$P(x) = (x-1)q(x) = x^k + (\lambda_0 - 2)x^{k-1} + (\lambda_0^2 - 2\lambda_0)x^{k-2}$$
$$+ \cdots + (\lambda_0^{k-1} - 2\lambda_0^{k-2})x - (\lambda_0^{k-1} - \cdots - 1),$$

from $p(\lambda_0) = 0$ we obtain $\lambda_0(\lambda_0^{k-1} - \lambda_0^{k-2} - \cdots - 1) = 1$, and so the constant term in P is $-\lambda_0^{-1}$, which is negative. Since $1 < \lambda_0 < 2$, the other coefficients are also negative, and we see that P is a primitive nonnegative polynomial, which implies that all roots of q must lie inside the unit circle. Since $p(x) = (x - \lambda_0)q(x)$, from this we obtain that all roots of p except λ_0 are within the unit circle.

D.1   Marden's Method

We now turn to Marden's method as applied to any polynomial f with real coefficients. The basic idea involves a kind of pairing between

$$f(x) = a_0 + a_1 x + \cdots + a_n x^n \text{ and its}$$

reciprocal polynomial $f^R(x) = a_0 x^n + a_1 x^{n-1} + \cdots + a_n$,

The relationship between f and f^R is such that for every non-zero root r of f inside the unit circle there's a corresponding root r^{-1} of $f^R(x)$ outside the unit circle, and conversely. This is true because

$$f(x) = a \cdot x^k \prod_{i=1}^{n-k}(x - r_i) \text{ implies } f^R(x) = A \cdot \prod_{i=1}^{n-k}\left(x - \frac{1}{r_i}\right),$$

where $A = a \cdot \prod_{i=1}^{n-k} r_i$.

For any f, Marden constructs the associated polynomial

$$G(x) = a_0 f(x) - a_n f^R(x)$$
$$= a_0(a_0 + a_1 x + \cdots + a_n x^n) - a_n(a_0 x^n + a_1 x^{n-1} + \cdots + a_n),$$

whose degree is at most $n-1$ and whose constant term is $a_0^2 - a_n^2$. He then streamlines some results of A. Cohn [27] to obtain his Lemma 42.1, which we reword as the following theorem.

Theorem D.1.1. *Suppose $f(x) = a_0 + a_1 x + \cdots + a_n x^n$ has p roots inside the unit circle. If $\delta(f) := a_0^2 - a_n^2$ is non-zero then $G(x) = a_0 f(x) - a_n f^R(x)$ has either p or $n-p$ roots inside the unit circle, and the sign of its constant term $\delta(f)$ determines which holds. Namely, G has p roots inside the unit circle iff $\delta(f) > 0$. Further, f and G have the same number of roots on the unit circle.*

Therefore, the number of roots of the constructed polynomial G that lie inside the unit circle is the same as the number of roots of either f or f^R within the unit circle, and the sign of $\delta(f)$ indicates which holds. For our current purposes we'll say that G **is equivalent to** f when G and f have the same number of roots within the unit circle. (Otherwise, G is equivalent to f^R.)

For example, let $f(x) = x^2 - \frac{1}{3}x - \frac{1}{3}$. Then f has two roots inside the unit circle, and its reciprocal polynomial $f^R(x) = 1 - \frac{1}{3}x - \frac{1}{3}x^2$ has no roots inside the unit circle. The constructed polynomial is

$$G(x) = a_0 f(x) - a_2 f^R(x)$$
$$= -\frac{1}{3}\left(x^2 - \frac{1}{3}x - \frac{1}{3}\right) - 1\left(-\frac{1}{3}x^2 - \frac{1}{3}x + 1\right)$$
$$= \frac{4}{9}x - \frac{8}{9},$$

and $\delta(f) = -\frac{8}{9} < 0$, which means that G is equivalent to f^R. Since the only root of G is $x = 2$, which lies outside the unit circle, G is indeed equivalent to f^R. Neatly enough, since $\delta(G) \neq 0$, in order to determine whether G has a root inside the unit circle we could have applied the same calculation to G. For this, we calculate $G^R(x) = \frac{4}{9} - \frac{8}{9}x$ and construct

$$-\frac{8}{9}G(x) - \frac{4}{9}G^R(x) = \left(\frac{8}{9}\right)^2 - \left(\frac{4}{9}\right)^2 = 3\left(\frac{4}{9}\right)^2,$$

a constant polynomial that has $\delta(G) > 0$ and so is equivalent to G. Therefore, G (and also f^R) has no roots inside the unit circle.

The iteration used in this last example is the basis for Marden's algorithm for counting the number of roots inside the unit circle. At the start of the procedure we set $f_0 = f$, which may have up to $\deg(f) = n$ roots within the unit circle. The degree of the constructed polynomial $f_1(x) = a_0 f_0(x) - a_n f_0^R(x)$ is at most $n - 1$. Provided $\delta(f_1) \neq 0$, we can continue to use f_1 to construct f_2, and so forth.

Notice how this technique can be used to count the number of roots. When the procedure starts we have $f(x)$, which may have up to n roots within the unit circle. The constructed polynomial $a_0 f - a_n f^R$ has at most $n - 1$ roots. If the constructed polynomial f_1 has the same number of roots as f^R within the unit circle, then f^R can have at most $n - 1$ roots within the circle, and hence must have at least one root outside the circle. But this root of f^R that is outside the circle must correspond to a root of f that is inside the circle. So, when f_1 is equivalent to f^R, we can increase the count of f's roots inside the circle by 1. On the other hand, if f_1 agrees with f, we know that f has at most $n - 1$ roots within the unit circle, and we can proceed with counting the roots of f_1 to get the count of the number of roots of f. One does not change the count in this situation.

As mentioned earlier, the sign of the constant term of f_1 indicates whether it is equivalent to f or equivalent to f^R. So the algorithm will have a switch, called DELTA in the version below, to keep track of whether the current polynomial is equivalent to f or f^R. This switch will be updated when a new polynomial is constructed and will also tell the algorithm when to increment the count.

The algorithm breaks down when some $\delta(f_i)$ is zero. In this situation we want to obtain an equivalent polynomial $\text{EQUIV}(f_i)$ [105, Section 45]. Suppose we encounter

$$g(x) := f_i(x) = a_0 + a_1 x + \cdots + a_m x^m \text{ with } \delta(g) = a_0^2 - a_m^2 = 0.$$

Then $a_m/a_0 = \pm 1$ which we call u. Construction of the equivalent polynomial $\text{EQUIV}(g)$ depends on whether or not $(a_m, \cdots, a_0) = (u a_0, \cdots, u a_m)$.

Case 1. If $(a_m, \cdots, a_0) = (u a_0, \cdots, u a_m)$, replace g with

$$\text{EQUIV}(g) = a_1 x^{m-1} + 2 a_2 x^{m-2} + \cdots + (m - 1) a_{m-1} x + m\, a_m,$$

the reciprocal polynomial of the derivative of g.

Case 2. If $(a_m, \cdots, a_0) \neq (u a_0, \cdots, u a_m)$, let $0 < q < m$ be the smallest subscript such that $u a_q \neq a_{m-q}$. Set $b = (a_{m-q} - u a_q)/a_m$ and

$$G(x) = (x^q + 2b/|b|) g(x) = \sum_{i=0}^{m+q} B_i x^i$$

with

$$\text{EQUIV}(g) = B_0 G(x) - B_{m+q} G^R(x),$$

a polynomial whose degree can be shown to be at most $\deg(g)$.

Marden's Algorithm

INPUT: $f(z) = a_0 + a_1 z + \cdots + a_n z^n$,
         a polynomial with real coefficients.

OUTPUT: COUNT, the number of roots of f
         within the unit circle.

<u>FOR</u> $s = 0$ <u>TO</u> n
$\quad a_s^{(0)} = a_s$
<u>ENDFOR</u>
DELTA $= 1$; COUNT $= 0$

<u>FOR</u> $j = 0$ <u>TO</u> $n - 1$
$\quad$ <u>IF</u> $|a_0^{(j)}| = |a_{n-j}^{(j)}|$
$\quad$ <u>THEN</u> EQUIV $(a_0^{(j)} + \cdots + a_{n-j}^{(j)} x^{n-j})$
$\qquad$ and update j if necessary.

$\quad$ <u>FOR</u> $k = 0$ <u>TO</u> $n - j - 1$
$\qquad a_k^{(j+1)} = a_0^{(j)} a_k^{(j)} - a_{n-j}^{(j)} a_{n-j-k}^{(j)}$
$\quad$ <u>ENDFOR</u>

$\quad$ DELTA $=$ DELTA $* \ sgn(a_0^{(j+1)})$
$\quad$ <u>IF</u> DELTA < 0 <u>THEN</u> COUNT $=$ COUNT $+ 1$
<u>ENDFOR</u>

<u>OUTPUT(COUNT)</u>

As written, the algorithm uses $O(n^2)$ time, and it uses $O(n^2)$ space for storing a two-dimensional array to hold the various a's. By slightly changing the order in which the a's are calculated and writing over the old a's with the new a's, the space usage can be cut to $O(n)$. The $O(n^2)$ run time assumes that products and differences can be computed in constant time. This may not be true. If one uses "real" computer arithmetic, then there may be catastrophic loss of accuracy, and extended precision may be necessary. Conversely, if rational or even integer arithmetic is used, the number of digits needed could double at each iteration, again forcing the use of extended precision.

Example D.1.1. Use Marden's algorithm to count the number of roots of $f(x) = -7x^3 + x^2 + 5x + 1$ inside the unit circle.
$n = 3$ DELTA $= 1$;   COUNT $= 0$

$j = 0$

$$k = 0 \qquad a_0^{(1)} = a_0^{(0)} a_0^{(0)} - a_3^{(0)} a_3^{(0)} = 1 \cdot 1 - (-7) \cdot (-7) = -48$$

$$k = 1 \qquad a_1^{(1)} = a_0^{(0)} a_1^{(0)} - a_3^{(0)} a_2^{(0)} = 1 \cdot 5 - (-7) \cdot 1 = 12$$

$$k = 2 \qquad a_2^{(1)} = a_0^{(0)} a_2^{(0)} - a_3^{(0)} a_1^{(0)} = 1 \cdot 1 - (-7) \cdot 5 = 36$$

$$\text{DELTA} = 1 \cdot (-1) = -1$$

$$\text{COUNT} = 0 + 1 = 1$$

$j = 1$

$$k = 0 \qquad a_0^{(2)} = a_0^{(1)} a_0^{(1)} - a_2^{(1)} a_2^{(1)} = (-48) \cdot (-48) - (36) \cdot (36) = 1008$$

$$k = 1 \qquad a_1^{(2)} = a_0^{(1)} a_1^{(1)} - a_2^{(1)} a_1^{(1)} = -48 \cdot 12 - 36 \cdot 12 = -1008$$

$$\text{DELTA} = -1 \cdot (1) = -1$$

$$\text{COUNT} = 1 + 1 = 2$$

$j = 2$

At this point, $|a_0^{(2)}| = |a_1^{(2)}|$, and EQUIV($f_2$) must be invoked. Continuing with our algorithm, Case 1 applies and gives EQUIV($1008 - 1008x$) $= \div 1008$. Updating to $j = 3$, the FOR loop terminates, and the algorithm outputs COUNT $= 2$ as the number of roots of $-7x^3 + x^2 + 5x + 1$ inside the unit circle. Notice that the root $x = 1$ is not *inside* the unit circle and is not counted.

Example D.1.2. Find the number of roots of $f(x) = x^2 + x + 1$ that lie inside the unit circle.

$n = 2 \qquad \text{DELTA} = 1; \qquad \text{COUNT} = 0$

$j = 0$

Since $a_0^{(0)} = a_2^{(0)}$, EQUIV(f) must be found. Case 1 again applies and

$$\text{EQUIV}(x^2 + x + 1) = a_1 x + 2a_2 = x + 2.$$

Since the degree of this polynomial is $\deg(f) - 1$, j must be increased by 1 and $a_0^{(1)} = 2$, $a_1^{(1)} = 1$.

$j = 1$

$$k = 0 \qquad a_0^{(2)} = a_0^{(1)} a_0^{(1)} - a_1^{(1)} a_1^{(1)} = 4 - 1 = 3 > 0.$$

DELTA remains positive and COUNT is not incremented. The FOR loop terminates, and 0 is returned as the number of roots inside the unit circle for the polynomial $x^2 + x + 1$. This is correct, because the two roots of this polynomial are the primitive third roots of unity, which lie on the unit circle.

Example D.1.3. We consider $f(x) = x^2 - x - 1$, the characteristic polynomial for the Fibonacci sequence. As we've noted many times, f has one root inside the unit circle and one root outside the unit circle. Applying Marden's algorithm, we obtain $a_0^{(0)} = -1$ and $a_2^{(0)} = 1$, giving $u = a_2/a_0 = -1$. But $(a_0, a_1, a_2) = (-1, -1, -1) \neq -(a_2, a_1, a_0)$, and Case 2 with $q = 1$ applies. Then $b = -2$,

$$G(x) = (x^q - 2)g(x) = (x - 2)(x^2 - x - 1) = x^3 - 3x^2 + x + 2 \,,$$

and

$$EQUIV(g) = 2G(x) - G^R(x) = -7x^2 + 5x + 3 \,.$$

Since $\deg(\mathrm{EQUIV}) = \deg(f)$, $j = 0$ is unchanged.

$j = 0$

$$
\begin{aligned}
k = 0 \quad & a_0^{(1)} = a_0^{(0)} a_0^{(0)} - a_2^{(0)} a_2^{(0)} = 9 - 49 = -40 < 0 \\
k = 1 \quad & a_1^{(1)} = a_0^{(0)} a_1^{(0)} - a_2^{(0)} a_1^{(0)} = 15 - (-35) = 50 \\
& \mathrm{DELTA} = 1 \cdot (-1) = -1 \\
& \mathrm{COUNT} = 0 + 1 = 1
\end{aligned}
$$

$j = 1$

$$
\begin{aligned}
k = 0 \quad & a_0^{(2)} = a_0^{(1)} a_0^{(1)} - a_1^{(1)} a_1^{(1)} = (-40)(-40) - (50)(50) < 0 \\
& \mathrm{DELTA} = -1 \cdot (-1) = 1 \\
& \mathrm{COUNT} = 1.
\end{aligned}
$$

The FOR loop terminates and the algorithm correctly outputs $\mathrm{COUNT} = 1$ for the number of roots of $x^2 - x - 1$ inside the unit circle.

Example D.1.4. As a final example we consider $f(x) = x^2 - 1$, which has the roots ± 1. Applying Marden's algorithm, we obtain $\delta(f) = 0$, which means that $\mathrm{EQUIV}(f)$ must be invoked. Since $(a_2, a_1, a_0) = -(a_0, a_1, a_2)$, Case 1 of the replacement algorithm applies, and $\mathrm{EQUIV}(f) = a_1 x + 2a_2 = -2$. This results in a decrement of the degree by 2, and j must be incremented by 2. But then the FOR loop terminates, and the algorithm correctly outputs $\mathrm{COUNT} = 0$.

D.2   Exercises

Ex D.1. Use Marden's algorithm to show that $5x^n - 1$ has n roots inside the unit circle.

Ex D.2. Show that $x^{n-1} + 2x^{n-2} + 3x^{n-3} + \cdots + (n-1)x + n$ has no roots inside the unit circle by applying Marden's algorithm to $x^n + x^{n-1} + \cdots + x + 1$.

Ex D.3. Use Marden's algorithm twice to determine the number of roots inside the unit circle and the number of roots inside the circle of radius 2 for the polynomial $2x^3 - 3x^2 + 2x + 2$. (The actual roots are $-\frac{1}{2}$, $1 + i$, $1 - i$.)

References

[1] F. Acton. *Numerical Methods That (Usually) Work.* Harper and Row, New York, NY, 1970.

[2] A. V. Aho, J. E. Hopcroft, and J. D. Ullman. *The Design and Analysis of Computer Algorithms.* Addison-Wesley, Reading, MA, 1974.

[3] W. R. Alford, A. Granville, and C. Pomerance. There are infinitely many Carmichael numbers. *Annals of Mathematics (2)*, 139:705–722, 1994.

[4] R. Askey and M. Ismail. *Recurrence Relations, Continued Fractions, and Orthogonal Polynomials.* American Mathematical Society, Providence, RI, 1984.

[5] W. W. Rouse Ball and H. M. S. Coxeter. *Mathematical Recreations and Essays.* Dover Publications, New York, NY, 1987.

[6] È. G. Belaga. Some problems involved in the calculation of polynomials. *Dokl. Akad. Nauk. SSSR*, 123:775–777, 1958.

[7] A. Berman and R. J. Plemmons. *Nonnegative Matrices in the Mathematical Sciences.* SIAM, Philadelphia, PA, 1994.

[8] H. Bernardelli. Population waves. *J. Burma Res. Soc.*, 31:3–18, 1941.

[9] W. Beyer, R. Roof, and D. Williamson. The lattice structure of multiplicative congruential pseudo-random vectors. *Mathematics of Computation*, 25:345–363, 1971.

370   References

[10] B. Bollobás. *Random Graphs*. Academic Press, London, 1985.

[11] B. Boncompagni. *Scritti di Leonardo Pisano: mathematico del secolo decimoterzo*. Rome, Italy, 1857–1862.

[12] C. B. Boyer. *A History of Mathematics*. Wiley, New York, NY, 1989.

[13] R. Brent and J. Pollard. Factorization of the eighth Fermat number. *Math. Comp.*, 36:627–630, 1981.

[14] E. F. Brickell. A fast modular multiplication algorithm with applications to two-key cryptography. In *Advances in Cryptology— Proceedings of CRYPTO '82*, pages 51–60. Plenum, 1983.

[15] Brother Alfred Brousseau. The relation of the zeros to periods in the Fibonacci sequence modulo a prime. *American Mathematical Monthly*, 71:897–899, 1964.

[16] Brother Alfred Brousseau. *Introduction to Fibonacci Discovery*. Fibonacci Association, San Jose, CA, 1965.

[17] Brother Alfred Brousseau. *Fibonacci and Related Number Theoretic Tables*. Fibonacci Association, San Jose, CA, 1972.

[18] W. G. Brown. Historical note on a recurrent combinatorial problem. *American Mathematical Monthly*, 72:973–977, 1965.

[19] P. Buneman and L. Levy. The Towers of Hanoi problem. *Information Processing Letters*, 10:243–244, 1980.

[20] R. L. Burden and J. D. Faires. *Numerical Analysis*. Brooks/Cole, Pacific Grove, CA, 2001.

[21] R. M. Capocelli, editor. *Sequences*. Springer-Verlag, New York, NY, 1990.

[22] R. M. Capocelli, G. Cerbone, P. Cull, and J. Holloway. Fibonacci facts and formulas. In *Sequences*, pages 123–137. Springer-Verlag, New York, NY, 1990.

[23] R. M. Capocelli and P. Cull. Rounding the solutions of Fibonacci-like difference equations. *Fibonacci Quarterly*, 41:133–141, 2003.

[24] R. D. Carmichael. On composite numbers P which satisfy the Fermat congruence $a^{P-1} \equiv 1 \pmod{P}$. *Amer. Math. Monthly*, 19:22–27, 1912.

[25] H. Caswell. *Matrix Population Models*. Sinauer Associates, Sunderland, Mass, 2001.

[26] E. Catalan. Note sur une équation aux différence finies. *J. Math. Pures Appl.*, 3:508–516, 1838.

[27] A. Cohn. Über die Anzahl der Wurzeln einer algebraischen Gleichung in einem Kreise. *Mathematische Zeitscrift*, 14:110–148, 1922.

[28] D. Coppersmith and S. Winograd. Matrix multiplication via arithmetic progression. *Journal of Symbolic Computation*, 9:251–280, 1990.

[29] T. Cormen, A. Leiserson, R. Rivest, and C. Stein. *Introduction to Algorithms (second edition)*. McGraw-Hill, Boston, MA, 2001.

[30] R. Crandall and C. Pomerance. *Prime Numbers: A computational approach*. Springer-Verlag, New York, NY, 2001.

[31] P. Cull. The problem of time unit in Leslie's population model. *Bulletin of Mathematical Biology*, 42:719–728, 1980.

[32] P. Cull. Global stability of population models. *Bulletin of Mathematical Biology*, 43:47–58, 1981.

[33] P. Cull. Local and global stability for population models. *Biological Cybernetics*, 54:141–149, 1986.

[34] P. Cull. Local and global stability of discrete one-dimensional population models. In L. M. Ricciardi, editor, *Biomathematics and Related Computational Problems*, pages 271–278. Kluwer, Dordrecht, 1988.

[35] P. Cull. Stability of discrete one-dimensional population models. *Bulletin of Mathematical Biology*, 50(1):67–75, 1988.

[36] P. Cull. Analysis of algorithms. In L. M. Ricciardi, editor, *Lectures in Applied Mathematics and Informatics*, pages 1–61. Manchester University Press, 1990.

[37] P. Cull. Linear fractionals — simple models with chaotic-like behavior. In D. M. Dubois, editor, *Computing Anticipatory Systems: CASYS 2001* , pages 170–181. Conference Proceedings 627, American Institute of Physics, Woodbury, N.Y., 2002.

[38] P. Cull and J. Chaffee. Stability in discrete population models. In D. M. Dubois, editor, *Computing Anticipatory Systems: CASYS'99*, pages 263–275. Conference Proceedings 517, American Institute of Physics, Woodbury, NY, 2000.

[39] P. Cull and J. Chaffee. Stability in simple population models. In *Cybernetics and Systems 2000*, pages 289–294. Austrian Society for Cybernetics Studies, 2000.

[40] P. Cull and J. Holloway. Computing Fibonacci numbers quickly. *Information Processing Letters*, 32:143–149, 1989.

[41] P. Cull and E. F. Ecklund Jr. Towers of Hanoi and analysis of algorithms. *American Mathematical Monthly*, 92(6):407–420, June–July 1985.

[42] P. Cull and A. Vogt. Mathematical analysis of the asymptotic behavior of the Leslie population matrix model. *Bulletin of Mathematical Biology*, 35:645–661, 1973.

[43] P. Cull and A. Vogt. The periodic limit for the Leslie model. *Mathematical Biosciences*, 21:39–54, 1974.

[44] P. Cull and A. Vogt. The period of total population. *Bulletin of Mathematical Biology*, 38:317–319, 1976.

[45] Paul Cull. Stability in one-dimensional models. *Scientiae Mathematicae Japonicae*, 58:349–357, 2003.

[46] G. Dalquist and A. Bjorck. *Numerical Methods*. Prentice-Hall, Englewood Cliffs, NJ, 1974.

[47] J. H. Davenport, Y. Siret, and E. Tournier. *Computer Algebra*. Chapman and Hall, San Diego, CA, 1988.

[48] A. de Moivre. *Miscellanea Analytica*. Londini, excudebant J. Tonson & J. Watts, London, 1730.

[49] R. Descartes. *Discourse on Method*. Penguin, New York, NY, 1968.

[50] R. Descartes. *Meditations*. Penguin, New York, NY, 1968.

[51] R. Devaney. *An Introduction to Chaotic Dynamical Systems*. Benjamin, Redwood City, CA, 1986.

[52] P. Diaconis, M. McGrath, and J. Pitman. Riffle shuffles, cycles, and descents. *Combinatorica*, 15:11–29, 1995.

[53] L. E. Dickson. *Linear Algebraic Groups and an Exposition of Galois Theory*. Dover Publications, New York, NY, 1958.

[54] D. M. Dubois, editor. *Computing Anticipatory Systems: CASYS'99 — Third International Conference* . Conference Proceedings 517, American Institute of Physics, Woodbury, N.Y., 2000.

[55] D. M. Dubois, editor. *Computing Anticipatory Systems: CASYS 2001 — Fifth International Conference* . Conference Proceedings 627, American Institute of Physics, Woodbury, N.Y., 2002.

[56] F. Dyson and H. Falk. Period of a discrete cat mapping. *American Mathematical Monthly*, 99:603–614, 1992.

[57] J. Eichenauer-Hermann. Inversive congruential pseudorandom numbers avoid the planes. *Mathematics of Computation*, 56:297–301, 1991.

[58] S. E. Eldridge and C. D. Walter. Hardware implementation of Montgomery's modular multiplication algorithm. *IEEE Transactions on Computers*, 42:693–699, 1993.

[59] P. Erdős. On almost primes. *American Mathematical Monthly*, 57:404–407, 1950.

[60] L. Euler. *Novi Commentarii Academiae Scientiarum Imperalis Petropolitanae*, 7:13–14, 1758–1759.

[61] L. Euler. *Introduction to Analysis of the Infinite*. Springer-Verlag, New York, NY, 1988. Translated by J. D. Blanton.

[62] W. Feller. *An Introduction to Probability Theory and its Applications*. John Wiley, New York, NY, 1968.

[63] M. E. Fisher, B. S. Goh, and T. L. Vincent. Some stability conditions for discrete-time single species models. *Bulletin of Mathematical Biology*, 41:861–875, 1979.

[64] M. Flahive and H. Niederreiter. On inversive congruential generators for pseudorandon numbers. *Lecture Notes in Pure and Applied Mathematics*, 141:75–80, 1993.

[65] H. Furstenberg. *Recurrence in Ergodic Theory and Combinatorial Number Theory*. Princeton University Press, Princeton, NJ, 1981.

[66] É. Galois. *Oeuvres Mathématiques*. Gauthier-Villars, Paris, France, 1897.

[67] F. R. Gantmacher. *The Theory of Matrices*. Chelsea Publishing Company, New York, NY, 1959.

[68] M. Garey and D. Johnson. *Computers and Intractability: A Guide to the Theory of NP-Completeness*. W. H. Freeman, San Francisco, 1979.

[69] F. Garvin. *The Maple Book*. Chapman and Hall, Boca Raton, FL, 2002.

[70] C. F. Gauss. *Disquisitiones Arithmeticae*. Springer-Verlag, New York, NY, 1986.

[71] B. S. Goh. *Management and Analysis of Biological Populations*. Elsevier, New York, NY, 1979.

[72] R. L. Graham, D. E. Knuth, and O. Patashnik. *Concrete Mathematics: A Foundation for Computer Science*. Addison-Wesley, Reading, MA, 1994.

[73] R. T. Gregory. *Methods and Applications of Error-Free Computation*. Springer-Verlag, New York, NY, 1984.

[74] R. Guy. How to factor a number. *Congressus Numerantium*, 16:49–89, 1976.

[75] M. A. Harrison. *Lectures on Linear Sequential Machines*. Academic Press, New York, NY, 1969.

[76] M. P. Hassel. Density dependence in single species populations. *Journal of Animal Ecology*, 44:283–296, 1974.

[77] R. Herken, editor. *The Universal Turing Machine*. Oxford University Press, Oxford, UK, 1988.

[78] K. Hoffman and R. Kunze. *Linear Algebra (second edition)*. Prentice-Hall, Englewood Cliffs, NJ, 1971.

[79] J. L. Holloway. *Algorithms for Computing Fibonacci Numbers Quickly*. MS thesis, Computer Science, Oregon State University, Corvallis, OR, 1989.

[80] C. Hooley. On Artin's conjecture. *J. Reine Angew. Math*, 226:209–220, 1967.

[81] Y. N. Huang. A counterexample for P. Cull's theorem. *Kexue Tongbao*, 31:1002–1003, 1986.

[82] D. Kalman. The generalized Vandermonde matrix. *Mathematics Magazine*, 57:15–21, 1984.

[83] J. Keller. How many shuffles to mix a deck? *SIAM Review*, 37:88–89, 1995.

[84] A. Knight. *Basics of MATLAB and Beyond*. Chapman and Hall, Boca Raton, FL, 2000.

[85] D. Knuth. Big Omicron and Big Omega and Big Theta. *SIGACT News*, 8:18–24, April-June 1976.

[86] D. Knuth. *Selected Articles on Analysis of Algorithms*, pages 35–42. Addison-Wesley, Reading, MA, 2000.

[87] D. Knuth. All questions answered. *Notices of the American Mathematical Society*, 49:318–324, 2002.

[88] D. E. Knuth. *The Art of Computer Programming*. Addison-Wesley, New York, NY, third edition, 1997.

[89] R. J. Kooman. *Convergence Properties of Recurrence Sequences*. Centrum voor Wiskunde en Informatica, Amsterdam, The Netherlands, 1991.

[90] T. Koshy. *Fibonacci and Lucas Numbers*. Wiley-Interscience, New York, NY, 2001.

[91] G. Lamé. Extrait d'une lettre de M. Lamé à M. Liouville sur cette question: un polygone convexe étant donné, de combien de manières peut-on le partager en triangles au moyen de diagonales? *J. Math. Pures Appl.*, 3:505–507, 1838.

[92] J. P. LaSalle. *The Stability of Dynamical Systems*. SIAM, Philadelphia, PA, 1976.

[93] H. Lebesgue. L'oeuvre mathématique de Vandermonde. *L'Enseignement mathématique (2)*, 1:203–223, 1956.

[94] D. H. Lehmer. Mathematical methods in large-scale computing units. In *Proceedings of the Second Symposium on Large-Scale Digital Computing Machinery*, pages 141–146. Harvard University Press, Cambridge, MA, 1951.

[95] P. H. Leslie. On the use of matrices in certain population mathematics. *Biometrika*, 33:183–212, 1945.

[96] H. Levy and F. Lessman. *Finite Difference Equations*. Dover, New York, NY, 1992.

[97] T-Y. Li and J. Yorke. Period three implies chaos. *American Mathematical Monthly*, 82:985–992, 1975.

[98] R. Lidl and H. Niederreiter. *Finite Fields*. Cambridge University Press, Cambridge, England, 1997.

[99] D. Lind and B. Marcus. *An Introduction to Symbolic Dynamics and Coding*. Cambridge University Press, New York, NY, 1995.

[100] E. N. Lorenz. Deterministic non-periodic flows. *J. Atmos. Sci.*, 20:130–141, 1963.

[101] A. J. Lotka. *Elements of Mathematical Biology*. Dover Publications, New York, NY, 1956.

[102] U. Manber. *Introduction to Algorithms: A Creative Approach.* Addison-Wesley, Reading, MA, 1989.

[103] M. Marcus and H. Minc. *A Survey of Matrix Theory and Matrix Inequalities.* Allyn and Bacon, Rockleigh, NJ, 1964.

[104] M. Marden. Much ado about nothing. *American Mathematical Monthly,* 83:788–798, 1976.

[105] M. Marden. *The Geometry of the Zeros of a Polynomial in a Complex Variable.* American Mathematical Society, New York, NY, 1989.

[106] G. Marsaglia. Random numbers fall mainly in the planes. *Proceedings of the National Academy of Sciences, U.S.A.,* 61:25–28, 1968.

[107] G. Marsaglia. The structure of linear congruential sequences. In *Applications of Number Theory to Numerical Analysis,* pages 249–285. Academic Press, New York, NY, 1972.

[108] Y. V. Matiyasevich. *Hilbert's Tenth Problem.* The MIT Press, Cambridge, MA, 1993.

[109] R. M. May. Biological populations with nonoverlapping generations: stable points, stable cycles, and chaos. *Science,* 186:645–647, 1974.

[110] R. M. May. Simple mathematical models with very complicated dynamics. *Nature,* 261:459–467, 1976.

[111] J. McCarthy. The Tower of Stanford (problem 10956). *American Mathematical Monthly,* 111:364–365, 2004.

[112] K. Mehlhorn. *Data Structures and Algorithms.* Springer-Verlag, New York, NY, 1984.

[113] L. M. Milne-Thomson. *The Calculus of Finite Differences.* Macmillan, London, 1933.

[114] L. Monier. Evaluation and comparison of two efficient probabilistic primality testing algorithms. *Theoret. Comput. Science,* 12:97–108, 1980.

[115] P. L. Montgomery. Modular multiplication without trial division. *Mathematics of Computation,* 44:519–522, 1985.

[116] P. A. P. Moran. Some remarks on animal population dynamics. *Biometrics,* 6:250–258, 1950.

[117] T. Muir. *The Theory of Determinants in the Historical Order of Development,* volume 3. Dover Publications, New York, NY, 1960.

[118] H. Neiderreiter. Quasi-Monte Carlo methods and pseudorandom numbers. *Bulletin of the American Mathematical Society*, 84:957–1041, 1978.

[119] H. Niederreiter. *Random Number Generation and Quasi-Monte Carlo Methods*. SIAM, Philadelphia, PA, 1992.

[120] I. Niven. *Diophantine Approximations*. Interscience Publishers, New York, NY, 1963.

[121] I. Niven. Formal power series. *American Mathematical Monthly*, 76:871–889, 1969.

[122] A. Nobile, L. M. Ricciardi, and L. Sacerdote. On Gompertz growth model and related difference equations. *Biological Cybernetics*, 42:221–229, 1982.

[123] V. Ya. Pan. On methods of computing polynomial values. *Russian Mathematical Surveys*, 21:105–137, 1966.

[124] F. Parker. Inverses of Vandermonde matrices. *American Mathematical Monthly*, 71:410–411, 1964.

[125] T. S. Parker and L. Chua. *Practical Numerical Algorithms for Chaotic Systems*. Springer-Verlag, New York, NY, 1989.

[126] C.J. Pennycuick, R.M. Compton, and L. Beckingham. A computer model for simulating the growth of a population, of two interacting populations. *Journal of Theoretical Biology*, 18:316–329, 1968.

[127] P. Petersen. *On computing maximal lattice dimensions of the inversive congruential generator*. MS thesis, Mathematics, Oregon State University, Corvallis, OR, 1998.

[128] J. M. Pollard. A Monte Carlo method for factorization. *BIT*, 15:331–334, 1975.

[129] G. Pólya. *How to Solve It*. Princeton University Press, Princeton, NJ, 1945.

[130] G. Pólya. On picture writing. *American Mathematical Monthly*, 63:689–697, 1956.

[131] W. H. Press. *Numerical Recipes: The Art of Scientific Computing*. Cambridge University Press, New York, NY, 1986.

[132] M. Rabin. Probabilistic algorithm for testing primality. *J. Number Theory*, 12:128–138, 1980.

[133] L. M. Ricciardi, editor. *Biomathematics and Related Computational Problems*. Kluwer, Dordrecht, 1988.

[134] L. M. Ricciardi, editor. *Lectures in Applied Mathematics and Informatics*. Manchester University Press, Manchester, UK, 1990.

[135] W. E. Ricker. Stock and recruitment. *Journal of the Fisheries Research Board of Canada*, 11:559–623, 1954.

[136] R. H. Risch. The problem of integration in finite terms. *Transactions of the American Mathematical Society*, 139:167–189, 1969.

[137] R. L. Rivest, A. Shamir, and L. M. Adelman. A method for obtaining digital signatures and public-key cryptosystems. *Communications of the ACM*, 21:145–152, 1978.

[138] H. Rogers. *Theory of Recursive Functions and Effective Computability*. McGraw-Hill, New York, NY, 1967.

[139] K. H. Rosen. *Elementary Number Theory and its Applications*. Addison-Wesley, Reading, MA, 1993.

[140] G. Rosenkranz. On global stability of discrete population models. *Mathematical Biosciences*, 64:227–231, 1983.

[141] J. J. Rotman. *Advanced Modern Algebra*. Prentice-Hall, Upper Saddle River, NJ, 2002.

[142] D. G. Saari and J. B. Urenko. Newton's method, circle maps, and chaotic motion. *The American Mathematical Monthly*, 91:3–18, 1984.

[143] A. Salomaa. *Automata–theoretic Aspects of Formal Power Series*. Springer-Verlag, New York, NY, 1978.

[144] A. Sarkovskii. Coexistence of cycles of a continuous map of a line to itself. *Ukr. Mat. Z.*, 16:61–71, 1964.

[145] A. Schonhage and V. Strassen. Schnelle Multiplikation grosser Zahlen. *Computing*, 7:281–292, 1971.

[146] E. Seneta. *Non-negative Matrices*. John Wiley & Sons, New York, NY, 1973.

[147] L. Sigler. *Fibonacci's Liber Abaci*. Springer-Verlag, New York, NY, 2002.

[148] D. Singer. Stable orbits and bifurcation of maps of the interval. *SIAM Journal on Applied Mathematics*, 35(2):260–267, Sept. 1978.

[149] D. Smith and M. Latham. *The Geometry of René Descartes*. Open Court Publishing Company, Chicago, IL, 1925.

[150] J. M. Smith. *Mathematical Ideas in Biology.* Cambridge University Press, Cambridge, 1968.

[151] J. M. Smith. *Models in Ecology.* Cambridge University Press, Cambridge, 1974.

[152] R. P. Stanley. Generating functions. In *Studies in Combinatorics*, pages 100–141. Mathematical Association of America, Washington, D.C., 1978.

[153] R. P. Stanley. *Enumerative Combinatorics*, volume 1. Wadsworth & Brooks/Cole, Monterey, CA, 1986.

[154] R. P. Stanley. *Enumerative Combinatorics*, volume 2. Cambridge University Press, New York, NY, 1999.

[155] J. Stoer and R. Bulirsch. *Introduction to Numerical Analysis.* Springer-Verlag, New York, NY, 1993.

[156] V. Strassen. Gaussian elimination is not optimal. *Numer. Math*, 13:354–356, 1969.

[157] A. Tarski. *Logic, Semantics, Metamathematics.* Oxford University Press, Oxford, England, 1956.

[158] R. Taylor and A. Wiles. Modular elliptic curves and Fermat's Last Theorem. *Annals of Mathematics (2)*, 141:443–551, 1995.

[159] A. Tucker. *Applied Combinatorics.* John Wiley, New York, NY, 2002.

[160] J. B. Urenko. Improbability of nonconvergent chaos in Newton's method. *Journal of Mathematical Analysis and Applications*, 117:42–47, 1986.

[161] S. Utida. Population fluctuation, an experimental and theoretical approach. *Cold Spring Harbor Symposium on Quantitative Biology*, 22:139–151, 1957.

[162] A. van der Poorten. *Notes on Fermat's Last Theorem.* John Wiley & Sons, New York, NY, 1996.

[163] J. Vinson. The relation of the period modulo m to the rank of apparition of m in the Fibonacci sequence. *Fibonacci Quarterly*, 1:37–45, 1963.

[164] N. N. Vorobev. *Fibonacci Numbers.* Blaisdell, New York, NY, 1961.

[165] D. D. Wall. Fibonacci series modulo m. *American Mathematical Monthly*, 67:525–532, 1960.

[166] C. D. Walter. Systolic modular multiplication. *IEEE Transactions on Computers*, 42:376–378, 1993.

[167] C. D. Walter. Space/time trade-offs for higher radix modular multiplication using repeated addition. *IEEE Transactions on Computers*, 46:139–141, 1997.

[168] A. Wiles. Ring-theoretic properties of certain Hecke algebras. *Annals of Mathematics (2)*, 141:553–572, 1995.

[169] H. S. Wilf. *generatingfunctionology*. Academic Press, New York, NY, 1990.

[170] J. H. Wilkinson. *The Algebraic Eigenvalue Problem*. Clarendon Press, Oxford, England, 1965.

[171] Jet Wimp. *Computation with Recurrence Relations*. Pitman, Boston, MA, 1984.

[172] N. Wirth. *Algorithms + Data Structures = Programs*. Prentice-Hall, Englewood Cliffs, NJ, 1976.

[173] A. Wiseman and T. P. Wiseman. *De bello Gallico*. English. D.R. Godine, Boston, MA, 1980.

[174] S. Wolfram. *Mathematica*. Addison-Wesley, Redwood City, CA, 1988.

[175] S. Wolfram. *A New Kind of Science*. Wolfram Media, Champaign, IL, 2002.

[176] O. Wyler. On second-order recurrences. *American Mathematical Monthly*, 72:500–506, 1965.

[177] N. Zierler. Linear recurring sequences and error-correcting codes. In *Error-Correcting Codes*, pages 47–59. Wiley, New York, NY, 1968.

Index